# Human-Centered AI

Artificial intelligence (AI) permeates our lives in a growing number of ways. Relying solely on traditional, technology-driven approaches won't suffice to develop and deploy that technology in a way that truly enhances human experience. A new concept is desperately needed to reach that goal. That concept is Human-Centered AI (HCAI).

With 29 captivating chapters, this book delves deep into the realm of HCAI. In Section I, it demystifies HCAI, exploring cutting-edge trends and approaches in its study, including the moral landscape of Large Language Models. Section II looks at how HCAI is viewed in different institutions—like the justice system, health system, and higher education—and how it could affect them. It examines how crafting HCAI could lead to better work. Section III offers practical insights and successful strategies to transform HCAI from theory to reality, for example, studying how using regulatory sandboxes could ensure the development of age-appropriate AI for kids. Finally, decision-makers and practitioners provide invaluable perspectives throughout the book, showcasing the real-world significance of its articles beyond academia.

Authored by experts from a variety of backgrounds, sectors, disciplines, and countries, this engaging book offers a fascinating exploration of Human-Centered AI. Whether you're new to the subject or not, a decision-maker, a practitioner, or simply an AI user, this book will help you gain a better understanding of HCAI's impact on our societies, and of why and how AI should really be developed and deployed in a human-centered future.

**Catherine Régis** is a specialist in health law and innovation law at Université de Montréal. Most of her work explores how to best regulate AI at the national and international levels. She holds a Canada Research Chair in Health Law and Policy and a Canada-CIFAR Chair in Artificial Intelligence and Human Rights (Mila). She has co-chaired (2021-2023) the Responsible AI Working Group of the Global Partnership on AI and is an associate academic member at MILA, the world's largest academic research center in machine learning.

**Jean-Louis Denis** is Professor of Health Policy and Management at the School of Public Health of Université de Montréal and a senior scientist the Research Center of Québec's largest hospital, CHUM. He holds the Canada Research Chair on Health System Design and Adaptation. He cochairs, with Catherine Régis, the HAICU Lab on human-centered AI of the U7+ Alliance of Global Universities.

**Maria Luciana Axente** is an award-winning AI ethics and public policy expert, a member of various advisory boards and an Intellectual Forum Senior Research Associate at the University of Cambridge, researching human-centric AI at the intersection between tech policy and ethics in the industry.

**Atsuo Kishimoto** is Professor at the Institute for Datability Science and Director of the Research Center on Ethical, Legal and Social Issues at the Osaka University. He is originally an economist by training and has worked for the National Institute of Advanced Industrial Science and Technology (AIST) as a risk assessor.

**Chapman & Hall/CRC Artificial Intelligence and Robotics Series**

*Series Editor: Roman Yampolskiy*

*Unity in Embedded System Design and Robotics: A Step-by-Step Guide*
Ata Jahangir Moshayedi, Amin Kolahdooz, Liao Liefa

*Meaningful Futures with Robots: Designing a New Coexistence*
Edited by Judith Dörrenbächer, Marc Hassenzahl, Robin Neuhaus, Ronda Ringfort-Felner

*Topological Dynamics in Metamodel Discovery with Artificial Intelligence: From Biomedical to Cosmological Technologies*
Ariel Fernández

*A Robotic Framework for the Mobile Manipulator: Theory and Application*
Nguyen Van Toan, Phan Bui Khoi

*AI in and for Africa: A Humanist Perspective*
Susan Brokensha, Eduan Kotzé, Burgert A. Senekal

*Artificial Intelligence on Dark Matter and Dark Energy: Reverse Engineering of the Big Bang*
Ariel Fernández

*Explainable Agency in Artificial Intelligence: Research and Practice*
Silvia Tulli, David W. Aha

*An Introduction to Universal Artificial Intelligence*
Marcus Hutter, Elliot Catt, David Quarel

*AI: Unpredictable, Unexplainable, Uncontrollable*
Roman V. Yampolskiy

*Transcending Imagination: Artificial Intelligence and the Future of Creativity*
Alexander Manu

*Human-Centered AI: A Multidisciplinary Perspective for Policy-Makers, Auditors, and Users*
Edited by Catherine Régis, Jean-Louis Denis, Maria Luciana Axente, Atsuo Kishimoto

For more information about this series please visit: https://www.routledge.com/Chapman--HallCRC-Artificial-Intelligence-and-Robotics-Series/book-series/ARTILRO

# Human-Centered AI

## A Multidisciplinary Perspective for Policy-Makers, Auditors, and Users

Edited by
Catherine Régis
Jean-Louis Denis
Maria Luciana Axente
Atsuo Kishimoto

CRC Press is an imprint of the
Taylor & Francis Group, an **informa** business
A CHAPMAN & HALL BOOK

Designed cover image: Shutterstock

First edition published 2024
by CRC Press
2385 Executive Center Drive, Suite 320, Boca Raton, FL 33431

and by CRC Press
4 Park Square, Milton Park, Abingdon, Oxon, OX14 4RN

*CRC Press is an imprint of Taylor & Francis Group, LLC*



ISBN: 978-1-032-34162-0 (hbk)
ISBN: 978-1-032-34161-3 (pbk)
ISBN: 978-1-003-32079-1 (ebk)

DOI: 10.1201/9781003320791

Typeset in Times
by SPi Technologies India Pvt Ltd (Straive)

# Contents

## SECTION II Sectoral Representations of HCAI

## *SECTION III LESSONS LEARNED AND PROMISING PRACTICES*

# Acknowledgments

This book's publication was made possible by the generous financial support of Université de Montréal (UdeM), Mila (Québec Artificial Intelligence Institute), and the French government through Université Côte d'Azur.[1]

Our gratitude extends to HAICU's editorial committee and the U7+ Alliance for their invaluable support and guidance throughout this project. We must highlight the distinctive role played by Annelise Riles, who was Executive Director of the Buffet Institute for Global Studies at Northwestern University at the time this book was written, in bolstering both HAICU's work and the U7+.

Special warmth and appreciation are directed to Réjean Roy, a strategic advisor at UdeM, whose expert guidance enabled the authors and our editorial team to complete a high-quality book on schedule. We also wish to express our thanks to Antoine Congost, a project manager at UdeM, for his invaluable aid in the final stages of the book's editing process.

Last but certainly not least, heartfelt thanks go to Daniel Jutras, Rector of UdeM, for his unwavering encouragement throughout this project. His support has been a cornerstone of this accomplishment.

## NOTE

1 As part of the UCAJEDI Investments in the Future Project, overseen by France's National Research Agency (reference number ANR-15-IDEX-01).

# Foreword

In July 2023, I testified before the US Senate Judiciary Subcommittee on Privacy, Technology, and the Law. My objective was to convince American officials and citizens of several key points: AI systems already outperform human intelligence in numerous areas; these tools might achieve broad human cognitive competence within decades or even years; we must proactively address the grave risks this poses to our societies and lives before they manifest; and this necessitates ensuring AI systems operate as intended, consistent with democracy, human values, and standards.

Designing, building, and deploying AI systems that adhere to human values and standards—that, to use the words of philosopher Shannon Valor[1], have goals that are centered around *us* and foster *our* flourishing—will be a massive undertaking. The ultimate success of this task will be influenced by various factors and levers of action, many highlighted in the book edited by Régis, Denis, Axente, and Kishimoto.

Firstly, we must decide to change the current trajectory leading to the dangerous concentration of extreme power derived from AI in the hands of a few in an uncontrolled way, because extreme power corrupts and threatens the very notion of democracy and human rights. Instead, as a collective, we must deeply reflect on the human values and standards we wish to instill in our AI systems and deliberate how and who determines them. Human rights are certainly a key element here, one that should have a stronger presence in AI debates. And we must combine this reflection with swift actions considering the pressing regulatory and governance needs that we face nationally and globally with respect to AI. We must act now because the commercial AI train is accelerating with the current massive investment in generative AI.

Secondly, we should ensure experts from diverse fields, including sociologists, legal experts, economists, and political scientists, join this endeavor. Interdisciplinary work is challenging, and producing results requires strong commitment, time, and deliberate strategies, but it is essential, as no single discipline can address all the complex governance issues raised by AI's development and deployment.

Thirdly, it is indispensable, as the United Nations' Secretary General recently remarked to the Security Council on Artificial Intelligence,[2] for academics to make their findings about the risks, safety, and social impact of AI more accessible to the broader public, to foster dialogues with AI developers, AI users, policymakers, and the civil society. This book was conceived with that idea in mind, which it achieves skillfully.

Régis, Denis, Axente, and Kishimoto's book is among the pioneering works emphasizing the importance of human-centered AI. It plays a vital role in addressing our common challenge: How do we envision our future with AI? And what will we do to activate this vision?

**Professor Yoshua Bengio**
*Full professor of Computer Science at Université de Montréal*
*Founder and Scientific Director of Mila—Québec AI Institute*
*2018 Co-recipient of the AM Turing Award*

## NOTES

1 See, chapter 1, “Defining Human-Centered AI: An Interview with Shannon Vallor,” p. 13.
2 See https://www.un.org/sg/en/content/sg/statement/2023-07-18/secretary-generals-remarks-the-security-council-artificial-intelligence-bilingual-delivered-scroll-down-for-all-english?_gl=1*1vit394*_ga*MjAxNTgyNDkwNC4xNjkxNTA3NjI5*_ga_TK9BQL5X7Z*MTY5MjczNDgwMi4zLjEuMTY5MjczNDg3Ny4wLjAuMA…

# Contributors

**Maria Axente**
University of Cambridge
Cambridge, UK

**Oshri Bar-Gil**
Behavioral Science Research Institute
Israel

**Cassandra Bowkett**
University of Manchester
Manchester, UK

**Garrick Cabour**
Polytechnique Montréal
Montréal, Québec, Canada

**Tapabrata Chakraborti**
Alan Turing Institute, University College London
London, UK

**Vicky Charisi**
European Commission, Joint Research Centre
Seville, Spain

**Allison Cohen**
Mila—Québec Artificial Intelligence Institute
Montréal, Québec, Canada

**Christina Colclough**
The Why Not Lab
Svendborg, Denmark

**Clementine Collett**
Oxford Internet Institute
University of Oxford
Oxford, UK

**Michael Da Silva**
University of Southampton
Southampton, UK

**Jean-Louis Denis**
Université de Montréal
Montréal, Québec, Canada

**Irene Di Bernardo**
University of Naples Federico II
Naples, Italy

**Virginia Dignum**
Department of Computing Science
Umeå University
Umeå, Sweden

**Philippe Doyon-Poulin**
Polytechnique Montréal
Montréal, Québec, Canada

**Rebecca Finlay**
Partnership on AI

**Jennifer Garard**
Sustainability in the Digital Age, Concordia University
Sherbrooke, Québec, Canada
and
Future Earth Canada
Montréal, Québec, Canada

**Julie (M.É.) Garneau**
Université du Québec en Outaouais
Gatineau, Québec, Canada

**Karine Gentelet**
Université du Québec en Outaouais
Gatineau, Québec, Canada

**Erin Gleeson**
Sustainability in the Digital Age, Concordia University
Sherbrooke, Québec, Canada
and
Future Earth Canada
Sherbrooke, Québec, Canada
and
McGill University
Montréal, Québec, Canada

**Ernest Habanabakize**
Sustainability in the Digital Age, Concordia University
Sherbrooke, Québec, Canada
and
Future Earth Canada
Sherbrooke, Québec, Canada
and
McGill University
Montréal, Québec, Canada

**Jakob Kappenberger**
University of Mannheim
Mannheim, Germany

**Haluna Kawashima**
Tohoku Fukushi University
Sendai, Japan

**Atsuo Kishimoto**
Osaka University
Suita, Japan

**Alistair Knott**
School of Engineering and Computer Science
Victoria University of Wellington
Wellington, New Zealand

**Kaspar Kundert**
Self-employed consultant

**Pierre Larouche**
Faculté de droit
Université de Montréal
Montréal, Québec, Canada

**Pierre-Majorique Léger**
HEC Montréal
Montréal, Québec, Canada

**Alexandre Lepage**
Université de Montréal
Montréal, Québec, Canada

**Christian Lévesque**
HEC Montréal
Montréal, Québec, Canada

**Rosette Lukonge Savanna**
Leapr Labs
Kigali, Rwanda

**Matt Malone**
Faculty of Law
Thompson Rivers University
Kamloops, British Columbia, Canada

**Gaétan Marceau Caron**
Mila—Québec Artificial Intelligence Institute
Montréal, Québec, Canada

**Marialuisa Marzullo**
University of Naples Federico II
Naples, Italy

**Sarit K. Mizrahi**
University of Ottawa
Ottawa, Ontario, Canada

**Cristina Mele**
University of Naples Federico II
Naples, Italy

**Gina Neff**
Minderoo Centre for Technology and Democracy
University of Cambridge
Cambridge, UK

**Tomoumi Nishimura**
Kyushu University
Fukuoka, Japan

**Joseph Nsengimana**
Mastercard Foundation Center for Innovative Teaching and Learning

**Takehiro Ohya**
Keio University
Tokyo, Japan

**Mario Passalacqua**
Polytechnique Montréal
Montréal, Québec, Canada

**Dino Pedreschi**
Dipartimento di Informatica
University of Pisa
Pisa, Italy

**Robert Pellerin**
Polytechnique Montréal
Montréal, Québec, Canada

**Sara Pérez-Lauzon**
HEC Montréal
Montréal, Québec, Canada

**Daoud Piracha**
Mila—Québec Artificial Intelligence Institute
Montréal, Québec, Canada

**Giada Pistilli**
Sorbonne University
Paris, France

**Bruno Poellhuber**
Université de Montréal
Montréal, Québec, Canada

**Benjamin Prud'homme**
Mila—Québec Artificial Intelligence Institute
Montréal, Québec, Canada

**Christopher D. Quintana**
Villanova University
Villanova, PA, USA

**Angelo Ranieri**
University of Naples Federico II
Naples, Italy

**Anand Rao**
PwC
Greater Boston, MA, USA

**Kinsie Rayburn**
Planet Labs
San Francisco, CA, USA

**Catherine Régis**
Université de Montréal
Montréal, Québec, Canada

**Tatiana Revilla**
Tecnológico de Monterrey
Monterrey, Mexico

**Sandra Rodriguez**
MIT lecturer, and Independent Creative Director and Producer

**Melissa Rosa**
Planet Labs
San Francisco, CA, USA

**Normand Roy**
Université de Montréal
Montréal, Québec, Canada

**Tiziana Russo Spena**
University of Naples Federico II
Naples, Italy

**Heiner Stuckenschmidt**
University of Mannheim
Mannheim, Germany

**Marina Teller**
Université Côte d'Azur
Nice, France

**Mélisande Teng**
Sustainability in the Digital Age, Concordia University
Sherbrooke, Québec, Canada
and
Future Earth Canada
Sherbrooke, Québec, Canada
and
Mila—Québec Artificial Intelligence Institute
Montréal, Québec,Canada

**Éliane Ubalijoro**
Sustainability in the Digital Age, Concordia University
Sherbrooke, Québec, Canada
and
Future Earth Canada
Sherbrooke, Québec, Canada

**Shannon Vallor**
University of Edinburgh
Edinburgh, Scotland

**Malwina Anna Wójcik**
Law, Science and Technology
University of Bologna
Bologna, Italy
and
University of Luxembourg
Esch-sur-Alzette, Luxembourg

**Tatsuhiko Yamamoto**
Keio University
Tokyo, Japan

**Kodai Zukeyama**
Kyushu Sangyo University
Fukuoka, Japan

# Introduction

*Atsuo Kishimoto*
Osaka University, Suita, Japan

*Catherine Régis and Jean-Louis Denis*
Université de Montréal, Montréal, Canada

*Maria Axente*
University of Cambridge, Cambridge, UK

## BACKGROUND

This book project was started by Catherine Régis and Jean-Louis Denis, two scholars from the University of Montreal and the co-leaders of the HAICU Lab, and Atsuo Kishimoto, an academic from Osaka University and one of the Lab members. From the onset, they were joined by Maria Axente, a pioneer practitioner, in the United Kingdom, in the field of responsible AI in industry. The HAICU Lab was founded by the U7+ Alliance of World Universities, an organization that includes more than 45 institutions from 20 countries across 6 continents. The lab's mission is to encourage collaboration between academia and public institutions, firms, and civil society organizations. Its aim is to develop human-centered responses to the significant challenges and opportunities brought about by the implementation of AI in our societies—in essence, to promote the creation and application of *human-centered AI (HCAI)*.

The journey of creating this book spanned across a time frame of 16 months. An international call for papers was launched in March 2022 and disseminated through various channels. Selected authors began work on their pieces in the autumn of 2022. Contributors were invited to participate in online seminars that took place in the winter of 2023. These forums offered attendees the chance to provide and receive peer feedback on their papers and exchange ideas on the role of HCAI from their perspective. In the subsequent months, the authors (who come from over 12 countries, like Canada, France, Italy, Japan, New Zealand, and the UK, and more than 12 disciplines, like computer science, education, the law, management, political science, and sociology) crafted revised drafts of their papers and received peer-reviewed comments. Different academic traditions and approaches to inquiry and argumentation are thus represented in the book.

We added to the book a series of interviews with key institutional players (like Rebecca Finlay, CEO at Partnership on AI; Joseph Nsengimana, Director of the Mastercard Foundation Center for Innovative Teaching and Learning; or Christian

DOI: 10.1201/9781003320791-1 

Colclough, Head of The Why Not Lab, a consultancy that serves trade unions and governments) to enrich our understanding of the many meanings of HCAI explored in the chapters as well as of future challenges and opportunities. This addition serves to add another layer of discussion on HCAI and encourage an interactive discussion between the emerging science of HCAI and views from regulators, developers, and users in various spheres of activity. The final manuscript was submitted to the publisher in July 2023.

## SIGNIFICANCE OF HUMAN-CENTERED AI

The information environment surrounding us is expanding rapidly due to digitalization and digital transformation. To navigate this new information landscape, AI has become indispensable for classifying, organizing, and quickly identifying what we need. However, debates persist about the degree to which tasks previously performed by humans should or can be entrusted to AI. It is crucial that we remain continually conscious and reflective about what responsibilities we delegate to AI and which ones we retain. In particular, we have been very cautious about high-stakes decisions, such as those involved in hiring, admissions, lending decisions, and court rulings, being left entirely to machine data processing. Instances have been observed where AI is applied merely because the data is available, without consideration of its benefits to the data subjects, In these circumstances, the "human-centered AI" concept emerges as a compelling approach.

Various strategies and guidelines have underscored the concept of HCAI as a crucial objective. The second of the five principles in the OECD AI Principles, formulated in 2019, is "human-centred values and fairness," which asserts that "AI actors should respect the rule of law, human rights and democratic values, throughout the AI system lifecycle." In Europe, the High-Level Expert Group on AI introduced the Ethics Guidelines for Trustworthy Artificial Intelligence in 2019, which declared that AI systems must be human-centric. The first among the seven requirements for "Trustworthy AI" is "human agency and oversight," encompassing fundamental rights, human agency, and human oversight. In addition, The Montreal Declaration for a Responsible Development of Artificial Intelligence, issued in 2018, anchors its ten principles in the notions of human well-being and respect for autonomy, which resonate deeply with the idea of human-centered AI.

By around 2020, numerous AI ethics guides and principles had been developed on global, national, and corporate scales. In recent years, the focus shifted toward translating these principles into daily practices for researchers and employees. The disparity between principles and practices, often referred to as the "implementation gap," is a topic addressed in several chapters of this book. As mentioned later in this introduction, the specific conditions required for actualizing HCAI in a practical sense are discussed at various moments in the book.

Since mid-2022, AI has become significantly more accessible due to the release of text-to-text, text-to-image, and text-to-music generative AI models from various companies. Now, even individuals without programming skills can utilize AI to generate text and images. This rapid proliferation of generative AI compels us to reevaluate the concept of HCAI. Until recently, AI technologies have enhanced the efficiency of human activities primarily through tasks like prediction and recognition, with the general public mainly being the recipient. However, generative AI differs from previous AI technologies in its ability to empower ordinary people to generate text and images. While this democratization of AI technology is a positive development, it also implies the "democratization" of potential misuse and abuse. In this regard, a more human-centered design is needed.

## DEFINITION AND SCOPE OF HUMAN-CENTERED AI

The concept of HCAI encompasses numerous facets, and as such, it defies a single, definitive interpretation. Generally, efforts to define HCAI can be categorized into process-focused and outcome-focused approaches.

The process-focused perspective centers on placing AI under human control. It stresses the importance of respecting human autonomy and accountability, particularly in relation to automated decision-making and the principle of "human in the loop" as set forth in the EU General Data Protection Regulation (GDPR). The emphasis lies on autonomy and responsibility: humans may err, but they alone can shoulder responsibility for mistakes.

The outcome-focused view appreciates the capacity of AI systems to satisfy human needs and values or to be user-friendly from an ergonomic standpoint. This perspective aligns with the human-centered design approach, which prioritizes the end user in the design of products and services. To create offerings that cater to user needs, it is advantageous to involve potential users from the early stages of product and service development. Consequently, an inclusive R&D process is imperative. Even with an outcome focus, human values related to the process are relevant, and an algorithm that can be unsafe to humans in unpredictable ways is not satisfactory from the point of view of HCAI. Therefore, the process and outcome approaches are closely interconnected; the key distinction lies in which is viewed as the goal and which is the means to attain that goal. Ultimately, the distinction between the two may not significantly alter what should be emphasized to achieve HCAI.

### THE POTENTIAL RISKS POSED BY AI TECHNOLOGY

The development process of machine-learning AI for prediction and recognition is depicted in Figure I.1. In order to create a human-centered approach, it is vital to properly gather training data (ensuring it is free of bias) and to scrutinize the models' algorithms for potential bias. It is also crucial to involve human oversight in high-stakes decisions, instead of blindly accepting the outputs as they are. For instance, issues arose with facial recognition technology when biases in the training data set

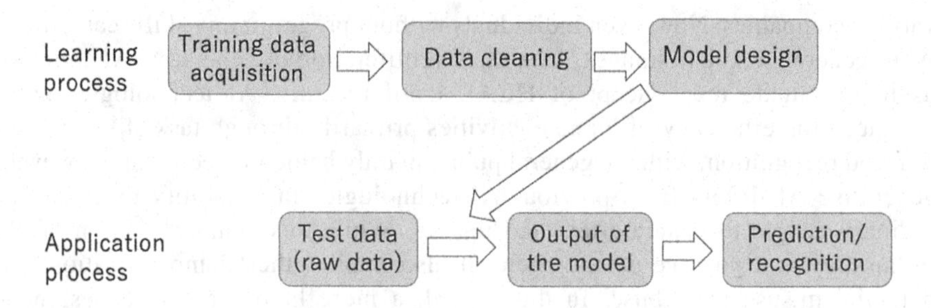


**FIGURE I.1** Flow diagram for the development of AI models used for prediction and recognition.

caused algorithmic biases, leading to accuracy disparities across genders and ethnicities and thereby resulting in discriminatory treatment. In response, there have been efforts to reassess the training data, audit the algorithms, and limit the use of certain facial recognition outputs.

For generative AI, the lower section of the figure depicts the generation process, which begins with prompt input and concludes with the creation of generated data. Text generation AIs, like ChatGPT, produce sentences that mimic human writing, but many "hallucinations" (i.e., errors) are known to occur in factual information. Numerous issues have already arisen because the output is used directly without human verification. It is also observed that such outputs tend to mirror and even amplify real-world stereotypes and biases.

## HCAI for Each Stakeholder

An inclusive approach is pivotal for the implementation of HCAI, necessitating governance where diverse stakeholders fulfill their respective roles. Simultaneously, the concept and approach of HCAI prove beneficial for a wide array of stakeholders. Through their agency, actors within the ecosystem influence how social expectations and prescriptions will be formed, promoted, or contested with consequences for decisions and choices made within the AI ecosystem (Hallett & Ventresca, 2006). The evolution of the AI ecosystem and the trajectory of innovations within that sector are an outcome of structures, activities, and actions at multiple levels of analysis, such as societal, field, organization, and individual (Scott, 2013). Within this ecosystem, power and countervailing power (e.g., healthy pressure being applied by various groups) are leading determinants. Here we consider how some representative stakeholders can use this book.

## POLICY-MAKERS

AI technology, including generative AI, will continue to advance and implement technologies that are unpredictable today. The concept of HCAI should apply to any advanced technologies and will aid in developing laws, regulations, and usage guidelines, and in identifying ethical, legal, and social issues. Concurrently, HCAI can also assist policy-makers in obtaining more accurate data and forecasts; AI analysis and predictions can provide a basis for policy decisions and help formulate more effective policies.

## INDUSTRY PRACTITIONERS

The rate of technological development often outpaces the amendments to laws and regulations. As such, compliance with laws and regulations alone is insufficient. It is necessary to constantly and self-reflectively assess whether business activities align with the norm of HCAI. To do so, a system and proper incentives need to be created that enables review throughout the entire data and technology lifecycle, from initial internal research and development reviews to post-implementation monitoring. The second part of this book serves as a reference since it provides individual discussions for each field.

## RESEARCHERS

With the spread of generative AI, AI is increasingly likely to have significant societal impacts, both positive and negative. The approach of HCAI can contribute to enhancing the positives and mitigating the negatives. Researchers must incorporate the concept of "human-centered" into technology development, as well as evaluate the impact of AI and propose suitable countermeasures and mitigation strategies. Research on social aspects and ethical issues can enhance the social acceptability and sustainability of AI technologies.

## GENERAL PUBLIC

With the advent of generative AI, members of the general public can become even more active users of AI, not just passive recipients of AI applications. Therefore, understanding more about AI, its mechanisms, and its characteristics is increasingly essential. It is important to know that while chatbots like ChatGPT can respond in complete sentences and were trained to imitate humans, they may behave in ways that are not like that of humans or that are unreliable. AI literacy is also required to minimize the dissemination of false information without malicious intent and avoid anthropomorphizing AI systems.

## THREE ANGLES OF DISCUSSION

This book is divided into three parts.

**Section I**, a foundational and definitional section, contains chapters that discuss and assess the many meanings of HCAI, as well as the main trends and approaches in the study of HCAI. Among other things, **Vallor** explains the difference between responsible AI and HCAI; presents the characteristics of what HCAI is or is not (e.g., discusses if ChatGPT is an HCAI technology or not); and explains why she is "cautiously optimistic" about the future of AI, while criticizing the recent engineering culture. **Wójcik** finds the roots of algorithmic inequality and social exclusion in Western notions of rationality and individualism, and argues that Ubuntu, a sub-Saharan philosophy that places human dignity in a person's ability to form harmonious relationships with others, may help build an ethical and legal framework for AI. **Bar Gil** focuses on recommendation engines—the machine-learning algorithms that help manage the flood of information. He uses Google's recommendation services as a subject to discuss the impact of HCAI on the key concepts related to the sense of self: autonomy, subjectivity, rationality, and memory. His chapter also briefly discusses generative AI. **Quintana** approaches HCAI from a user experience perspective. He examines the ethical challenges at the "intersection" between humans and machines and between ethics and user experience design. **Larouche** focuses on the aspect of translating AI ethics into legal regulations. He first points out the shortcomings in the conceptual examination phase and then addresses the challenges in the legislative action phase. Recent discussions on AI are organized in terms of the contrast between permissive and protective approaches in relation to innovation. Finally, these arguments are applied to recent legislation in Europe and Canada. **Malone** examines HCAI from the perspective of privacy, an elusive concept. Building on the historical development of privacy concepts, the author examines how the latter will change in the future as we increasingly give AI more time, attention, and trust, and as demands on HCAI increase. Finally, **Pistilli** discusses General-Purpose Large Language Models (LLM), citing several ethical challenges and exploring potential solutions for LLMs to become HCAI. The need for collaboration between engineers and philosophers is emphasized in this process. In summary, Section I of the book discusses the cross-cutting topics that are necessary to consider HCAI practices in various fields. It covers topics such as algorithmic inequality, recommendation engines, ethical challenges in human-machine interaction, translating AI ethics into legal regulations, privacy concepts, and the recent development of large language models (LLMs).

**Section II** contains chapters that probe the potential and representations of HCAI among different groups and stakeholders, within various institutions (justice systems, health systems, higher education institutions, etc.). In other words, it dives into the role HCAI could play, positively and negatively, in specific sectors of societal activity. An institutional lens on the AI ecosystem emphasizes that technological developments and the way they become embedded in society, organizations, and practices are not purely rational or responses to efficiency considerations, but are shaped by a wide set of social expectations and prescriptions (Hinings et al., 2018). Incentive structures, social expectations, and prescriptions are important shapers of the emerging AI ecosystem. **DaSilva, Denis, and Regis** explore the healthcare sector, where AI is increasingly being developed and deployed across all facets of healthcare systems. They argue that HCAI requires two interdependent and interactive governance approaches: (1) a normative approach that ensures the interests of healthcare actors are protected at a regulatory level, and (2) a contextual approach that allows for understanding, responding, and adapting to such interests at a practical level. They discuss how to practically implement this two-step governance strategy. **Garard et al.** discuss what HCAI could mean for the domain of sustainability and agriculture, a field of high importance for many countries. They propose four guiding principles for HCAI in this context: inclusive data stewardship, robust AI applications, trust in partners and technology, and systems of accountability. They present the Data-Driven Insights for Sustainable Agriculture project as a concrete illustration of their proposal. **Ohya** tackles the challenge of implementing HCAI in the field of justice, specifically judiciary systems. Through a creative analytical lens, the author explains how AI could change the basis of judicial reasoning and what we should require from AI systems to be implemented in judiciary-related decisions, ensuring respect for democratic imperatives and fundamental rights. **Lévesque et al.** explore the sector of workspaces, where the transformative impact of AI is uncertain and disputed. They offer perspectives on how HCAI could be approached for better work, including increasing workers' autonomy. They argue, among other points, that all stakeholders need to seriously examine the link between better work and a better society. Disempowerment through increasing economic and social insecurity, deskilling, and meaningless jobs may have destructive consequences for social solidarity and democracy. Specific policies that change incentives and encourage socially responsible innovation are therefore needed. **Poellhuber and his colleagues** discuss the topic of AI in higher education, a sector which has been at the forefront of many public debates with the recent release of generative AI that demonstrates how such technology can transform informational skills, methodology skills, and writing skills. Specifically, the author examines three areas where AI has significant potential in higher education: learning analytics, adaptive learning, and generative artificial intelligence. For each of these areas, The authors analyze their potential to enhance the educational experience, as well as their associated limitations. Moving to the specific reality of universities, **Mele and her colleagues** explore how HCAI could assist universities in advancing three critical missions, which they identify as engaged education, engaged research, and engaged society. This "engaged university" would benefit from AI-based technologies in addressing societal challenges through research and innovation while enhancing the quality of education. **Knott, Chakraborti, and Pedreschi** delve into the highly debated and influential sector of social media platforms, where AI brings numerous opportunities and challenges. They explain that AI systems operate in two main types on social media platforms: recommender systems and content classifiers. After such an explanation, they ultimately target the lack of transparency as the main issue arising from such AI usage, a problem social media platforms must address as a priority to ensure they can be regarded as human-centered. Finally, **Rodriguez** takes on the challenge of reflecting on the interplay between cultural and artistic work and HCAI, drawing lessons from a research-creation project called CHOM5KY vs CHOMSKY. In this project, visitors converse with an AI entity to better grasp how AI systems function and rethink their pitfalls and opportunities. This original chapter approaches the role of artistic work in better communicating AI realities and for developing human skills to inquire, collaborate, and create. In summary, Section II of the book offers a sectorial and cross-sectorial overview of the challenges and opportunities AI can play in some key domains of activity and showcase how HCAI is not necessarily a one-size-fits-all lens and approach, despite some strong common features like alignment to end users' needs, rights, and values. It can be context sensitive and therefore will need to be adapted to the unique reality of the institutions they are embedded in and their related actors.

**Section III** comprises chapters on lessons learned and promising practices that should inform policy-making and policy innovation. It deals with the public policy implications of the proliferation of AI in various sectors. Policy-makers play a crucial role in guiding AI development and ensuring that AI applications align with societal objectives. Additionally, AI can enhance public services and amplify the policy capacity of governments and public bodies. Through public policies,

AI's potential can be realized. **Dignum and Charisi** explore an often-overlooked topic: how AI can be child-friendly and respectful, or what they call an age-appropriate system. Using frameworks and approaches developed or promoted by UNICEF, they propose relying on the generative potential of regulatory sandboxes to support the development and deployment of HCAI for children. Regulatory sandboxes blend evidence-informed approaches to issues with the inherent pluralism of society and sectors in the creation of HCAI. This chapter highlights the importance of having relevant and effective tools and methodologies to assist policy-makers in their work. **Gentelet and Mizrahi** propose to shift away from current widespread normative approaches. They suggest discarding approaches that focus solely on improving the technologies themselves. In their view, debiasing AI requires the recognition that biases and discriminations are socially constructed. They see citizen participation as the antidote to ensure that AI remains human-centered. Citizen participation, in their terms, is based on the need to protect and nurture citizen entitlements to capabilities. Thus, the right policies for regulating AI will be based on strategies and processes that ensure genuine and effective participation. Once again, it is crucial that policy-makers have frameworks and methodologies at their disposal to facilitate active and generative citizen participation. **Teller** promotes a similar approach in her chapter. Analyzing emerging regulations within the European Commission context, the author argues that their risk-based approach results in a top-down regulatory dynamic that is ineffective against the forces of algocracy. Public participation is proposed to inform the development of regulations that benefit from a variety of perspectives and viewpoints developed by interested public groups. **Passalaqua and his colleagues** discuss the role of AI in shaping the next industrial revolution. They suggest taking the effect of AI on work seriously and assessing AI within the context of the complex socio-technical environments that characterize industries and manufacturing. They identify three design properties of HCAI at work: worker or operator well-being, their engagement with the technology, and the contribution of system performance. They propose a human-centric design methodology to support the development of trust and human agency in the context of significant technological changes and transitions. The application of this framework and methodology in two sectors is suggested. **Zukeyama et al**. examine the approach used by the Japanese government to govern AI for the public and human goods. They suggest that HCAI can be achieved through a renewed approach to governance and regulations, which they term agile governance. Three characteristics of agile governance are proposed to promote HCAI: goal-based regulations as opposed to rule-based ones, multi-stakeholder dialogues, and a focus on realizing value through AI in relation to the development of human rights. Ultimately, this chapter suggests an approach that policy-makers can use to promote HCAI across various sectors. **Kappenberger and Stuckenschmidt** also develop a policy-oriented framework to regulate AI. The framework is based on identifying desirable attributes of AI and its use within the context of public policies. The framework is structured around three dimensions: prerequisite level, process level, and output level. The application of this framework could support the achievement of three desirable public policy goals: improving efficiency, increasing decision-making capacities, and nurturing citizen-government interactions. An application of the framework in the domain of traffic regulation is provided. Lastly, **Axente and her colleagues** demonstrate that three implementation gaps explain why the adoption of inclusive AI is not more widespread in organizations: the engagement, translation, and dialogue gaps. Their research indicates that to actualize HCAI, adopting companies will need to build and support diverse and inclusive teams; collaborate with a broad and diverse group of stakeholders; and share best practices with other organizations. In summary, Section III of the book offers resources that could aid policy-makers as they navigate the largely unexplored landscape of AI. The increasing proliferation and spread of AI present both a chance and a challenge for policy-makers. On one hand, AI can enhance policy capacity. On the other hand, AI can present issues that require proper regulation. Regardless of the scenario, chapters in this section reveal that principles alone are insufficient. Frameworks, tools, and methodologies are necessary to properly regulate AI, maximize its benefits, and minimize potential risks.

We hope the readers will enjoy the book.

**The editors,**
*Atsuo, Catherine, Jean-Louis, Maria*

## REFERENCES

Hallett, T., & Ventresca, M. J. (2006). Inhabited institutions: Social interactions and organizational forms in Gouldner's Patterns of Industrial Bureaucracy. *Theory and Society*, *35*, 213–236.

Hinings, B., Gegenhuber, T., & Greenwood, R. (2018). Digital innovation and transformation: An institutional perspective. *Information and Organization*, *28*(1), 52–61.

Scott, W.R. (2013). *Institutions and Organizations: Ideas, Interests, and Identities*, SAGE.

# *Section I*

## *Meanings, Trends and Approaches in the Study of HCAI*

# 1 Defining Human-Centered AI

## *An Interview with Shannon Vallor*

*Shannon Vallor*

University of Edinburgh, Edinburgh, Scotland

Shannon Vallor holds the Baillie Gifford Chair in the Ethics of Data and Artificial Intelligence at the University of Edinburgh's Edinburgh Futures Institute, where she directs the Centre for Technomoral Futures. A Fellow of the Alan Turing Institute, her research explores the ethics of emerging science and technologies. Her current project focuses on the impact of emerging technologies—particularly those involving automation and AI—on the moral and intellectual habits, skills, and virtues of human beings: our character. Prof. Vallor is the author of a 2016 book, *Technology and the Virtues: A Philosophical Guide to a Future Worth Wanting*, and the forthcoming *The AI Mirror: Reclaiming Our Humanity in an Age of Machine Thinking*. She currently codirects the UK's BRAID (Bridging Responsible AI Divides) research program and regularly advises the government and industry on responsible AI design and use.

**You're one of the world's foremost experts on human-centered AI. How would you define that concept? What does it mean for AI to be human-centered?**

I would define human-centered AI as systems that are designed by people, for people, and with people, in such a way that the ultimate design aim is the promotion of human flourishing. There are multiple components in that definition, right? First, the aim of the technology has to be centered around humans, specifically the *flourishing* of people, and the interests of people as humans. But the process matters as well. The design, development, and deployment of AI need to be shaped and guided by humans that have in mind—and are in conversation with—the humans that this technology will interact with.

DOI: 10.1201/9781003320791-3

Of course, there are no technologies that aren't designed and developed by humans, but often humans are developing an AI system in order for that system to serve an institution, an economy, an industrial process, and a market. And that's not in itself human-centered, right? So just having humans involved in the design and deployment of AI is trivial because all technology is human-centered in that way. Human-centered AI really is about the alignment between the humans driving the technology and the humans standing at the center of where the technology is being driven.

**How does the concept of human-centered AI differ from other concepts that are often used in the AI world? For example, how is it different from "responsible AI"?**

There's obviously considerable overlap between those ideals. But I think where they differ is that the notion of responsibility (which is something I also work on right now) is largely about the specific human relationships that underpin, for example, social trust. And so, responsibility, and the notion of responsible AI, is about the way that humans' use of the power they have in their hands (or of the powers they create) impacts the vulnerabilities of other people—the relationships between individual humans, human organizations, and communities and even the broader social trust.

I think responsibility then is that intrinsic social practice of earning the trust of others that licenses the use of our power in ways that can affect them. That's something that's really important to talk about because of the new kind of power AI represents in human hands. It is a power that is affecting the people who currently are left out of the process of shaping the technology—people who, often, are not empowered to push back on, question, or reform technologies that are harmful to them. That's, summarized, the responsible AI piece.

The human-centered AI piece, I think, is less about things like social trust and power, and more about the aim of the technology, its ultimate ends. So, human-centered technology is about aligning the entire technology ecosystem with the health and well-being of the human person. The contrast is with technology that's designed to replace humans, compete with humans, or devalue humans as opposed to technology that's designed to support, empower, enrich, and strengthen humans.

The important thing, here, is also to have a concept of the human that's appropriately rich. So, there are technologies and there are AI systems that we might think are human-centered because their goal is to improve human performance in a given context. But if the humans at the other end of that performance, the persons we're trying to augment with AI, are just treated as other machines whose performance we want to boost and if we're indifferent to their full human personality and the full range of their interests, then we're not really talking about human-centered technology. Even if the technology deployed was designed explicitly for the purpose of augmenting human capability.

If all I'm doing is trying to augment human capabilities so that I can extract more value from that human as part of an economic system, or as part of a set of physical operations, and I'm indifferent to everything else about the person that makes them a person, then that's not human-centered technology and that's not human-centered AI.

That's the other important piece of this. It's to understand that human-centered AI has to have a robust and comprehensive conception of the human person and human flourishing at the heart of it.

**Would it have been better, in your opinion, to center public discussions about the concept of human-centered AI instead of the concept of responsible AI? Would conversations have been richer?**

I think which lens you choose depends upon what the context is and what the aims are. So, are you trying to, for example, address the political legitimacy of a certain kind of technology and the safety of that tool? In that case, you know, a responsible AI framework might be more appropriate than a human-centered one, because you're really concerned about the appropriate regulation and governance of that technology. Responsible AI is the right lens when what you're dealing with primarily is a governance challenge.

The concept of human-centered AI, to me, is a little bit above the responsible AI lens, and it's broader. I think you should use it when what you want to talk about is: what is this technology for? And what kind of future do we want to create with it?

That's when the human-centered AI lens, I think, is more illuminating. Because I don't think you can answer that question simply by pointing to the responsible AI concept. To say that I develop an AI system responsibly doesn't tell me anything about what kind of future I'm trying to shape with it or enable with that. So, I think the bar for responsible AI is high—certainly higher than the point we're currently at—but I think the bar for human-centered AI … I don't know if I would say the bar is higher, but the ambition certainly is bigger.

**Do we have examples of products or services that were developed using the human-centered lens, or are we basically talking about something that does not really exist?**

I'll say that there are technologies, AI technologies, that are compatible with a human-centered approach to AI. For example, think about the kinds of applications of AI that are used to enhance the accessibility and tools for real-time transcription for people who are a part of the deaf community or people who might not speak the native language of the speaker. The kinds of tools we have for automatic translation and captioning can be seen as part of a broader impulse to enrich and broaden access to knowledge and culture for a wider swath of humanity. It's certainly possible to consider that those kinds of technologies that are accessibility-driven are compatible with a human-centered approach.

But I think that answering the question of whether that's accidental or purposeful is important, right? Whether they're designed with a human-centered philosophy, that's a question you can't really answer easily from the outside. You have to really be able to have a good view into how that process came to be, how that technology came to be, what assumptions were baked into it from the beginning, what incentives and goals determined its final form, and to what extent the wider human interests were kept at the center of the design process and the design thinking.

So, I think you can have technologies that are clearly compatible with human-centered AI that might have ended up being compatible with it through a process that wasn't particularly human-centered. But you can also have an AI technology that's human-centered in the way it was conceived, designed, and deployed, but that unfortunately fails—a system that, in the end, is neither safe nor beneficial. That's a perfectly conceivable scenario.

So, I don't think human-centered AI guarantees that every artifact that we build in the AI domain will be ethical, responsible, safe, and beneficial. But I think if we don't have human-centered as the AI Gold Standard, then the ecosystem as a whole will be much more likely to be misaligned with human flourishing and unsustainable in the long run.

**Can you think of an example of an AI tool that definitely was not conceived in a human-centered fashion?**

Sure! Unfortunately, too many AI systems are primarily motivated by the need to manipulate, survey, and control people or treat them as objects to be analyzed and constrained without any clear story about whom this technology benefits or how it improves society.

For example, there's a great deal of effort to design technologies for things like emotion recognition, to enable machines to mimic human emotions in speech or gesture, but, also, to classify humans according to emotional state, so that they can be more effectively classified, ranked, and predicted. Think about the use of emotion recognition features in a video interview application for hiring. There was a lot of concern early on about companies that were advertising that their automated video interview software would be analyzing candidates' facial expressions for cues of different kinds of emotional states. Or think about the way that, reportedly, the Chinese government has deployed technology in schools that monitors children's attention and gaze and emotional displays, presumably in order to facilitate the management of those students' response to material.

Even if you can identify particular benefits for these technologies, it's very clear that the humans that are the object of these tools are not their intended beneficiaries, that these tools were designed to classify, manipulate, and control humans for the sake of some *other* institution or process. That's the opposite of what human-centered AI is.

It's very clear, also, that ChatGPT is not a human-centered technology. Think about the way it's transforming classrooms right now. Now, we might find ways to make good use of it in the classroom—in part because we'll be forced to. But that's precisely the point: ChatGPT wasn't developed in an effort to answer a question like, "What does humanity, what do people, need most from technology right now?" Instead, it was developed by saying, "What will we get if we just scale up these technologies as big as we can get them, and pour as much computing power into them as we could? And what we get is something that we then have to cope with as opposed to something designed by us, for us, and to benefit us.

As a result, we now have educators who have to throw out the way of teaching that they've been developing over centuries and *remake* education in a way that won't be frustrated by the existence of this tool. And it's an open question whether what we'll get will be better or worse than what we had before. But the point is nobody asked educators, "What do you need from AI to allow you to teach better?" ChatGPT is not what they would have asked for. It's not the technology we needed. So there's all this talk about the upsides of ChatGPT, but most of the upsides are ways in which we can *accommodate* that technology, which was dropped on people, not chosen by them.

**You mentioned that, very often, human-centered AI will not be the end result, but the process should be human-centered. Can you tell me a bit more about how this process should be implemented? What does it look like?**

I think there are a lot of different methodologies and practices that have been developed that can support human-centered AI. Value-sensitive design is one of them. It consists of asking things like, "What are the values shaping this particular design architecture?", "Who will it serve?", and "What existing values might it press upon or conflict with? So, value-sensitive design, I think, is part of the answer.

Participatory design is also part of the solution. There's a well-known phrase, "nothing about us without us," which arose in the disability community in response to proposals that were made to design things like exoskeletons for people in wheelchairs. When you ask most people in wheelchairs, they would say, "I don't want to stand. I don't want an exoskeleton, that's clunky." A lot of wheelchair users were saying instead, "What we want is what we've been asking for a long time: curb cuts on sidewalks, counters that are an appropriate height, airlines that handle our wheelchairs without breaking them. These are the things we need. Give us these things!" Instead, they were given exoskeletons, which didn't take off in part because they weren't designed with the community that they were intended to benefit.

"Nothing about us without us." It's not just about designing a tech that you imagine will serve disabled folks. It's about tech that's going to transform the lives of parents or change the impact or the experience of care in a hospital setting. If you're not designing it with the people who the technology is about, then you've already failed. It won't be human-centered. If the people these technologies are going to land on are not part of the process of conceiving, designing, testing, deploying, and iterating these technologies, if they're not part of that process all the way through, then you're very likely to end up with something that isn't human-centered.

Implementing a human-centered process is also about incentivizing the design, development, and deployment of AI, and its regulation, in a way where the incentives chosen represent this kind of broader interest of human flourishing. For example, if all I'm trying to do is get an AI system to be more accurate or to be faster at reaching a decision, if that's my only metric for success, then it's not a human-centered technology. Why? Because I haven't answered the question, "Faster, for what?" or "More accurate, so what?" or "What is the thing that speed or accuracy will enable?" If you can't tell a story that ends with greater human flourishing, if it's just about allowing for a more efficient extraction of value that flows to already powerful and wealthy people at the expense of millions, then speed and accuracy have no particular humane value in that context. They actually have negative value. They're actually allowing you to do something that *undermines* human flourishing faster and more efficiently than you were able to before. So, putting in place a human-centered process is about making sure that the end metrics of success are directly tied to reliable and rich measures of human flourishing.

**What kind of a team do you need to build human-centered AI?**

That's the really important piece here. If you're going to design technologies that have the totality of the human person at their center, then you need the knowledge behind those technologies—their design, development, and deployment—to represent that fullness. That means that you need not just computer scientists, ethicists, and philosophers. You need people with expertise in anthropology, sociology, psychology, history, media, and the arts, depending upon what you're doing.

It's not the case that every technology has to be designed by an entire university, but depending upon what you're doing and what aspects of the human experience an AI tool is going to affect, you have to be able to selectively draw upon the range of knowledge of the human experience, and the human person, that is relevant to that application in that context. To give another example, there's a lot of discussion about building robots to support care for the elderly in care homes.

There's a whole bucket of issues with that, but one of the first questions you would need to ask is, "What kind of knowledge is needed in order to do this well?" It definitely isn't just knowledge of machine learning, robotics, and healthcare regulations. You will need people who are experts in geriatric medicine, geriatric psychology, and the flourishing and well-being of the elderly. And you'll need to have people who have experience of what that means in the particular culture and region that you're expecting this robot to be deployed in. So, if I'm designing a tool that's going to be deployed in elder care homes in South Korea, I can't design it simply with people who have knowledge of the experience of elder care in Scandinavia or of what elderly people in Scandinavia generally value, or what forms of activity are most conducive to the happiness of people in that cultural context.

But we've actually seen concrete examples where this wasn't done—actually, in the case I'm aware of, the reverse was done: robots that were designed in South Korea were brought to Scandinavia to be tested in care homes. And the residents in these care homes reacted much more negatively to the affective design of these social robots. Whereas in South Korea the affective displays of the robot—the way it spoke, the way it moved, and the way it interacted—were seen as warm and charming, in this particular Scandinavian test environment, the robots were seen as childish, infantilizing, and annoying. And there had obviously been no sort of preparation for that different cultural set of norms and values in the design.

So, it's really important that when we do human-centered AI, we understand the "Who?" and the "Where?" and the "How?" of where this technology is going to be deployed. And then, we must identify the diverse forms of knowledge that are relevant to understanding what human flourishing looks like *there*, in *that context*. And then, we have to sort of reverse engineer human flourishing by going back and saying, "Alright, how do we reach that goal?" and "How do we build a technology that enables that flourishing more fully, or that makes it accessible to a wider range of individuals, or that makes it more sustainable?"

The usual process, though, is quite different. The usual process is: we design a technology, and then we try to find somewhere to make it work, and then we ask people in that context to reshape *themselves* to accommodate what we've built. This is something that has been identified by many scholars (Langdon Winner might have been the first person to talk about this). It's the idea of reverse adaptation.

Instead of adapting technologies to our needs, we adapt ourselves to technology's needs. The technology needs more data or the technology needs a safe physical envelope around it so that it doesn't get pushed over or the technology needs to be powered all the time. And then we have to reshape our context in order for the technology to function. Human-centered AI involves the reverse: it's about, "What are the human needs and opportunities that we want to enable?" and then "How do we adapt our technological capacity to meet those needs?"

**Do you think we'll be able to design this human-centered AI that we're talking about?**

In the long term, I am cautiously optimistic … but I'm less optimistic than I was five years ago. And the reason is, I think we have sort of normalized what was really an aberrant culture of engineering, initially.

I mean, the traditional models of engineering elaborated in the 20th century—prior to the software revolution—were not perfect by any means, but, certainly, they incorporated notions of ethics, responsible design, safety engineering, monitoring and auditing of performance, understanding the user, and user-centered design. These were part of any Technology 101 course!

But a lot of the lessons that traditional engineering disciplines learned over the centuries and formalized into professional codes of practice in the 20th century—a lot of those lessons were just thrown out the window, or just never learned, by software developers and computer scientists, people who grew up in a very different kind of culture.

What worries me, then, is how much, over the last ten years, we've just normalized that culture in such a way that people now talk about these aberrant practices as *the culture of engineering*. I taught engineering ethics for years before Facebook appeared. My mind explodes when someone tells me some of the things we see nowadays are just part of the culture of engineering. This is *not* the culture of engineering. This is a *departure* from the culture of engineering. We've forgotten that history. We've forgotten what human-centered engineering was; we've forgotten that baseline.

My point is that we have farther to go than we did ten years ago because we've had ten years of normalizing this aberration, something which is profoundly harmful to our democracy, the health of our information ecosystem, and human well-being and flourishing. We see that harm every day, and yet we shrug and say, "Well, what can we do? This is tech, this is engineering, this is how things are."

What I really think is we need to be more aggressive on things like regulation, which, fortunately, is moving forward now (I think that's where I get some optimism, as I think it's clear that there's a mandate to govern AI more effectively than we have governed social media platforms, for example). But that push for regulation won't be enough if we don't really get to the roots of engineering culture and realign it with the long historical mission of engineering and technology more broadly.

Technology began as a way to meet human needs—to comfort and cure, feed and shelter, and strengthen people and communities. That's what we use technology for. *Technology is humane at its heart*. And this idea that technology is that inhumane thing that doesn't care about humanity, that just develops on its own, and that we have to adapt to, that's a bizarre fiction that we have to change.

If we change that narrative, if we can recover the history and meaning of technology and engineering and creative practice and apply that to AI, then I think the sky's the limit. If we could realign AI with those norms and values, the possibilities would be inconceivably beneficial to the human family. That's where my optimism lies.

I'm not a determinist. I do not believe that technology develops independently of human will. If no one's steering the bus, it's because we took our hands off the wheel. All we have to do to regain control is put our hands back on the wheel and think about where we're going. And because that's still possible for us, and because that's in our interest, I remain cautiously optimistic we will wake up and begin to steer again.

# 2 Toward Addressing Inequality and Social Exclusion by Algorithms

## *Human-Centric AI through the Lens of Ubuntu*

*Malwina Anna Wójcik**
University of Bologna, Bologna, Italy
University of Luxembourg, Luxembourg City, Luxembourg

### INTRODUCTION

The perspective of human-centric AI (HCAI) encompasses the development of technologies that not only "augment, amplify, empower, and enhance human performance" (Shneiderman, 2022: 4) but also embrace human values. In its Ethics Guidelines for Trustworthy AI, the Independent High-Level Expert Group on AI set up by the European Commission (AI HLEG) underlines that the human-centric approach to AI should be rooted in "respect for human dignity, in which the human being enjoys a unique and inalienable moral status" (European Commission, 2019: 37). Human dignity is indeed considered the cornerstone of both the values on which the EU is founded[1] and the human rights enshrined in the European Charter of Fundamental Rights[2] (European Union, 1997; European Union, 2010). Thus, in EU law, dignity is a multilayered concept—a value, a fundamental right, as well as an ethical rationale for principles, rights, and obligations. In this sense, dignity constitutes "the bridge between the two disciplines of law and practical philosophy" (Frischhut, 2019: 130).

Dignity is closely tied with equality, another value enshrined in Art. 2 of the Treaty on European Union (TEU) (European Union, 1997). As underlined by the European Court of Justice, the equality of human beings stems from their inherent dignity—the concept often referred to as *égale dignité*.[3] Thus, the HCAI project requires an effort to achieve a fair and inclusive society, aiming to eradicate algorithmic discrimination and social exclusion. The AI HLEG guidelines list "equality, non-discrimination and solidarity" (European Commission, 2019: 11) among the fundamental rights that should constitute the basis of trustworthy AI, underlining the need to go beyond strictly construed nondiscrimination, to protect vulnerable groups

* PhD candidate in Law, Science and Technology, University of Bologna and University of Luxembourg.

DOI: 10.1201/9781003320791-4

from social exclusion and to "foster equality in the distribution of economic, social and political opportunity" (European Commission, 2019: 9). The proposed AI Act regulating the use of high-risk AI systems acknowledges that algorithmic harms "contradict Union values of respect for human dignity, freedom, equality, democracy and the rule of law and Union fundamental rights, including the right to non-discrimination, data protection and privacy and the rights of the child" (European Parliament and Council, 2021: Recital 15).

The rise of algorithmic decision-making (ADM) in key areas of public life, including the justice system, employment, healthcare, social benefits, and credit scoring, poses new challenges for the achievement of *égale dignité*. Firstly, historically marginalized communities do not play a meaningful role in the design and development of AI technologies. The lack of diversity in technological teams, as well as scarce mechanisms for the involvement of minority stakeholders in the heavily value-laden algorithmic design process, has led to the entrenchment of prevailing power narratives. Secondly, as widely documented, ADM systems tend to reflect and exacerbate existing societal biases (Wachter et al., 2021). The unavailability of data concerning minorities or their stereotypical, inaccurate representation in datasets leads to unlawful discrimination by algorithms. Thirdly, unequal deployment and availability of ADM systems widen power gaps. Algorithmic feedback loops result in vulnerable groups being subjected to excessive digital surveillance, for instance, through the use of predictive policing tools (EFRA, 2022). Additionally, the benefits of AI technologies are often withheld from them and concentrated in the hands of the privileged.

Sabelo Mhlambi, a scholar writing on the intersection of human rights, ethics, culture, and technology, argues that overcoming these challenges requires departing from the Western conception of personhood which underpins the development of AI (Mhlambi, 2020). From Aristotle who described humans as "rational animals" to Descartes' famous "I think therefore I am" declaration, the European philosophy grounds personhood in the idea of rationality, informing the quest for developing machines that mimic human intelligence. In this narrative, reason is construed as an individual quality, indispensable for making human beings "complete, ruler and autonomous." However, according to Mhlambi, the individualist conception of personhood suffers from inherent flaws, including justifying and worsening inequality, as evidenced by the historical weaponization of rationality as an exclusion tool. The application of the Enlightenment ideas of liberty, equality, and human rights was limited to those deemed rational enough to be considered a person, justifying the enslavement and subjugation of non-Europeans and the inferior treatment of women.

Thus, feminist philosophers were among the first to criticize rationality as the basis of human dignity, arguing for a relational concept of autonomy (Stoljar, 2013).

In a similar manner, Mhlambi proposes an alternative approach to AI ethics building on the South African philosophy of Ubuntu, which construes personhood and human dignity in relational terms, based on the premise that individual well-being can only be achieved by society as a whole flourishing (Mhlambi, 2020). As underlined by Tom Zwart, Ubuntu urges humanity to look beyond individual interest, requiring "cooperation, interdependence and collective responsibility" (Zwart, 2012: 555).

Thus, the Ubuntu perspective can shed new light on the challenges posed by ADM in the context of discrimination and social exclusion. The purpose of this chapter is to contribute to this effort by showing how Ubuntu can be operationalized as an ethical and legal principle in the context of HCAI, informing intercultural dialogue concerning the regulation of AI.

## UBUNTU AS AN ETHICAL AND LEGAL CONCEPT

Stemming from the Nguni philosophy, the tradition of Ubuntu constitutes a cornerstone of the sociopolitical identity of postapartheid South Africa, affecting the structure of its government, corporations, and organizations (Reviglio et Alunge, 2020). Moreover, the influence of Ubuntu extends beyond South Africa, constituting the basis of moral philosophy present in many sub-Saharan societies.

Elaborating a comprehensive definition of Ubuntu is a challenging, perhaps even impossible task. This is because its meaning is highly contextual and open-ended (Bohler-Muller, 2005). Chuma Himonga argues that as a metanorm Ubuntu is best conceived in abstract terms, with reference to its underlying attributes of "community, interdependence, dignity, solidarity, responsibility and ideal" (Himonga, 2013: 173). The former Justice of the South African Constitutional Court, Yvonne Mokgoro, describes Ubuntu as

> a philosophy of life, which in its most fundamental sense represents personhood, humanity, humaneness, and morality; a metaphor that describes group solidarity where such group solidarity is central to the survival of communities with a scarcity of resources, and the fundamental belief is that [...] a human being is a human being because of other human beings.
>
> (Mokgoro, 2012: 317)

This description sheds some light on Ubuntu's contribution to the debate about human-centric AI, underlining that, in essence, Ubuntu is preoccupied precisely with what it means to be human in the (digital) society. The very etymology of the word evokes the process of acquiring personhood—in isiZulu, one of the languages belonging to the Nguni family, the root "Ntu" indicates a person or people, while the prefix "ubu" signifies being or becoming (Mhlambi, 2020: 13). In this sense, personhood in Ubuntu is different from a mere biological distinctiveness of the human species, going "beyond such requirements as the presence of consciousness, memory, will, soul, rationality, or mental function" (Cornell, 2014: 160). As underlined by a prominent African philosopher, Ifeanyi Menkiti, the achievement of full personhood is not something individuals can attain on their own because it requires the involvement of the whole community (Menkiti, 2004).

The relational notion of personhood has implications on how social relationships are regulated by the law, whose essential function is to "secure human beings in their being" (Murungi, 2004: 525). Therefore, in a normative sense, although not explicitly articulated in the South African constitutional text, Ubuntu is considered a value that "permeates the Constitution generally and more particularly chapter three which embodies the entrenched fundamental rights."[4] According to the Constitutional Court of South Africa, Ubuntu "regulates the exercise of rights by the emphasis it lays on sharing and co-responsibility and the mutual enjoyment of rights by all."[5]

## THE CONCEPT OF *ÉGALE DIGNITÉ* THROUGH THE LENS OF UBUNTU—LESSONS FROM THE SOUTH AFRICAN CONSTITUTIONAL COURT

Ubuntu points toward a sense of kinship between the people (Ramose, 1999), and is inextricably linked to the value of human dignity, "with emphasis on the virtues of that dignity in social relationships and practices" (Mokgoro, 2012: 318). However, there are notable differences between the understanding of dignity in Ubuntu and European philosophical thought. This section illustrates how the South African Constitutional Court construes dignity through the lens of Ubuntu, shielding vulnerable groups from social exclusion.

### Social Exclusion as a Violation of Relational Dignity

While the father of the modern conception of human dignity, Immanuel Kant, associates dignity with the capacity for practical reason which allows human beings to exercise their autonomy, in Ubuntu dignity is not rooted in reason but rather in the singularity and uniqueness of human beings and their capacity to build harmonious relationships with others (Metz, 2020). As Tom Bennett contends, it is precisely the notion of being "embedded in the community" that makes Ubuntu different from the Western idea of dignity (Bennett, 2011). Thus, the spirit of Ubuntu is often expressed in the maxim: "I am, because we are and since we are, therefore I am" (Mbiti, 1990).

In the jurisprudence of the South African Constitutional Court, discrimination and social exclusion are perceived as a violation of Ubuntu and an attack on dignity because they prevent human beings from thriving as members of the community. Thus, in *Mahlangu*,[6] Justice Victor contended that the exclusion of domestic workers from the occupational injury compensation scheme infringed on their dignity through the devaluation of their work and, thus, their contribution to society. She explicitly argued that the marginalization of care workers goes against "the values of our newly constituted society namely human dignity, the achievement of equality and Ubuntu."[7] Similarly, in *Khosa*,[8] Justice Mokgoro opined that the exclusion of permanent residents from social security has a detrimental impact on their dignity, as they are "cast in the role of supplicants," whose survival is dependent on the goodwill of others.[9] In other words, the violation of their dignity is based on the fact that they are placed on the margins of society and deprived of a chance to engage in harmonious relationships. For Drucilla Cornell and Karin van Marle, the essence of Ubuntu in Justice Mokgoro's argument is found in her assertion that permanent residents deserve the same constitutional guarantees as citizens, because through their social engagements they have become a part of the South African society (Cornell et van Marle, 2022).

According to the Court in *Khosa*, the well-being of the poor is connected with the well-being of the other members of the community and the community as a whole. This approach illustrates a contrast between the Western notion of social contract and Ubuntu. The former construes social bonds as a result of an individual decision to concede one's natural liberty. The latter is based on the premise that one is in fact born into the social bond (Cornell, 2014). An important consequence of this reasoning is that by virtue of being human one acquires both rights and duties. Therefore, in Ubuntu, freedom and responsibility are interconnected, creating a bridge between the individual and the community (Cornell et van Marle, 2022). Belonging to a community, requires both identity as a group member and solidarity with the group, understood as a genuine will to help others for their sake (Metz, 2014). The exclusion of vulnerable groups is therefore incompatible with Ubuntu, which requires that they are treated with compassion and understanding.[10]

### Relational Autonomy—Balancing Individual and Group Rights through Meaningful Engagement

At the heart of Ubuntu lies the balance between the inalienable rights of individuals and the universal rights of the community. Kwame Gyekye attempts to explain their interdependence through an Akan proverb: "The clan is like a cluster of trees which, when seen from afar, appear huddled together, but which would be seen to stand individually when closely approached" (Gyekye, 1997: 40). Thus, while Ubuntu does not discard the notion of individual autonomy, it necessarily construes it in a social matrix. Under Ubuntu, autonomy is best conceived in its relational form, according to which a self-governing individual is "socially constituted and possibly defines her value commitments in terms of interpersonal relations and mutual dependencies" (Christman, 2004: 143). Thus, because personal rights are shaped by the rights of others, "strictly speaking, from the perspective of Ubuntu there can be no absolute individual rights" (Chuwa, 2014: 36). Therefore, the achievement of Ubuntu necessarily involves ongoing dialogue and the negotiation of the differences between the members of the community. Equality is the necessary prerequisite of this dialogue, which cannot be reduced to the conceptions of social cohesion or consensus, as it requires a deliberative effort aimed at understanding the position of others (Bohler-Müller, 2012).

Perhaps, the operation of relational autonomy is best illustrated by the Constitutional Court's judgment in *Port Elizabeth*.[11] This eviction case, brought by Port Elizabeth Municipality, involved balancing the right to private property and the unlawful occupants' right to protection against illegal eviction. The judgment, delivered by Justice Sachs, made explicit reference to Ubuntu, which the Court perceived to be "a part of the deep cultural heritage of the majority of the population" which "suffuses the whole constitutional order."[12] According to Justice Sachs, Ubuntu combines "individual right with communitarian philosophy" and supports the "need for human interdependence, respect and concern."[13] Thus, using Ubuntu as a foundational value behind the Constitution, Justice Sachs highlighted the need to remedy the injustices of the past and restore the dignity of the homeless. Reading the two conflicting rights together, the Court embraced relational autonomy, mitigating the individualistic character of the right to private property. In this way, balancing competing interests, the Court held that the eviction was unlawful. One of the arguments against the lawfulness of the eviction used in *Port Elizabeth* was the lack of any attempt on the municipality's side to consider the problems of squatters and engage in mediation with them.[14]

Building on this reasoning, the Constitutional Court in *Olivia Road*,[15] another eviction case, set a requirement according to which the authority seeking eviction must "meaningfully engage" with the occupants, in an attempt to resolve the dispute "in the light of the values of the Constitution, the constitutional and statutory duties of the municipality and the rights and duties of the citizens concerned."[16] This formulation reflects the spirit of Ubuntu, which fosters dialogue, reconciliation, and cooperation between human beings, acknowledging both their rights and duties. Meaningful engagement, as a transformative tool designed to give voice to the vulnerable, is broader than a simple duty to hear the other party, as it requires genuine participation of civil society (Liebenberg, 2012).

## BEING HUMAN IN A (DIGITAL) SOCIETY—UBUNTU'S CONTRIBUTION TO THE HCAI DEBATE

The communitarian view of human dignity offered by Ubuntu could bring a new perspective to the HCAI debate, framing the questions of algorithmic bias and digital exclusion as a primarily relational problem. Thus, building on the findings of the previous section, this section explores the implications of Ubuntu for the AI fairness research community, AI programmers and computer scientists, and the interaction between AI systems and their users.

### The AI Fairness Research Community

The relational approach of Ubuntu leads to a departure from the narrow definition of algorithmic bias as a "bug in the system" that can be permanently fixed, favoring a more contextual understanding of the problem, and exploring why certain patterns of algorithmic discrimination emerge. This involves uncovering asymmetrical power dynamics and structural inequalities that shape our digital society, as well as questioning "assumptions regarding knowledge, justice, and technology itself" (Birhane, 2021: 8). Therefore, instead of focusing solely on rectifying the symptoms of algorithmic exclusion, for instance, by resampling imbalanced datasets, the research community should aim to understand its causes, inquiring about how data are generated and collected. To foster this aim, researchers should engage in interdisciplinary and collaborative projects on the intersection of computer science, ethics, law, and social sciences. Moreover, promoting diversity in the research community, with a specific focus on vulnerable and underrepresented groups, is indispensable for understanding the causes of algorithmic bias. An example of good practice in this field is provided by Queer in AI, an initiative created with the purpose of remedying harms stemming from the underrepresentation of queer persons in the AI community (Organizers of Queer in AI et al., 2023). The key principles of the organization comprise decentralized structure, community-led initiatives, and intersectionality. Queer in AI encourages the broad participation of its members in discussion and program planning, bringing to light lived experiences of exclusion based on intersecting identities, including ethnicity, gender, class, disability, or caste. These experiences are carved into concrete steps aimed at increasing the representation of queer persons, such as graduate school application financial aid scheme, inclusive conference guides, workshops and socials, and trans-inclusive publishing advocacy.

Moreover, the perspective of Ubuntu could contribute to the debate surrounding competing notions of fairness. The conflict between group fairness and individual fairness metrics permeates machine learning literature (Kleinberg et al., 2016; Zehlike et al., 2022). The core tenet of group fairness is equality of opportunity. However, group fairness constraints can lead to two individuals of seemingly identical characteristics being classified differently, breaching individual fairness, which requires that similar cases should be treated alike. While much of the AI fairness literature focuses on how to balance these inherent trade-offs, some researchers view the dichotomy between individual and group fairness as artificial, because at the abstract level the two notions "are in fact different aspects of the same consistent set of fundamental moral and political concerns" (Binns, 2019: 517). Ubuntu aligns with this criticism. The notion of relational personhood implies that individual fairness cannot be construed in isolation from the group fairness considerations, because individuals share possible patterns of privilege or disadvantage with members of the group they belong to. To argue otherwise would mean to refuse to view individual experiences in their entirety. As put by Murungi, "to discount what one has in common with other human beings is to discount oneself as a human being" (Murungi, 2004: 525). Therefore, operationalizing fairness should not be constrained by the individual vs. group dichotomy. Rather, as argued by Binns, the starting point for choosing fairness metrics should be mapping the probable kinds of injustices that might have an impact on the outcome of the automated decision in a given context (Binns, 2019).

## AI Programmers and Computer Scientists

Freedom from negative biases is accounted for in value-sensitive design frameworks for automated systems (Friedman et al., 2013). According to Batya Friedman and others, value-sensitive design should encompass a tripartite methodology: conceptual investigations, empirical investigations, and technical investigations. Ubuntu impacts each of these elements (Friedman et al., 2013).

The conceptual investigations' aim is to determine the relevant stakeholders and values, and to explore how they are affected by the technology and how competing values should be balanced (Friedman et al., 2013). At this stage, Ubuntu's contribution is to underline the role of indirect stakeholders, who, although not directly interacting with the system, can be severely affected by it. Thus, the implementation of ADM technologies should always be accompanied by an assessment of their main beneficiaries, ensuring that vulnerable populations do not suffer because of an unequal distribution of the burdens and benefits of the technology. A solution for algorithmic bias should therefore be centered around those that are disproportionately affected by the problem, underscoring the importance of lived experiences of discrimination (Birhane, 2021). Thus, the HCAI perspective, imagined through the lens of Ubuntu, requires new technologies not to be deployed based on utilitarian arguments of benefiting the majority. Instead, augmenting human dignity requires their effect to be tested against the experiences of those most vulnerable.

The empirical investigations' aim is to determine how stakeholders apprehend and prioritize the values in question (Friedman et al., 2013). As echoed in the jurisprudence of the South African Constitutional Court, Ubuntu plays an important role in reaching a trade-off between values, requiring that primacy be given to protecting the most vulnerable and disadvantaged. For instance, the judgments in *Port Elizabeth* and *Olivia Road* underlined that the right to property cannot be perceived as an absolute right, as it has to be balanced against the rights of those living in poverty. Thus, it can be argued that in the tradition of Ubuntu, value-sensitive design should prioritize the rights of the community's most vulnerable members. In this sense, the societal ideal of Ubuntu evokes restorative justice, demanding that the technology contributes to the achievement of a more just and equitable society, rectifying historical patterns of exclusion.

Finally, technical investigations' purpose is to ascertain what values the technology fosters and how it can be redesigned to support values identified and conceptualized in the preceding investigations (Friedman et al., 2013). Discussing the application of Ubuntu tradition to algorithmic design, the IEEE Guidelines on the ethically aligned design of autonomous and intelligent systems highlight the impact of AI on the value of community, the central tenet of Ubuntu (The IEEE Global Initiative on Ethics of Autonomous and Intelligent Systems: 55-56, n.d.). The guidelines acknowledge AI's potential for both community-disruptive and supplementing effects, recommending designers and programmers to "work closely with the end-users and target communities to ensure their design objectives, products, and services are aligned" with the needs (The IEEE Global Initiative on Ethics of Autonomous and Intelligent Systems: 56, n.d.). Thus, Ubuntu mandates inclusion in the AI discourse. This encompasses fostering participatory design processes for ADM technologies. The relational ethics of Ubuntu requires that civil society stakeholders are directly involved in making value-laden decisions concerning the design of algorithmic technologies.

The doctrine of meaningful engagement developed by the South African Constitutional Court could serve as a model for community-based decision-making about AI technologies. As underscored in the Court's jurisprudence, meaningful engagement is not limited to participation rights but involves a genuine dialogue between the stakeholders. In other words, it entails a deliberative effort aimed at the negotiation of differences in the design process. As stated by the Constitutional Court in *Olivia Road*, the process should encourage a proactive approach, as opposed to treating the vulnerable communities as a "disempowered mass."[17] Thus, the elimination of significant power imbalances between the ADM developers and the target communities is a prerequisite for a successful dialogue. For instance, the intellectual property rights of ADM developers should not be used to justify algorithmic opacity and hinder algorithmic audits by civil society. Moreover, meaningful engagement in ADM system design should involve appropriate education and training for stakeholders without technical experience to allow meaningful deliberation.

The case of the Māori indigenous community's data governance in New Zealand provides an interesting example of how prioritization of historically underserved communities' values in ADM participatory design can lead to rectifying power asymmetries. In 2018, Te Hiku Media, a small nonprofit radio station, initiated a project inviting the Māori community to record and annotate audio data in their native language (Birhane et al., 2022). Notably, the community decided that the data, which was subsequently used to create natural language processing tools, should be shielded from commercial use by Western corporations. Thus, data sovereignty frameworks were put in place to ensure that the technologies developed with the use of community-sourced data directly benefit the Māori population and respect its right to self-determination. Similarly, participatory efforts have been made to operationalize Māori values in data collection by the government (Sporle et al., 2020).

## The Interaction between AI Systems and Their Users

The ethics of Ubuntu affects the relationship between the AI systems and their users, having implications for fair ADM practices.

First and foremost, Ubuntu stands in opposition to a technology-centric perspective that foresees that artificial intelligence can outperform human intelligence and possibly replace human decision-making. Dorine Eva Van Norren argues that Ubuntu rejects the assumption that the mind is the sole locus of intelligence. Instead, intelligence in Ubuntu is also found in intuition and "the heart," that is, the ability to "feel for" and sympathize with others (van Norren, 2022: 112). Thus, even if AI is able to mimic the human mind, it will never replace human intelligence, as it is incapable of empathy and compassion. Therefore, ADM systems should not be deployed without meaningful human oversight.

In terms of guidelines on how the interaction between the human and the machine should be shaped, Ubuntu's ideal of social personhood points toward the idea of hybrid collective intelligence. Collective intelligence can be understood as the human ability to "connect in a way that allows us to collectively act more intelligently than any individual person" (Peeters et al., 2021: 218). It is a core element of Ubuntu which believes that the community, including the ancestors and the future generations, is the true locus of intelligence. In recent years, collective intelligence has attracted growing interest from the AI research community and developers who explore how the paradigm could be applied to the interaction between the human and AI agent, leading to the development of hybrid collective intelligence. As underlined by Justice Sachs in *Port Elizabeth*, the spirit of Ubuntu reminds the people that they are not "islands onto themselves,"[18] and neither is AI, as no autonomous system operates in complete isolation from society (Johnson et Vera, 2019). Thus, the future of intelligence lies in the interaction between the collective of humans and machines, each retaining its distinctive features. As argued by Peeters and others, the perspective of collective hybrid intelligence is particularly well suited to analyze properties that emerge on the societal level, including equality and fairness (Peeters et al., 2021). It also helps to ensure that "misunderstanding or misalignment between stakeholder groups" does not lead to injustice and discrimination at the collective level (Peeters et al., 2021: 233).

## CONCLUSION—TOWARD INTERCULTURAL DIALOGUE FOR HCAI

If stimulating a truly global discussion on trustworthy and ethical AI is to be taken seriously, the EU's international policy initiatives, such as the International Outreach for Human-Centric Artificial Intelligence (European Commission, 2021), should involve intercultural dialogue with non-Western traditions. Such dialogue is crucial to eliminate power imbalances and avoid what Charles Ess calls "imperialistic homogenisation" of the ethics of information (Ess, 2006). He argues that the perspective of ethical pluralism, while acknowledging irreducible differences, can help to stimulate engagement across them in order to enrich both sides of the discussion. The cross-cultural dialogue between the European ethical and legal tradition and Ubuntu could lead to finding points of convergence between the two cultures, compatible with their respective identities. For instance, some parallels could also be drawn between the communitarian and relational perspectives of Ubuntu and the EU value of solidarity, which entails an equal share of burdens and benefits. In the HCAI framework, solidarity implies an assessment of the long-term implications of AI systems, making sure that their burdens and benefits are equally distributed and that they do not entrench inequality (Luengo-Oroz, 2019). Regrettably, while the draft AI Act evokes other Art. 2 TEU values in its recitals, it does not mention solidarity at all (European Parliament and Council, 2021). A dialogue with the HCAI perspective offered by Ubuntu could lead to revisiting the EU value of solidarity and engaging it in the ethical and regulatory AI framework.

## NOTES

1 Art. 2 of the Treaty on European Union (TEU) provides: "The Union is founded on the values of respect for human dignity, freedom, democracy, equality, the rule of law and respect for human rights, including the rights of persons belonging to minorities. These values are common to the Member States in a society in which pluralism, non-discrimination, tolerance, justice, solidarity and equality between women and men prevail."
2 Art. 1 of the Charter of Fundamental Rights (CFR) provides: "Human dignity is inviolable. It must be respected and protected."
3 Case C-36/02 *Omega Spielhallen- und Automatenaufstellungs-GmbH v Oberbürgermeisterin der Bundesstadt Bonn* ECLI:EU:C:2004:162, para 80.
4 *S v Makwanyane* 1995 (3) SA 391 (CC), para 237.
5 Ibid, para 224.
6 *Mahlangu and Another v Minister of Labour and Others* 2021 (2) SA 54 (CC).
7 Ibid, para 65.
8 *Khosa and Others v Minister of Social Development and Others* 2004 (6) SA 505 (CC).
9 Ibid, para 76.
10 *Hoffmann v. South African Airways* 2000 (11) BCLR 1211 (CC), para 38.
11 *Port Elizabeth Municipality v Various Occupiers* 2005 (1) SA 217 (CC).
12 Ibid, para 37.
13 Ibid.
14 Ibid.
15 *Occupiers of 51 Olivia Road, Berea Township, and 197 Main Street, Johannesburg v City of Johannesburg* 2008 (3) SA 208 (CC).
16 Ibid, para 16.
17 *Olivia Road*, para 20.
18 *Port Elizabeth*, para 37.

## REFERENCES

Bennett, T. (2011). Ubuntu: An African Equity. *Potchefstroom Electronic Law Journal, 14*, 30.

Binns, R. (2019). On the apparent conflict between individual and group fairness. In Conference on Fairness, Accountability, and Transparency (FAT*'20), January 27–30, 2020, 514-524. New York, NY: Association for Computer Machinery. https://doi.org/10.1145/3351095.3372864.

Birhane, A. (2021). Algorithmic injustice: A relational ethics approach. *Patterns*, *2*(2), 100205.

Birhane, A., et al. (2022). Power to the People? Opportunities and Challenges for Participatory AI. In *Equity and Access in Algorithms, Mechanisms, and Optimization (EAAMO '22)*, Article 6, 1–8. New York, NY: Association for Computing Machinery, https://doi.org/10.1145/3551624.3555290

Bohler-Muller, N. (2005). The Story of an African Value: Focus: Ten Years after Makwanyane. *SA Public Law, 20*, 266.

Bohler-Müller, N. (2012). Some Thoughts on the Ubuntu Jurisprudence of the Constitutional Court. In D. Cornell & N. Muvangua (Eds.), *UBuntu and the Law: African Ideals and Post-Apartheid Jurisprudence* (pp. 367–376). Fordham University Press.

Christman, J. (2004). Relational autonomy, liberal individualism, and the social constitution of selves. *Philosophical Studies: An International Journal for Philosophy in the Analytic Tradition, 177*, 143.

Chuwa, L. T. (2014). *African indigenous ethics in global bioethics: Interpreting Ubuntu*. Springer.

Cornell, D. (2014). Is There a Difference That Makes a Difference between Dignity and Ubuntu? In *Law and revolution in South Africa: UBuntu, Dignity, and the struggle for constitutional transformation* (pp. 149–168). Fordham University Press.

Cornell, D., & van Marle, K. (2022). Exploring UBuntu: Tentative Reflections. In *Exploring UBuntu: Tentative Reflections* (pp. 344–366). Fordham University Press.

EFRA. (2022). *Bias in Algorithms: Artificial Intelligence and Discrimination*. LU: Publications Office.

Ess, C. (2006). Ethical Pluralism and Global Information Ethics. *Ethics and Information Technology, 8*, 215.

European Commission, Directorate-General for Communications Networks, Content and Technology. (2019). *Ethics guidelines for trustworthy AI*, Publications Office. https://data.europa.eu/doi/10.2759/346720

European Commission. (2021). *International Outreach for Human-Centric Artificial Intelligence Initiative|Shaping Europe's Digital Future*. Retrieved January 25, 2023, from https://digital-strategy.ec.europa.eu/en/policies/international-outreach-ai

European Court of Justice. (2004). Case C-36/02, *Omega Spielhallen- und Automatenaufstellungs-GmbH v Oberbürgermeisterin der Bundesstadt Bonn*. ECLI: EU:C:2004:162, para 80.

European Parliament and Council. (2021). Proposal for a Regulation of the European Parliament and the Council Laying Down Harmonised Rules on Artificial Intelligence (Artificial Intelligence Act) and Amending Certain Union Acts, COM/2021/206 final, Recital 15.

European Union. (1997). *Consolidated versions of the treaty on European Union and of the treaty establishing the European community*. Office for Official Publications of the European Communities.

European Union. (2010). Charter of Fundamental Rights of the European Union. *Official Journal of the European Union C83, 53*. European Union.

Friedman, B., et al. (2013). Value Sensitive Design and Information Systems. In N. Doorn et al. (Eds.), *Early Engagement and New Technologies: Opening up the Laboratory* (pp. 55–95). Springer.

Frischhut, M. (2019). *The Ethical Spirit of EU Law*. Springer International Publishing.

Gyekye, K. (1997). *Tradition and Modernity: Philosophical Reflections on the African Experience*. Oxford University Press.

Himonga, C. (2013). The Right to Health in an African Cultural Context: The Role of Ubuntu in the Realization of the Right to Health with Special Reference to South Africa. *Journal of African Law*, *57*, 165–195.

Hoffmann v. South African Airways, 2000 (11) BCLR 1211 (CC).

The IEEE Global Initiative on Ethics of Autonomous and Intelligent Systems. (n.d.). *Classical Ethics in A/IS*. Retrieved from https://standards.ieee.org/wp-content/uploads/import/documents/other/ead1e_classical_ethics.pdf (Accessed May 25, 2023).

Johnson, M., & Vera, A. (2019). No AI Is an Island: The Case for Teaming Intelligence. *AI Magazine*, *40*(1), 16–28.

Khosa and Others v Minister of Social Development and Others, 2004 (6) SA 505 (CC).

Kleinberg, J., Mullainathan, S., & Raghavan, M. (2016). Inherent trade-offs in the fair determination of risk scores. *arXiv:1609.05807*.

Liebenberg, S. (2012). Engaging the Paradoxes of the Universal and Particular in Human Rights Adjudication: The Possibilities and Pitfalls of "Meaningful Engagement." *African Human Rights Law Journal*, *12*, 1.

Luengo-Oroz, M. (2019). Solidarity Should Be a Core Ethical Principle of AI. *Nature Machine Intelligence*, *1*, 494.

Mahlangu and Another v Minister of Labour and Others, 2021 (2) SA 54 (CC).

Makwanyane, S. V. 1995 (3) SA 391 (CC).

Mbiti, J. S. (1990). *African Religions & Philosophy*. Heinemann.

Menkiti, I. (2004). On the Normative Conception of a Person. In K. Wiredu (Ed.), *A Companion to African Philosophy* (pp. 324–331). Blackwell.

Metz, T. (2014). Dignity in the Ubuntu Tradition. In D. Mieth et al. (Eds.), *The Cambridge Handbook of Human Dignity: Interdisciplinary Perspectives* (pp. 310–318). Cambridge University Press.

Metz, T. (2020). Human Dignity, Capital Punishment, and an African Moral Theory: Toward a New Philosophy of Human Rights. *Journal of Human Rights*, *9*, 81–99.

Mhlambi, S. (2020). *From Rationality to Relationality: Ubuntu as an Ethical and Human Rights Framework for Artificial Intelligence Governance*. Carr Center Discussion Paper Series, no. 2020–009.

Mokgoro, Y. (2012). UBuntu and the Law in South Africa. In D. Cornell & N. Muvangua (Eds.), *UBuntu and the Law: African Ideals and Post-Apartheid Jurisprudence* (pp. 317–323). Fordham University Press.

Murungi, J. (2004). The Question of an African Jurisprudence: Some Hermeneutic Reflections. In K. Wiredu (Ed.), *A Companion to African Philosophy* (pp. 519–526). Blackwell.

Occupiers of 51 Olivia Road, Berea Township, and 197 Main Street, Johannesburg v City of Johannesburg, 2008 (3) SA 208 (CC).

Organizers of Queer in AI et al. (2023). Queer In AI: A Case Study in Community-Led Participatory AI. In *Proceedings of the 2023 ACM Conference on Fairness, Accountability, and Transparency (FAccT '23)* (pp. 1882–1895). Association for Computing Machinery, New York, NY. https://doi.org/10.1145/3593013.3594134

Peeters, M. M. M., et al. (2021). Hybrid Collective Intelligence in a Human–AI Society. *AI & Society*, *36*, 218.

Port Elizabeth Municipality v Various Occupiers, 2005 (1) SA 217 (CC).

Ramose, M. B. (1999). *African Philosophy Through Ubuntu*. Mond Books.

Reviglio, U., & Alunge, R. (2020). "I Am Datafied Because We Are Datafied": An Ubuntu Perspective on (Relational) Privacy. *Philosophy & Technology*, *33*, 595–612.

Shneiderman, B. (2022). *Human-Centered AI*. Oxford University Press.

Sporle, A., Hudson, M., & West, K. (2020). Indigenous data and policy in Aotearoa New Zealand. In M. Walter et al. (Eds.), *Indigenous Data Sovereignty and Policy* (pp. 62–80). Routledge.

Stoljar, N. (2013). Feminist Perspectives on Autonomy. In *Stanford Encyclopedia of Philosophy*. Retrieved May 23, 2023, from https://plato.stanford.edu/entries/feminism-autonomy/

van Norren, D. E. (2022). The Ethics of Artificial Intelligence, UNESCO and the African Ubuntu Perspective. *Journal of Information, Communication and Ethics in Society*, *21*, 112.

Wachter, S., Mittelstadt, B., & Russell, C. (2021). *Bias Preservation in Machine Learning: The Legality of Fairness Metrics Under EU Non-Discrimination Law*. 123 W. Va. L. Rev. 735.

Zehlike, M., et al. (2022). Beyond Incompatibility: Interpolation between Mutually Exclusive Fairness Criteria in Classification Problems. arXiv:2212.00469.

Zwart, T. (2012). Using Local Culture to Further the Implementation of International Human Rights: The Receptor Approach. *Human Rights Quarterly*, *34*, 546–569.

# 3 Redefining Human-Centered AI

## *The Human Impact of AI-Based Recommendation Engines*

*Oshri Bar-Gil*

Behavioral Science Research Institute, Israel

## INTRODUCTION

In recent years, the stream of information has grown, and we are constantly bombarded with nonstop information for our day-to-day decision-making. As the digital culture advocate Kevin Kelly says in his book, *The Inevitable* (2016), the fastest-increasing quantity on this planet is the amount of information we are generating. This information is not only present on the servers of tech companies but also surrounds us in every direction. Whether it's the news, social media feeds, or the agenda and timing of our meetings, there is just too much information that inundates us and urges our attention (Andrejevic, 2013).

Evans et al. (2017, p. 36) defined affordances as possibilities for action between an object/technology and the user that enables or constrains potential behavioral outcomes in a particular context. Some social-technological affordances, notably machine learning (ML) algorithms and big-data analytics, promise to enhance our use of information and "help" us in information processing and decision-making to make life "faster," less stressful, and "frictionless" (Rosa, 2013; Virilio, 2005). Typically, these affordances are organized in services offered by platforms incorporating ML algorithms applied to the user's digital data-doppelgänger as curated and preserved carefully by the platforms themselves. The platforms use it to recommend, and potentially influence, the user's future action by matching ads, directly or indirectly, to support better profiling of users (Zuboff, 2018). These recommendations can be about the information that will be most appropriate to our search, the next news piece we would like to read, and include other fields such as music, commerce, and so on (Karakayali et al., 2018).

This chapter is based on a large research project that used user reviews of Google products and services to conduct a phenomenological inquiry on changes to self and self-concept as a consequence of utilizing platformized services (Bar-Gil, 2021). The platform selected for the research was Google and its multitude of personalized


DOI: 10.1201/9781003320791-5

AI-based services that function as recommendation engines which, over the last decade, have become our personal thinking assistants.

Current efforts to create personalized AI recommendation engine services use personal data accumulated in assorted services as our data doppelgänger and as a prototype for our information processing and decision-making. These augmented self-services could be seen as our contemporary superpower to handle information overload—a partner for discussions, thoughts, actions, and decisions.

We allow search engines to remember the links and pathways to online content, the right spelling for words, and the queries we wouldn't dare to ask anybody else (Stephens-Davidowitz, 2017). However, Google is more than just a search engine; among its AI-based services is an email service that recommends automatic yet personalized responses and filters the mail information stream in order to direct our attention and decisions (Bullock, 2017). Google's recommendation engines also suggest the next video on YouTube (Airoldi et al., 2016), filter the restaurants that "might be a good match for us" (Duong, 2017), determine the news that fits our interests (Nayak, 2020), and more. We use Google's Maps to recommend which route is currently the best one (Wang, 2021), and its integration with Google's calendar recommends what time to set off, considering traffic congestion, and reminds us of our grocery list just as we pass by the shopping center (Umapathy, 2015).

The average Google user communicates with Google's servers over 100,000 times daily (Hill, 2019).[1] This implies our close connection with them and the tight coupling between us and our cloud selves. Is such tight coupling enabling us to preserve our human control, autonomy, or agency in these processes? As the *Wired* editor and technology researcher Kevin Kelly (2016, p. 127) asked pointedly, where does my "I" end and the cloud start?

While the social effects of Google as a company (Alphabet), and as a service, have been the topic of numerous books and studies, the question of how it affects us, the individual human users of its services, has not been addressed fully. It is a critical question if we would like to make a Human-Centered AI (HCAI) that not only does our tasks but helps us to flourish and realize deeper human goals. In this chapter, we will dive into this topic, exploring the human impact of using Google's recommendation engines in their platformized application, as a foundation for comprehending how AI affects our lives in the hyper-technological world of the information age.

The chapter will start by describing the contemporary information environment and then describe different technological efforts to adapt to this environment, which include the use of external "cognitive technologies" such as recommendation engines and detail their potential impact on the user. The potential impact of those cognitive technologies will be explained and illustrated by adding a recommendation engine user figure to a classic extended cognition thought experiment and exploring its potential effects on behavior, cognition, and decision-making. From there, the chapter will elaborate on the possible changes in intention, rationality, and memory patterns to form the discussion about ways to make AI more human-centered, considering its effect on us as humans.

## NEW OPPORTUNITIES

Today's technologically saturated environment has been given a variety of names, such as the fourth industrial revolution (Schwab, 2017), the second machine age (Brynjolfsson & McAfee, 2016), or the late information age (Fuchs, 2008). They all point to a widespread increase in information transfer with high-speed communication networks, increased storage capabilities, and advanced algorithms with powerful computing capabilities (McAfee & Brynjolfsson, 2017). These affordances offer new opportunities for users of digital services, but they also create pressure to make decisions more frequently and rapidly. According to German philosopher Hartmut Rosa (2013), as the rate of technological development increases, humans enter a cycle of acceleration. Technological acceleration causes a hastening of lifestyle, which requires new technologies to deal with that acceleration, and so on.

According to French philosopher Bernard Stiegler (1998), since the beginning of history, humans have used various artificial aids, or prostheses, to deal with the inherent limitations of human cognition. The French phenomenologist Maurice Merleau-Ponty (1945/2002) proposed that external expansion can sometimes be realized on a physical or perceptual level. A blind man's cane, for example, demonstrates how the cane, a technological object, serves as an extension of his body's sensory organs, providing an additional sense for the blind man. Other researchers, from the cognitive sciences field, suggest viewing various technological aids as an extension of the self beyond the actual body (Menary, 2010). Following them, Don Ihde (2009), the American philosopher of technology, proposed that like the blind man's cane, technology not only allows us a different sensory perception of the environment but also changes our perception of the world, including how we think about ourselves. The microscope is an example he frequently used. Aside from allowing us to see tiny particles, the microscope made us realize that we are made up of those particles, cells, bacteria, and more. Following those thinkers and others, we can see that AI, as a technology, might serve to augment our cognition but also change our perceptions of ourselves and the world.

Currently, HCAI is a concept mainly used in reference to realizing human and fundamental rights (Sigfrids et al., 2023). However, the potential influences described above call for a broadening of the current definitions of what HCAI is. For example, Shneiderman (2022) defines it as using AI to "serve the collective needs of humanity" by understanding "human language, feelings, intentions and behaviors." Others, such as Wang et al. (2021, p. 1) define it as designing and implementing "AI techniques to support various human tasks, while taking human needs into consideration and preserving human control."

### AI AS AN EXTENDED COGNITIVE AID

The philosophers of consciousness, Andy Clark and David Chalmers (1998), developed the "extended mind" hypothesis. They argued we cannot simply point to the boundary of our skin and skull as a limit for our mind or the self. They presented a thought experiment meant to examine how one's cognition can be extended to assess the validity of their idea. Let us imagine two characters, Otto and Inga, who wish to visit the museum. Inga can find her way from looking at a map and memorizing her path. Otto has Alzheimer's and so must write down the directions on paper and then follow them to reach his destination. This thought experiment ends happily, with both characters arriving at the museum. Analyzing their thinking process reveals Inga relied only on her "biological" cognition to remember her planned route and make judgments and decisions about which roads to take. In comparison, Otto's cognition was "extended." His journey to the museum involved the use of an external "memory technology," a sheet of paper, to aid in his deliberation and decision-making on the turns along the way. Clark and Chalmers argue that Otto's use of a pencil and paper to replace his biological memory demonstrates that humans are capable of thinking with, and through, external technologies. According to their findings, the same cognitive decision-making process underlies both sets of outcomes, even if different components of the mind (even external) participated in the process. For the sake of this chapter's central theme, which is to understand the effect of using recommendation engines on us as humans, let's add Nadia as a character to this thought experiment. Nadia relies on the Google Maps application to get her to the museum. When Nadia puts the museum's name into the app, it will show her the institution's location, offer multiple routes, and prompt her to pick one, maybe while subtly suggesting the optimal route. Then it will use her phone's GPS to lead the way, make path corrections as they are needed, and even think ahead of potential "better" routes.

What is the difference between Otto's usage of memory extension and Nadia's decision-making based on Google Maps? First, Otto and Inga had to put in varying amounts of focused attention and intentional cognitive effort while navigating and orienting themselves. It would appear that Nadia has an easier task ahead of her in terms of attention and intention. It's not hard to picture her strolling along, with her mind focused on something other than finding her way or orienting herself—maybe even browsing social media. As Google's recommendations for when and where to turn mediate her trip, she might lose some degree of agency or autonomy.

The most significant loss of autonomy can be explained by the change in intention, rationality, and memory. Inga deliberated on all her decisions, first to go to the museum and then on each subsequent decision (to take the detour, turn right, and go straight). Otto was content with his preconceived intention to arrive at the museum by the path he had picked and written down ahead of time, and he kept to it the entire way. Nadia, on the other hand, had to "synchronize" the app with her prior intention to visit the museum, but after that she could delegate the navigational decisions, or action intentions, to the app and follow its directions—take the advised route and turn at the appropriate moment. Moreover, if Nadia is an advanced user of the app, it may have notified her she needed to leave early due to traffic on the road or a longer-than-usual line at the museum. It is even possible that when Nadia established her prior intention to visit a specific museum, Google presented her with alternative museums and even suggested that she modify her intention since other museums had a higher rating or were more suitable to her former interests. This further undermines her intentional decisions.

Another distinction may be the level of rationality assigned to Nadia's behavior. According to Google Maps' algorithms, the path determined for her is the most accurate one regarding her prior experiences, and preferences—saving distance, walking time, or even avoiding stair climbing. These types of optimized decisions, based on algorithmic planning which incorporates near-real-time situational awareness, may be regarded as more "rational" by current social standards (Fisher, 2020), and maybe by Nadia herself.

The last distinction that we shall consider for the time being is about memory—Inga used only her biological memory; Otto utilized a sheet of paper as an external memory. Nadia, on the other hand, relied on a different type of external memory, Google Maps' spatial "memory," to determine which routes and pathways she should take. The path she selected to walk was extensively examined in real-time by Google, using the GPS on her cellphone, with an accuracy of less than one meter (Milner, 2016). It was added to both her personal digital data in Google's archive and all the route choices of all the individuals who have walked in this region. Nadia will actively contribute to Google users' shared, connective memory, the same memory that will guide others' actions as it guided hers.

The example in this thought experiment shows how a spatial recommendation engine might affect its users by allowing them to delegate parts of their decision-making processes. However, these changes occur not only in the spatial domain but in a wide range of judgments and decision-making that use platformized recommendation engines in similar ways. To provide these personalized services to users, Google and other technology platforms create a "data self" about the user that aims to represent the user in the platform through a massive amount of data (Van Dijck et al., 2018; Zuboff, 2018). For example, my personal data file in Google services contains over 100 gigabytes of data.[2] If this information were printed on a stack of paper, it would be taller than the Burj Khalifa, the world's tallest building. This representation, combined with the platforms' algorithms, enables them to make decisions for and about the users. For example, which route to take to the museum or which restaurant to recommend in the Google Maps application, given all the places the user has visited and enjoyed in the past (Lardinois, 2018), which news item will pique his interest most at the time and place he enters the application (Newton, 2017), which automatically generate responses for emails to suggest (Deahl, 2018) and more.

This causes a snowball effect when it comes to delegating more tasks over information-related judgments, decisions, and interactions. Increased reliance on technology and the development of self-representation-databased service platforms have altered not only how we acquire knowledge but also how we evaluate it, choose among options, and process it in our daily lives. The ability of a person to act in the real or virtual world is increasingly mediated, almost controlled, by such services. Consider how much of the decision to select a route is yours or algorithmic. How much more challenging it would be to perform a wide range of tasks such as finding information, getting around, and more if Google services weren't around! Frequently, someone who was late to a meeting blames his delay on the fact that "the navigation app told him he would be on time,"[3] or someone who did something because "it was written on Google," treating these sources as infallible imperatives or predictions. Using those self-databased recommendation engines follows a predictable cycle: first, we create our data doppelgänger using Google or the other platforms, then it gets data from our service usage, then it uses the data and preferences we provide to improve its own performance; finally, it provides us with even more utility in a variety of ways, so we keep on and continue using it. These efficient and beneficial recommendation engines have a hidden cost—their influence on us as humans. To paraphrase Winston Churchill, first we shape our profiles; thereafter, they shape us.[4] Instead of making us more autonomous, independent, and free, the recommendation engines may make us heteronomous (Castoriadis, 2011), conforming to the rules of others enforced upon us algorithmically—not just any other, but sophisticated technological corporations.

## INTENTION

Simplifying the philosophical debate about autonomy, one can consider it as an individual's capacity for self-determination or self-governance (Pham et al., 2021). From this view, agency is the ability to choose for ourselves, without any undue influence from others, what is right or wrong for us. Intention refers to the purpose or aim behind an action. In this context, intention can be seen as closely related to agency and autonomy because it reflects the individual's ability to make choices and act on them (Smith, 2017). Ihde (2009), mentioned above, claims that technology does more than just help us fulfill our plans or the way we do things. It also influences the very process by which these plans are conceived—our intention. Thus, it diminishes the autonomy and agency of its users, as seen from the thought experiment. Another example might be when a user starts a search for information about a particular concept but is redirected to a different concept mid-search by the autocomplete service of Google's search engine.

The philosopher John Searle (1983, p. 1) defined intention as "that feature of certain mental states and events that consists in their being directed at, being of, or representing certain other entities and state of affairs." Following him, Bratman (1987) focused on intention as a means by which the self regulates its interactions and long-term activity. According to these definitions, we can see that using recommendation engines might change our intention and our mental state, as well as our self-regulation.

At the very least, we can see the delegation of a particular mental state from the intention in action—of doing, deciding, or thinking about something to the intentional action of using an engine that does it for us.

This makes those services very useful for us—we don't even have to explicitly tell it what our intention is; it'll come up with its "own" ideas to fulfill our wishes as it understands them, becoming an "intermediary agency," diminishing the human agency in the process (Pickering, 2010).

## RATIONALITY ENGINES

Our inclination to forget, lack of awareness, restricted processing speed, inadequate attention, and weak planning abilities are some qualities that mark human cognition. Herbert Simon (1971) coined the term "bounded rationality," referring to those limitations that result in using mental shortcuts rather than thorough calculations and preparations while solving problems and making decisions. It refers to the thinking process itself, in which we attempt to employ as few mental resources as possible, and to the decision-making, which may or may not be obtained in a calculated, optimal, and accurate manner—thus reflecting the limitations of human rationality.

In contrast to human-bounded rationality, the use of recommendation engines, such as Google's navigation service, which Nadia uses to choose the route to the museum, enables optimal and precise timing, down to the minute level of departure. The route is calculated while analyzing a variety of data and preferences, creating alternatives, and choosing the most suitable ones under a clear and reproducible algorithmic logic that analyzes a significantly higher amount of digital information in order to recommend a decision (Doneson, 2019).

From the social perspective, recommendations made by algorithms are perceived as more rational and efficient as they become more pervasive in our daily lives (Fisher, 2020). Human users cannot do it in a more justifiable or logical manner than Google, but they can support the algorithmic decision-making process in a better way by feeding the platforms' recommendation engines with more data and by delegating autonomy to inform their decisions. All of this is done in the hope that it will make decisions similar to theirs, considering all the historical data that has been given and analyzed by sophisticated algorithms to be more rational. Using these products also impacts how rational users perceive themselves to be (Bar-Gil, 2020). Users believe that using recommendation engines makes them more rational, efficient, and sophisticated. They also believe that it allows them to accomplish "more with less" by saving cognitive resources for decision-making and even utilizing effective algorithms to realize their intentions. This is especially evident in the management of a resource that is always in short supply: time. Consider, for example, how Google's calendar services can automatically coordinate meetings (Asara, 2016), travel plans, and even the achievement of personal objectives, as suggested in the launch of Google's new goal-based-rescheduling feature:

> Whether it's reading more books, learning a new language, or working out regularly, achieving your goals can be really hard ... That's why starting today, we're introducing Goals in Google Calendar. Just add a personal goal—like "run 3 times a week"—and Calendar will help you find the time and stick to it.
>
> (Ramnath, 2016)

However, it is not just about being able to coordinate everything without "wasting" time planning and organizing; Google's calendar even encourages its users to complete their tasks more actively, using behavioral nudges and reminders to direct attention and decision-making into realizing prior intentions, which begs the question, whose agenda is set for whom? And is algorithmic rationality and efficiency a good thing for us as humans? Or maybe, to respect our bounded rationality and allow for human autonomy, do we need our HCAI system to do something different?

## REMEMBERING IN THE CLOUD

Along with changes in intention and rationality, we can also observe changes in memory patterns. The emergence of complex, dialogic, and flexible modes of communication enabled by digital and cloud platforms, as well as the undermining of previous configurations of individual–group–societal relations, necessitates a new way of thinking about individual and social memory patterns. Memory researcher Andrew Hoskins (2017) called it "connective memory"—the patterns of memory emerging from increased connections and entanglements with others in real-time through an assortment of digital apps, platforms, and networks. The extended memory-in-the-cloud offered by Google and other platforms enables users to not only contribute to the formation of a shared collective memory but also process this memory content for their own individual benefit (Ward & Wegner, 2013). Consider our version of the extended mind thought experiment—when a user of the navigation application refers to a section they have traveled, their "memory" of that section is converted into a shared, connected memory of specific geographic locations, which affects how they and others who have not yet "experienced" that section of travel choose to move. The spatial decision-making, which we emphasized in this example, is only a small portion of Google's entire ecosystem of decision-making services, which also contains visual selection (Google Photos), verbal ones (the search engine, Google Books, and more), and video ones (YouTube). Although Google's global memory is varied, dispersed, and rich, it can also be retrieved, searched for, filtered, examined, and algorithmically processed. Those affordances provide users the option to process information themselves or let Google services or other recommendation engines do it for them. To use it efficiently, we save larger portions of our memory on cloud platforms, letting go of these memories for the sake of more information-processing and decision-making, until they become deeply ingrained in the way we remember and think (Sparrow et al., 2011).

## CONCLUSION: THE EFFECTS OF RECOMMENDATION ENGINES ON US—HOW CAN RECOMMENDATION ENGINES BE HUMAN-CENTERED?

While the progress of technology opens exciting new opportunities, it also creates a flood of data that can be difficult to wade through. To make our lives easier and achieve our daily goals with less mental energy and time spent on them, we turn to technology to help us think. We delegate some cognitive processes and efforts to those cognitive technologies, organized in platforms (Clowes, 2015). As the technological platform ecosystem grows and serves more functions, we delegate more and more cognitive processes to help us cope with information overload and daily churns. The possible benefits of this process are numerous, including increased leisure time, improved capacity for foresight and organization, enhanced orientation and navigation skills, answers to our most intimate questions at the tip of our fingers, and many more. But those have a price. They change our thinking processes, altering some of our core human aspects of intentionality, rationality, and memory in the digital sphere and the real world, diminishing our agency and autonomy.

### The Next Frontier—Delegating Ourselves to Generative AI Services

While writing this chapter, a new technology emerged—generative AI models for texts (chat), images, sound, and video that can create content on behalf of the human user who initially prompted them (Browne, 2022).

It seems that combining those with our data doppelgängers can create a powerful personal assistant that will be very helpful in various tasks of content creation and save us a lot of time and "friction." If we follow the argument of this chapter, it has the potential to escalate the delegation process described, manipulate our intention, and diminish human agency and autonomy, even in tasks that were formerly considered "creative."

### Can We Consider These Effects as Human-Centered Artificial Intelligence?

The ecology of recommendation engines grants users powers that would have been regarded as superpowers at any other time in history, a kind of realization of the idea of the merging of man (or, at the very least, his data doppelgänger) and machine. However, is it challenging to perceive this empowerment and extension as making us "superhuman," as transhumanists claim? To paraphrase Nietzsche, one of their favorite philosophers (Sorgner, 2009), does it truly improve the human experience in a way that we can say that AI is human-centered? It appears that AI-based recommendation engines emphasize other aspects of Nietzsche's ideas, such as those of the "last man" (Nietzsche, 1883/2006), particularly those dealing with more prevalent human qualities such as laziness, desire for comfort, and low effort.

HCAI cannot be considered as a technological design. It must include ethical, psychological, social, political, and legal aspects to analyze its profound impact on individuals and society. Challenging the concept of HCAI from these fields is necessary to continue harnessing artificial intelligence's wonderful powers, whilst augmenting the human parts of us, or at least minimizing the potential harm to our most crucial aspects of humanity, such as our ability to be autonomous subjects, conscious, developing, and responsible for ourselves and our actions. How might such thinking be fostered? This will be covered in greater detail in the subsequent chapters of this book.

## NOTES

1 Technology reporter Kashmir Hill conducted an experiment in which she blocked a different technology company every week. The 104,000 daily access requests for Google can be compared to 15,880 requests for Facebook, 15,600 for Microsoft, and about 292,000 access requests for Amazon, which stores a large number of servers in its cloud storage service (AWS).

2 It is a very simple procedure. See: https://support.google.com/accounts/answer/3024190?hl=en

3 For an even weirder story about reliance on navigation systems see: https://www.telegraph.co.uk/technology/5081143/When-satnav-systems-go-awry.html

4 We shape our buildings; thereafter they shape us. See: http://automatedbuildings.com/news/aug20/articles/lynxspring/200721102909lynxspring.html

## REFERENCES

Airoldi, M., Beraldo, D., & Gandini, A. (2016). Follow the algorithm: An exploratory investigation of music on YouTube. *Poetics*, *57*, 1–13. https://doi.org/10.1016/j.poetic.2016.05.001

Andrejevic, M. (2013). *Infoglut: How too much information is changing the way we think and know*. Routledge.
Asara, F. (2016, September 29). *Save time with smart scheduling in Google Calendar*. Google. https://blog.google/products/calendar/save-time-with-smart-scheduling-in-google-calendar/
Bar-Gil, O. (2020). Clipping us together: The case of the Google Clips camera. *NECSUS European Journal of Media Studies*, *9*, 215–236. https://doi.org/10.25969/mediarep/14308
Bar-Gil, O. (2021). *Google's self: Self-perception at the age of information*. [Doctoral Thesis, Bar-Ilan University].
Bratman, M. (1987). *Intention, plans, and practical reason*. Center for the Study of Language and Information.
Browne, K. (2022). Who (or what) is an AI artist? *Leonardo*, *55*(2), 130–134. https://doi.org/10.1162/leon_a_02092
Brynjolfsson, E., & McAfee, A. (2016). *The second machine age: Work, progress, and prosperity in a time of brilliant technologies* (1st ed.). W. W. Norton & Company.
Bullock, G. (2017, May 17). *Save time with Smart Reply in Gmail*. Google. https://blog.google/products/gmail/save-time-with-smart-reply-in-gmail/
Castoriadis, C. (2011). *Postscript on insignificance: Dialogues with Cornelius Castoriadis* (English language ed). Continuum.
Clark, A., & Chalmers, D. J. (1998). The extended mind. *Analysis*, *58*(1), 7–19.
Clowes, R. (2015). Thinking in the cloud: The cognitive incorporation of cloud-based technology. *Philosophy & Technology*, *28*(2), 261–296. https://doi.org/10.1007/s13347-014-0153-z
Deahl, D. (2018, May 10). *Here's how to use Gmail's new Smart Compose* [Blog]. The Verge. https://www.theverge.com/2018/5/10/17340224/google-gmail-how-to-use-smart-compose-io-2018
Doneson, D. (2019). The conquest of fortune: On the Machiavellian character of algorithmic judgment. *Social Research: An International Quarterly*, *86*(4), 871–883.
Duong, Q. (2017, November 7). Skip the line: Restaurant wait times on Search and Maps. *Google*. https://www.blog.google/products/maps/skip-line-restaurant-wait-times-search-and-maps/
Evans, S. K., Pearce, K. E., Vitak, J., & Treem, J. W. (2017). Explicating affordances: A conceptual framework for understanding affordances in communication research. *Journal of Computer-Mediated Communication*, *22*(1), 35–52. https://doi.org/10.1111/jcc4.12180
Fisher, E. (2020). The ledger and the diary: Algorithmic knowledge and subjectivity. *Continuum*, 1–20. https://doi.org/10.1080/10304312.2020.1717445
Fuchs, C. (2008). *Internet and society: Social theory in the information age*. Routledge.
Hill, K. (2019, January 29). I Cut Google Out Of My Life. It Screwed Up Everything. *Gizmodo*. https://gizmodo.com/i-cut-google-out-of-my-life-it-screwed-up-everything-1830565500
Hoskins, A. (2017). Memory of the multitude: The end of collective memory. In A. Hoskins (Ed.), *Digital memory studies: Media pasts in transition*. Routledge.
Ihde, D. (2009). *Postphenomenology and technoscience: The Peking University lectures*. SUNY Press.
Karakayali, N., Kostem, B., & Galip, I. (2018). Recommendation systems as technologies of the self: Algorithmic control and the formation of music taste. *Theory Culture & Society*, *35*, 3–24. https://doi.org/10.1177/0263276417722391
Kelly, K. (2016). *The inevitable: Understanding the 12 technological forces that will shape our future*. Viking.
Lardinois, F. (2018, June 26). The new Google Maps with personalized recommendations is now live. *TechCrunch*. https://techcrunch.com/2018/06/26/the-new-google-maps-with-personalized-recommendations-is-now-live/

McAfee, A., & Brynjolfsson, E. (2017). *Machine, platform, crowd: Harnessing our digital future*. W. W. Norton & Company.

Menary, R. (Ed.). (2010). *The extended mind*. MIT Press.

Merleau-Ponty, M. (2002). *Phenomenology of perception* (C. Smith, Trans.). Routledge. (Original work published 1945).

Milner, G. (2016). *Pinpoint: How GPS is changing technology, culture, and our minds*. W. W. Norton & Company.

Nayak, P. (2020, September 10). *Our latest investments in information quality in Search and News*. Google. https://blog.google/products/search/our-latest-investments-information-quality-search-and-news/

Newton, C. (2017, July 19). *Google introduces the feed, a personalized stream of news on iOS and Android*. The Verge. https://www.theverge.com/2017/7/19/15994156/google-feed-personalized-news-stream-android-ios-app

Nietzsche, F. W. (2006). *Thus spoke Zarathustra: A book for all and none* (A. Del Caro & R. B. Pippin, Trans.). Cambridge University Press. http://dx.doi.org/10.1017/CBO9780511812095 (Original work published 1883)

Pham, A., Rubel, A., & Castro, C. (Eds.). (2021). Autonomy, agency, and responsibility. In *Algorithms and autonomy: The ethics of automated decision systems* (pp. 21–42). Cambridge University Press. https://doi.org/10.1017/9781108895057.002

Pickering, A. (2010). Material culture and the dance of agency. In *The Oxford handbook of material culture studies*. https://doi.org/10.1093/oxfordhb/9780199218714.013.0007

Ramnath, J. (2016, April 12). Find time for your goals with Google Calendar. *The Keyword*. https://www.blog.google/products/calendar/find-time-goals-google-calendar/

Rosa, H. (2013). *Social acceleration: A new theory of modernity* (J. Trejo-Mathys, Trans.). Columbia University Press.

Schwab, K. (2017). *The Fourth Industrial Revolution*. Crown Business.

Searle, J. R. (1983). *Intentionality, an essay in the philosophy of mind*. Cambridge University Press.

Shneiderman, B. (2022). *Human-centered AI*. Oxford University Press.

Sigfrids, A., Leikas, J., Salo-Pöntinen, H., & Koskimies, E. (2023). Human-centricity in AI governance: A systemic approach. *Frontiers in Artificial Intelligence*, *6*. https://www.frontiersin.org/articles/10.3389/frai.2023.976887

Simon, H. A. (1971). Designing organizations for an information-rich world. In M. Greenberger (Ed.), *Computers, communication, and the public interest* (pp. 37–72). The Johns Hopkins Press. https://digitalcollections.library.cmu.edu/awweb/awarchive?type=file&item=33748

Smith, M. N. (2017). Intentions: Past, present, future. *Philosophical Explorations*, *20*(sup2), 1–12. https://doi.org/10.1080/13869795.2017.1356360

Sorgner, S. L. (2009). Nietzche, the overhuman, and transhumanism. *Journal of Evolution and Technology*, *20*(1), 29–42.

Sparrow, B., Liu, J., & Wegner, D. M. (2011). Google effects on memory: Cognitive consequences of having information at our fingertips. *Science*, *333*(6043), 776–778. https://doi.org/10.1126/science.1207745

Stephens-Davidowitz, S. (2017). *Everybody lies: Big data, new data, and what the internet can tell us about who we really are*. Dey St.

Stiegler, B. (1998). *Technics and time: The fault of Epimetheus*. Stanford University Press.

Umapathy, V. (2015, December 7). *Add to-dos to your Google Calendar using Reminders*. The Keyword. https://blog.google/products/calendar/add-to-dos-to-your-google-calendar/

Van Dijck, J., Poell, T., & de Waal, M. (2018). *The platform society: Public values in a connective world*. Oxford University Press.

Virilio, P. (2005). *The information bomb*. Verso.

Wang, D., Ma, X., & Wang, A. Y. (2021). *Human-Centered AI for Data Science: A Systematic Approach* (arXiv:2110.01108). arXiv. https://doi.org/10.48550/arXiv.2110.01108

Wang, J. (2021, November 4). *Google Maps navigates its way to 10 billion installs*. Android Police. https://www.androidpolice.com/google-maps-navigates-its-way-to-10-billion-installs/
Ward, A. F., & Wegner, D. M. (2013). Mind-blanking: When the mind goes away. *Frontiers in Psychology, 4*. https://doi.org/10.3389/fpsyg.2013.00650
Zuboff, S. (2018). *The age of surveillance capitalism: The fight for a human future at the new frontier of power*. PublicAffairs.

# 4 Ethics at the Intersection
## *Human-Centered AI and User Experience Design*

*Christopher D. Quintana*
Villanova University, Villanova, PA, USA

In attempts to understand and develop the notion of human-centered artificial intelligence, there is an increasing call to recognize the rich collaborative possibilities between the theory and practices of human–computer interaction, AI development, and moral concerns often examined within moral and social philosophies. Work that intersects with one or all of these areas often emphasizes the interaction between users' context of use, the nature of the digital environments users navigate, and the wide range of design practices that are corralled into creating a user's experience. In this chapter, I aim to highlight emerging moral and social concerns that potentially arise throughout these points of interactions between users and differing forms of artificial intelligence. I suggest that the point of interaction between humans and machines is a rich vein to draw from for examining both the nature of the ethical problem and the mechanisms that give rise to these problems. Thus, the titular intersection refers to both the user interactions and the point where ethics and philosophy can meet insights from user experience design. I begin with a brief overview of some influential approaches to philosophical AI ethics and then turn to exploring the intersections alluded to above in order to enable a broader understanding of what goes into designing human-centered AI systems and the moral questions this raises.

## DOMINANT CURRENTS IN THE ETHICS OF ARTIFICIAL INTELLIGENCE

In this chapter, I follow existing AI Ethics literature in suggesting that "it makes little sense to consider the ethics of algorithms independent of how they are implemented and executed in computer programs, software and information systems" and for finding much to gain from ethicists of technology focusing on decision-making algorithms that make "generally reliable … decisions based upon complex rules that challenge or confound human capacities for action and comprehension … algorithms whose actions are difficult for humans to predict or whose decision-making logic is difficult to explain after the fact" (Mittelstadt et al., 2016) or highlight the important questions or problems that are not being addressed at all. The reality of the current deployment of artificial intelligence, often referred to as narrow AI,[1] shows us there is plenty to work on already. As we shall see, work has been and is being carried out to account for this situation. Nevertheless, I highlight a complementary alternative to the current


DOI: 10.1201/9781003320791-6

analysis of narrow AI deployment, namely the nature of the digital environments and hardware necessary for those environments to function. Such an emphasis is ecumenical with the call to consider the ethics of AI within specific implementations.

What then are the problems being taken into consideration by influential philosophical AI ethicists interested in the application of narrow AI? Two sibling papers (Mittelstadt et al., 2016; Tsamados et al., 2022) do us the service of an extensive surveying of the literature and outlining the key areas of concern in the following ways:

I. Inconclusive Evidence Leading to Unjustified Actions
II. Inscrutable Evidence Leading to Opacity
III. Misguided Evidence Leading to Bias
IV. Unfair Outcomes Leading to Discrimination
V. Transformative Effects Leading to Challenges for Autonomy
VI. Transformative Effects Leading to Challenges for Informational Privacy
VII. Traceability Leading to Moral Responsibility

The topics can be clustered into predominately epistemic (I–III) and predominately normative (IV–VI) concerns. I begin here with an overview of the papers on the key epistemic concerns and then turn to the normative issues, particularly the transformative effects on autonomy. Epistemic issues I and III address the nature of the data processed. Issue I involves the reliance on probabilistic yet uncertain knowledge, while III highlights the "garbage in, garbage out" principle, which implies that algorithmic conclusions are only as reliable and neutral as their input data (Mittelstadt et al., 2016). Issue II addresses the inaccessibility or "opacity" of data and algorithms, which often defy comprehensibility and accessibility, both of which are key components of transparency. These issues result in an asymmetry of power favoring those who process and hold data and data subjects. In such a situation, it is difficult for data subjects to make informed choices about the exposure of their data due to the limited awareness of its use. At the same time, these issues in opacity and transparency are a barrier to oversight, as the swaths of data are so large that it becomes difficult for human actors to analyze and comb through the mass of data.

Normative issues, such as unfair outcomes leading to bias and discrimination, are nevertheless closely linked to the epistemic concerns mentioned earlier. Biased data and profiling algorithms are implicated in perpetuating these issues. Personalization algorithms, for example, can reinforce existing social disadvantages and contribute to self-fulling prophecies and stigmatization of certain groups, ultimately undermining their potential participation in society. Furthermore, profiling and aggregation of individuals within broader groups, along with the opaque processes and data used, raise questions about informational privacy. The concept of informational privacy is typically understood as the right to shield personal data from third parties, control one's information, and require effort from others to obtain it. Researchers note that the specificity of an individual's identity is not necessary for informational privacy concerns to arise, as algorithms tend to aggregate people based on shared characteristics.[2] Mittedelstadt et al. consequently critique the 2012 European Commission's data protection laws for only protecting data usage connected to identifiable individuals. They argue that even within anonymized aggregated data, an individual's informational identity can still be breached, as profiling can occur in relation to surrounding data points despite an obscured identity.

Another form of traceability relevant concerns moral responsibility (VII). When a technology fails, there is generally a desire to have blame, sanctions, or accountability traced to an actor with control. Computer programmers have historically had a general sense of program effects and potential failures. However, with rapid advancements in information communication technologies and programming, tracing responsibility is increasingly difficult due to complexity and volume. Ethicists and researchers (Bozdag, 2013; Kramer et al., 2011) argue that with nonlearning algorithms, programmers retain some responsibility. Still, issues arise with tracing responsibility in the context of massive and opaque datasets that defy transparency and human oversight. The issue is further complicated with algorithms possessing learning capacities, as their indeterminate nature makes it difficult to pinpoint responsibility on specific programmers or groups, who cannot predict how the algorithm will learn and respond to the new data inputs. Whether self-learning algorithms can be ascribed agency, moral standing, and ethical decision-making remain contested. Similarly, the responsibility of developers in this case is also up for debate:

> Assigning moral agency to artificial agents can allow human stakeholders to shift blame to algorithms. Denying agency to artificial agents makes designers responsible for unethical behavior of their semi-autonomous creations; bad consequences reflect bad design. Neither extreme is entirely satisfactory due to the complexity of oversight and the volatility of decision-making structures.
>
> (Mittelstadt et al., 2016, p. 11)

The final theme focuses on the transformative effects of algorithms on autonomy, with researchers primarily analyzing personalization algorithms. Tsamados et al. (2022) identify three sources of concern regarding user autonomy: (i) pervasive distribution of proactive learning algorithms informing user choices, (ii) users' limited understanding of algorithms, and (iii) lack of appeals over algorithmic outcomes. Many users lack a deep understanding of how learning algorithms work. The lack of understanding can hinder users from fully utilizing algorithms to achieve their goals. However, since the publication of Tsamados et al., some progress has been made, with many laypersons developing tacit knowledge of algorithms' reliance on explicit data inputs for curation and a move toward limited access to appeals on algorithmic outcomes. This awareness is nevertheless insufficient to encapsulate full comprehension and the ability to exert control over outcomes. Examining the threat to autonomous choice posed by learning and personalization algorithms thus reveals the complex intersection of technical, ethical, and political problems. Public fascination and concern with algorithms mirror scholarly worries over echo chambers, epistemic bubbles, misinformation, lack of informational diversity, and the overall shaping of user personalities and ideologies.

## TECHNO-SOCIAL AI ETHICS

The normative and epistemic issues addressed above are justifiably major areas of inquiry and action. Nevertheless, a broader vision of the system reimagines our sense of how something like autonomy is impacted. For example, one way to understand the normative and epistemic issues addressed above is to imagine, on the one hand, a stable "subject," a user equipped with some stable set of capabilities (such as rationality and choice) as well as some right or other (such as autonomy or privacy). On the other hand, you have a technology or algorithm *in medias res*, seemingly without history or regard for the many hands which went into designing it. However, such a framing risks obscuring the social-productive processes that enable some technology or another.[3] In a recent call to reimagine AI ethics policy debates, Emma Ruttkamp-Bloem critiques the "false Cartesian perceptions of what it means to be human in a technologically mediated socio-cultural world" (Ruttkamp-Bloem, 2022, p. 3).[4] Contrasting with this understanding of Cartesianism, Ruttkamp-Bloem's draws on Andy Clark's (2004) philosophical anthropology of the "natural-born cyborg" featuring an extended mind and body through technologies. This perspective views humans as co-extensive with their environments and technologies, constantly interacting with them, shaping, and being shaped by them: technologies, society, and individuals coconstruct each other from the outset. This approach emphasizes the distributed interplay of humans, technology, and society to better understand the source, spread, and impact of technology and identify potential solutions to the issues it raises. From this point of departure, the following sections target the forms of habituation and interaction encouraged by affective, embodied, and choice architectures of AI-laden applications, interfaces, and machines. I conclude by briefly examining the tensions that arise out of design in an industry context.

## Techno-Social Engineering

Scholars Brett Frischmann and Evan Selinger express concern over what they call the techno-social reengineering of human beings. For Frischmann and Selinger (2018), information communications technologies (ICTs) provide affordances (i.e., capabilities enabled by a relationship between tool and user) that ultimately undermine the freedom afforded by human practical agency.[5] In their view, ICTs do this by offering routes to outsourcing a wide variety of physical, cognitive, emotional, and ethical tasks to technical devices, systems, or applications. One example is GPS devices and the capacity for navigation (Dreyfus & Kelly, 2011). Furthermore, the theory of techno-social engineering argues that reliance on these affordances shape and in certain cases program human behaviors. In a search for empirical support for the theory, Haenschen et al. (2021) conducted experiments that manipulated the birthday notification system on Facebook to show that the notifcation of a (false) birthday of a friend was enough to nudge users into wishing someone a happy birthday—Even when users reportedly knew the actual birthday of the friend. In the words of the authors:

> Facebook users have been programmed to respond to a stimulus without stopping to think about its veracity, even when they might have reason to do so. The Facebook platform has changed the social practice of wishing friends a happy birthday by providing an automated system to nudge individuals to send a greeting, resulting in a dramatically higher number of birthday wishes than one would give or receive otherwise.
>
> (Haenschen et al., 2021), p. 1479)

Important to note is that what occurs when someone is caught up in the user experience of Facebook, what is happening here is not simply brute informational nudging, i.e., a nudge toward an action based on the availability of information. No doubt the raw information of a birthday happening is crucial, but the interface design engages the user beyond just textual information: the interfaces deploy the use of color theory (e.g., the use of red for notification), affective and symbolic design (e.g., the use of relevant birthday imagery), and tailored presentation (e.g., displaying how others are actively posting on the profile of the person whose birthday it is). A groundbreaking study of machine gambling in Las Vegas provides a helpful parallel (Schüll, 2014). It is not simply the information that gambling some relatively small money could lead to big gains which keep casino slot machines sections populated with eager users—however indispensable this might be. Casino designers long understood the benefits of the interior design and layout of a casino: the architecture, ambiance, and affective responses, produced by walking the casino floor in keeping users engaged—not to mention the satisfying visual spectacle, tactile feedback, and auditory engagement provided by the machine. Designers and users of a variety of digital platforms similarly ought to recognize how their designs and the narrow forms of artificial intelligence deployed therein create a digital environment capable of fundamentally altering the behaviors and habits of users.

## Moral De-Skilling and Habituation

Within the philosophical ethics of technology, there is an increasing concern that technologies not only alter the behaviors and habits of users but potentially cause atrophy of key moral skills or abilities. One of the first attempts to address the impact information communication technologies could have on the cultivation of moral skills, as understood in theories of virtue, is found in Vallor (2015). In the landmark paper, Vallor attempts to "adapt the conceptual apparatus of sociological debates over economic deskilling to illuminate a different potential for technological deskilling/upskilling, namely the ability of ICTs to contribute to the moral deskilling of human users."[6]

To carry out such an examination, Vallor begins with the role of moral skills in the cultivation of character. Building on a Neo-Aristotelian foundation, she highlights how in the virtue theory literature, virtues are cultivated rather than innate states of character. Under such a framework, the likelihood that someone develops a particular virtue will depend on whether they "engage repeatedly in the kinds of practices that cultivate it," specifically "practices that successfully engender certain skills of acting rightly in particular moral contexts" (Vallor, 2015, p. 109). Take the practice of skillful chess playing. Recent controversies over cheating at high-level competitions emphasize the importance of a virtue such as honesty. In a competitive setting, especially one where significant prestige and financial incentives are in place, there is a lurking incentive to undermine fair competition. This form of undermining fair competition ultimately threatens the very practice of skillfully excelling in the practice of chess. In participating in this activity, players are thus put in a position to evaluate whether they should undermine their practice and the expectations of someone developing and practicing the virtue of honesty. Naturally, competitive chess is not the only route to exercising honesty, but it is these sorts of opportunities for exercising virtues that are engendered by practices such as competitive chess playing. In contrast, participating and developing oneself within, say, a guild of skilled thieves, misses the mark since virtues are aimed at some conception of human flourishing, rather than deceit, robbery, and undermining of trust in the human community.

What the example illustrates is the importance of what the context of a user's experience or interaction with a particular piece of technology does for enabling the capacity of the user to cultivate moral skills and habits. Vallor's framework admits to wide applicability, including autonomous or "smart" weapons systems and digital platforms. But Vallor offers a particularly instructive example of embodied artificial intelligence in the form of social care robots. On the moral skill model, the concern would be that social care robots might undermine some of the key contexts where practices such as care are exercised. To call care or caring a virtue is to distinguish between a simple attitude and an "an activity of personally meeting another's need" through the "skillful, attentive, responsible and emotionally responsive disposition to personally meet the needs of others who share our techno social environment" (Vallor, 2018, p. 221). Of particular concern here is the process of automating or outsourcing away the "burdens" of care, in favor of designing a "robot which will support the skillful carrying-out of holistic caring practices. In this way, a carebot may not only benefit patients but also help meet the moral needs of caregivers, by allowing them to become more skillful carers" (Vallor, 2015, p. 120). I will not posture to have the design blueprints for such a complex robotics design and would defer to the exponentially more skillful designers and engineers in this domain. Nevertheless, it is worth noting some recent meta-analyses of user experience research in the context of social care robots.[7] Although the authors cite Dautenhahn (2013)'s call for social robotics meeting the social and emotional needs of their individual users as well as respecting human values, there is nearly no analysis of the ethical stakes in the design of user experience or in the research program of researchers examining the nature of the human–robot interaction. The meta-analyses instead show a strong commitment to what some HCI scholars call the usability paradigm, by which they suggest that the positive associations of usable and useful products are the paradigmatic drive of user experience design.[8] Furthermore, there appears to be less emphasis on the interactions between the human caregivers already at work—the coworkers of these care robots, so to speak. Following Vallor's framing, what ought to be an object of study in order to have human-centered AI in the form of social robotics would then be not only the capacities of care robots to interact with the receivers of care with moral skill but also the extent to which carebots better enable caregivers to deliver on the moral goods of their practice.

## Design in Context

The recognition that designers are doing "ethics by other means" (Verbeek, 2005) and the embeddedness of design itself requires a look at designers and their contexts. In turning to designers and their context, I will draw on the notion of a practice and the goods related to a practice, from philosopher Alasdair MacIntyre. For MacIntyre, a practice is a coherent and complex form of social human activity with its own internal goods and standards of excellence. To be worthy of the name, participants in a practice must develop or exercise skills, capacities, or moral and epistemic virtues. In short, the practice must contribute to participants' flourishing as individuals and not just as practitioners or workers (MacIntyre, 1981, pp. 181–203). Furthermore, the practice itself must have what MacIntyre calls an internal good. An internal good is the end goal of the activity which practitioners aim to achieve in their activity. So, for example, we would not call someone a proper practitioner in medical care if they intentionally poisoned patients rather than treating their ailments or alleviating their symptoms. And within that practice, there are standards of criteria of acceptability and at the other end excellence which must be met for that good to be truly achieved. A dentist who is haphazard and negligent fails to achieve the goal of good oral health, in a way that a methodical and careful dentist does not.

There is one other characteristic of practices besides internal goods, standards of excellence, and the cultivation of human capabilities, namely the complementary notion of external goods. External goods include money, fame, prestige, etc., that could come because of excelling in a practice. Indeed, such goods often are important aids in the pursuit of a practice, as in the case of professional associations which foster and sponsor opportunities for a practice to be carried out or further theorized. Nevertheless, while internal and external goods can be complementary, it is often the case that they run in tension with each other, as when an organization jettisons some of the internal goods of a practice for the sake of profit motive, efficiency, or the prestige associated with being first to market (Moore, 2019, pgs. 55-74). Finally, while many of us tend to participate in a variety of practices, the practices we engage in at work are likely to be, as business ethicist Geoff Moore puts it, "one of the most significant practices given the amount of time and energy." Furthermore, work is one of the most significant ways that workers can contribute "to the common good: the products of services which we are involved in supplying, and the way in which our engagement in the practice develops us as people" (Moore, 2019, p. 84).

Designers, especially those within user experience and human-centered implementation of AI, can benefit from viewing their activities as contributing to these two distinct goods. In the case of pursuing human-centered design—understood here as ethical design—this is especially important. Take, for example, the crucial role that user experience has played in advocating and designing for accessibility in the case of the deaf and blind's online experiences. In such a situation, a design is good not only because it meets certain usability or legal criteria but also for its broader social contribution to the common good in expanding who has access to something as important as the digital ecosystems that often influence the offline world—to the extent such a distinction is even still tenable. Design concerns like the one above contribute to the internal good of design by contributing to products that improve the lives of users and open avenues to new experiences. Fostering this practice helps meet worker's expectations of their work activity, the growing desire for more ethical consumption,[9] and the instrumental value of this form of thinking.[10] As should be clear to those of us interested in the effect of technologies on humanity, the reduction of deceitful or outright harmful design is increasingly a subject of scrutiny across a wide variety of groups from policy-makers and professional organizations to parents and users themselves. The 2022 record-setting $520 million (USD) Federal Trade Commission settlement with Epic Games offers an instructive example. The company was charged for issues with one of its video games, which pertained to two broad categories: privacy violations and dark or deceptive patterns. For our purposes, we can focus on the decisions that led to the design of deceptive patterns, that is, designs that trick users, obscure the processes they are engaged in, and/or undermine the autonomy of the user. In the case of Epic Games, they were charged with using dark patterns to trick users into making purchases, made default settings that harmed children and teens, and consciously obfuscated the existence of refund features. Indeed, the designer who helped design the refund request path in testing reported that not a single player found the option in their testing. When the designers submitted their findings, their superior reportedly was told that the obfuscated refund feature was perfect where it was.[11]

The example underscores that my portrayal of designers' roles in this context is undoubtedly limited and only part of the bigger picture. By focusing on user experience designers, I do not intend to place the burden of responsibility on a part of product development often lacking in investments or stage of maturity.[12] The realization of design's internal goods are also a responsibility for managers and organizations in general. Rather, my aim is to provide ways of framing design practice as value-laden and resources for contemplating the ethical impacts arising from experience and interface design. The integration of AI in experiences, interfaces, and machines that humans interact with further intensifies the issue. Unsurprisingly, AI ethics as is carried out in the virtue ethics tradition often emphasizes the need for designers and organizations to cultivate virtues like honesty, empathy, care, justice, and prudence for putting AI ethics into practice (Hagendorff, 2020; Vallor, 2018).

## DISCUSSION

The pervasiveness of dynamic and adaptive AI systems, such as content recommenders (e.g., news, video, audio), predictive models (e.g., financial modeling), decision-making (e.g., mortgage lending), and natural language understanding systems (e.g., chatbots), presents challenges to human-centered design. Given the societal implications of AI, researchers have argued that explainability exceeds legal requirements and instead "support[s] users in taking control" and helps "designers enhance correctness, identify improvements in training data, account for changing realities" (Shneiderman, 2020, pg. 26:9). For Shneiderman (2020), this aspect of development is part of the technical practices of developing explainable and exploratory UIs, an aspect creating reliable software engineering systems. Here "explainable and exploratory UIs" refer to interfaces equipped with components that allow users to comprehend and control these processes.

A case in point is the case study on the Google Flights service, which uses machine learning to predict flight prices (Polonski, 2020). Recognizing that customers continuously compare prices, its designers aimed to promote "price intelligence" by providing clear and understandable information. This was done by incorporating design elements that showed how good a current price is, simple text explanations, predictive information about potential price changes, and visual elements. For example, the price visualization bar indicates good vs. average vs. bad pricing with numbers and colors green, yellow, and red, on a slider. Below the bar, users can click to access a graph showing price history. In addition to these elements, there was an easy way to access the data sources used to make calculations. The designers thus avoided cognitive overload while enabling the user to be informed and explore their options. This approach fosters informed decisions, avoiding cognitive overload and aligning with Shneiderman's (2020) concept of explainable and exploratory interfaces.

One of the key considerations, from the perspective of the "deskilling and habituation" model described in the "Moral De-Skilling and Habituation" section of this chapter, is the nature of skills and habits developed through these interactions. It is not simply about becoming skillful in purchasing flights, helpful as that may be, but about fostering skillful engagement with digital environments users frequent by facilitating exploration, feedback, and understanding. Conversely, interfaces that obfuscate AI interaction risk leading users to become habituated into misplacing trust and flawed expectations of capabilities. The implications of this habituation extend far beyond consumer choice since digital interfaces encompass many aspects of social, political, aesthetic, and educational practices. From the perspective of virtue ethics and epistemology, there is a risk to both epistemic and moral virtues such as intellectual humility (Heersmink, 2018) or empathy and appropriate or discerning attention (Vallor, 2018). However, this virtue theoretical perspective can inform product planning stages and product analysis by identifying the types of habits and skills that a product might cultivate and their moral implications. Such an approach attempts to transcend calls for fairness, transparency, and accountability, and aids in calls to promote "practices that raise self-efficacy, encourage creativity, clarify responsibility, and facilitate social participation" (Shneiderman, 2020).

While this chapter has focused on the techno-social engineering embedded in human–AI interactions, the case of FTC vs. Epic Games in the section "Design in Context" underscores that decision-making surrounding explainable and exploratory interfaces is often complicated by managerial or commercial pressure. Consequently, Shneiderman argues that the creation of reliable, human-centered systems will require not just good design practice but some combination of a safety culture at the organizational level (e.g., leadership commitment, hiring and training, and internal review) and trustworthy external review (e.g., auditing firms, professional organizations, and governance). These are complex and contested zones of action at both organizational and political levels. Furthermore, if human-centered design necessitates clarifying what value the AI system provides to the user, then organizations ensure AI implementation coheres with user needs, capabilities, and expectations. If the benefits of an AI system are unclear, or its performance questionable, there is a risk of deploying it unnecessarily. Such a reflexivity aims at curbing the development of a technological culture that threatens to deploy AI-powered products unreflectively and at the cost of human capabilities and practices.

## ACKNOWLEDGMENT

Thanks to Sally Scholz for her comments on earlier versions of this work. As the work developed, my thanks go to fellow writers in this volume as well as editors of this volume for their comments and suggestions. Any shortcomings are mine alone.

## NOTES

1 I follow Tasioulas (2022) in understanding narrow AI as "AI-powered technology that can perform limited tasks (such as facial recognition or medical diagnosis)." In contrast, artificial generalized intelligence would, theoretically, be capable of a range of human cognitive abilities, and perhaps learn to leverage capabilities to tasks that were not part of predetermined training.

2 Van Otterlo's metaphor is particularly helpful here: "Even if I would—as an individual—replaced all the glass in my own [glass] house by wood (i.e., protect my data) it would still be possible to build profiles of all my neighbors and derive information about me." (Van Otterlo, 2013, p. 3), as quoted in Mittelstadt et al. (2016).

3 Important philosophers of technology in this alternative tradition include Jeroen van den Hoven, Peter-Paul Verbeek, Albert Borgmann, and Bruno Latour, among others. See especially Verbeek (2005).

4 Cf. the preface to Paul Dourish (2004), *Where the Action Is: The Foundations of Embodied Interaction*, "Much of contemporary cognitive science is based on a rigorous Cartesian separation between mind and matter, cognition and action … [in contrast a] new approach abandoned the idea of disembodied rationality and replaced it with a model of situated agents, at large in the world, and acting and interacting with it." Note that I am here more invested in explaining what "Cartesianism" means in the context being explored, rather than the history of philosophy project of attributing a position to a canonical thinker.

5 For a broad overview of philosophical and ethical approaches to ICTs, see Quintana (2023).

6 (Vallor, 2015, p. 108). Vallor also highlights the possibility of upskilling or reskilling. For example, she notes the ambiguous character of the phenomena, citing how the computer revolution freed white-collar workers from a slew of tasks such as filing or copying. However, for the moment I focus on aspects of the paper which take a critical approach to skillful action.

7 Shourmasti et al. (2021). See also Alenljung et al. (2017).

8 Fallman (2011). In contrast, Fallman argues that HCI needs to move toward a new and moral philosophy-inspired account of what constitutes a "good"—in the moral sense of good—user experience.

9 (Papaoikonomou et al., 2011) offers a glimpse at how the desire for ethical consumption has warranted extensive empirical work on the subject.

10 Nevertheless, it must be stressed that from the perspective of virtue ethics, it is not sufficient for motivations for virtuous action to be solely based on instrumental value. An individual or organization who does so falls short of virtue. On this, see "The Virtuous Agent's Reasons For Action" in Hursthouse (2010), Chapter 9 of MacIntyre (2006), and chapter 7 of Moore (2019).

11 See Lesley Fair, "$245 million FTC settlement alleges Fortnite owner Epic Games used digital dark patterns to charge players for unwanted in-game purchases," 2022: https://www.ftc.gov/business-guidance/blog/2022/12/245-million-ftc-settlement-alleges-fortnite-owner-epic-games-used-digital-dark-patterns-charge

12 See Pernice et al. (2021): https://www.nngroup.com/articles/ux-maturity-model/ on this notion of UX maturity.

## REFERENCES

Alenljung, B., Lindblom, J., Andreasson, R., & Ziemke, T. (2017). User Experience in Social Human-Robot Interaction. *International Journal of Ambient Computing and Intelligence*, *8*(2), 12–31. https://doi.org/10.4018/ijaci.2017040102

Bozdag, E. (2013). Bias in algorithmic filtering and personalization. *Ethics and Information Technology*, *15*, 209–227. https://doi.org/10.1007/s10676-013-9321-6

Clark, A. (2004). *Natural-Born Cyborgs: Minds, technologies, and the future of human intelligence*. Oxford University Press.

Dautenhahn, K. (2013). *Human-robot interaction*. The Interaction Design Foundation. https://www.interaction-design.org/literature/book/the-encyclopedia-of-human-computer-interaction-2nd-ed/human-robot-interaction

Dourish, P. (2004). *Where the action is: The foundations of embodied interaction*. MIT Press.

Dreyfus, H., & Kelly, S. D. (2011). *All things shining: Reading the Western classics to find meaning in a secular age*. Amsterdam University Press.

Fallman, D. (2011). The New Good. *Proceedings of the SIGCHI Conference on Human Factors in Computing Systems*. https://doi.org/10.1145/1978942.1979099

Frischmann, B., & Selinger, E. (2018). *Re-engineering humanity*. Cambridge University Press.

Haenschen, K., Frischmann, B. M., & Ellenbogen, P. (2021). Manipulating Facebook's notification system to provide evidence of techno-social engineering. *Social Science Computer Review*, *40*(6), 1478–1495. https://doi.org/10.1177/08944393211008855

Hagendorff, T. (2020). AI Virtues – The Missing Link in Putting AI Ethics into Practice. *ArXiv* http://arxiv.org/pdf/2011.12750.pdf

Heersmink, R. (2018). A virtue epistemology of the internet: Search engines, intellectual virtues and education. *Social Epistemology*, *32*(1). https://doi.org/10.1080/02691728.2017.1383530

Hursthouse, R. (2010). *On virtue ethics*. Oxford University Press.

Kraemer, F., van Overveld, K. & Peterson, M. (2011). Is there an ethics of algorithms? *Ethics and Information Technology*, *13*, 251–260. https://doi.org/10.1007/s10676-010-9233-7

MacIntyre, A. (1981). *After virtue: A study in moral theory*. University of Notre Dame Press.

MacIntyre, A. (2006). *Dependent rational animals: Why human beings need the virtues*. Open Court.

Mittelstadt, B. D., Allo, P., Taddeo, M., Wachter, S., & Floridi, L. (2016). The Ethics of Algorithms: Mapping the debate. *Big Data & Society*, *3*(2). https://doi.org/10.1177/2053951716679679

Moore, G. (2019). *Virtue at work: Ethics for individuals, managers, and organizations*. Oxford University Press.

Papaoikonomou, E., Ryan, G., & Valverde, M. (2011). Mapping ethical consumer behavior: Integrating the empirical research and identifying future directions. *Ethics & Behavior*, *21*(3), 197–221. https://doi.org/10.1080/10508422.2011.570165

Pernice, K., Gibbons, S., Moran, K., & Whitenton, K. (2021). *The 6 levels of UX Maturity*. Nielsan Norman Group. https://www.nngroup.com/articles/ux-maturity-model/

Polonski, S. (2020). *When to book and when to fly? explaining prices in Google Flights*. Medium. https://medium.com/people-ai-research/pair-guidebook-google-flights-case-study-1ba8c7352141

Quintana, C. D. (2023). Information Communication Technology. In M. Sellers & S. Kirste (Eds.), *Encyclopedia of the philosophy of law and social philosophy*. Springer. https://doi.org/10.1007/978-94-007-6730-0_1037-2

Ruttkamp-Bloem, E. (2022). Re-imagining current AI Ethics policy debates: A view from the ethics of technology. *Artificial Intelligence Research*, 319–334. https://doi.org/10.1007/978-3-030-95070-5_21

Schüll, N. D. (2014). *Addiction by design: Machine gambling in Las Vegas*. Princeton University Press.

Shneiderman, B. (2020). Bridging the gap between ethics and Practice. *ACM Transactions on Interactive Intelligent Systems*, *10*(4), 1–31. https://doi.org/10.1145/3419764

Shourmasti, E. S., Colomo-Palacios, R., Holone, H., & Demi, S. (2021). User experience in Social Robots. *Sensors*, *21*(15), 5052. https://doi.org/10.3390/s21155052

Tasioulas, J. (2022). Artificial Intelligence, humanistic ethics. *Daedalus*, *151*(2), 232–243. https://doi.org/10.1162/daed_a_01912

Tsamados, A., Aggarwal, N., Cowls, J., Morley, J., Roberts, H., Taddeo, M., & Floridi, L. (2022). The ethics of algorithms: Key problems and solutions. *AI & Society* https://doi.org/10.2139/ssrn.3662302

Vallor, S. (2015). Moral deskilling and upskilling in a new machine age: Reflections on the ambiguous future of character. *Philosophy & Technology 28*(1) (February 21, 2014): 107–124. https://doi.org/10.1007/s13347-014-0156-9

Vallor, S. (2018). *Technology and the virtues: A philosophical guide to a future worth wanting*. Oxford University Press.

Verbeek, P.-P. (2005). *What things do: Philosophical reflections on technology, agency, and design*. Pennsylvania State University Press.

# 5 Human-Centered Artificial Intelligence (HCAI)

## *From Conceptual Examination to Legislative Action*

*Pierre Larouche*

Université de Montréal, Montréal, Québec, Canada

When ChatGPT burst into public awareness in the fall of 2022, it drew attention and gave salience to a debate about the legal approach toward the current AI wave that had been going on for many years already in specialist circles. Current calls, in the media, for intervention by public authorities come at the very time when the specialist debate is moving from a prolonged phase of conceptual examination toward concrete legislative action. That action is spearheaded by the EU, where a proposal for an "AI Act"[1] is now making its way through the EU legislative process and is expected to be enacted in the course of 2023 (European Union, 2022d). Other jurisdictions, including Canada, are following.

This chapter will focus on the opportunities and challenges arising out of this transition from conceptual examination into concrete legal action. The first part points to shortcomings in the conceptual examination phase that must be overcome in the transition to concrete legal action. The second part introduces the main challenge at the legislative action stage, namely the choice of approach toward innovation. The third part illustrates how the AI Act addresses these shortcomings and deals with that challenge. The fourth puts the AI Act in a broader context in order to highlight its full impact.

## SHORTCOMINGS IN THE CONCEPTUAL EXAMINATION PHASE

In the conceptual examination phase, hundreds of groups and committees around the world delved into what became known under the broad heading of "AI and ethics." They produced many reports and studies that typically resulted in a set of ethical principles intended to govern AI (Algorithm Watch, 2020).[2] Most notably, the OECD and the G20 endorsed one such set of ethical principles, comprising (i) inclusive

 DOI: 10.1201/9781003320791-7

growth, sustainable development, and well-being, (ii) human-centered values and fairness, (iii) transparency and explainability, (iv) robustness, security, and safety, and (v) accountability (OECD, 2019; G20, 2019). In parallel, the AI community also looked into how to design AI systems with a view to making them work with humans (Human-Centered AI or HCAI).[3] This massive endeavor definitely paved the way for the substance of any legal approach to AI. The aim of this chapter is not to summarize or synthesize this body of work, but rather to point to some weaknesses that affect many of these "AI and ethics" initiatives and hinder the transition to concrete legal action. Two shortcomings stand out: the focus on AI as a standalone object of law and regulation (a form of "shiny object syndrome"), and the assumption that there is no law currently applicable to AI (the "blank page" fallacy).

## AI as a "Shiny Object"

Conceivably, there is some "shiny object syndrome" at work as regards AI. AI and ethics endeavors tend to treat AI as a standalone object for law or regulation to attach to, hence the framing as "AI regulation" and the calls to "regulate AI." Yet AI is not an autonomous entity to which legal obligations can be attached[4]; certainly, engagement and compliance with the various "AI and ethics" principles described above is not something that AI as such can undertake. Rather, like most technologies, AI will be implemented in various products and services offered by firms—profit or nonprofit—to customers or users, as the case may be ("AI Systems").[5] Some of these AI Systems will be offered directly to end-users (B2C). Other AI Systems will rather be purchased on a B2B basis by other firms to be included in their own products and services. Others still will be procured by public authorities to be used in the provision of services to citizens. For the purposes of the present discussion, the precise mechanisms by which AI will be introduced and used in practice are not material. What matters is that AI, as a technology, will be integrated into a well-known framework, comprising actors—firms, individuals, and public authorities—and mechanisms—sale contracts, employment contracts, public services, and markets, to name but the main ones. In that sense, "regulating AI" is really regulating the development, supply, and use of AI by these actors within these mechanisms. In legal terms, this involves having legal rights and regulations—reflecting the "AI and ethics" principles in substance—bear on these actors when they interact within these mechanisms.

The preceding paragraph might seem trite, but its implications are not. Once it is agreed that regulating AI means regulating firms, individuals, and public authorities as they deal with AI Systems, then the sheen of novelty enshrouding "AI regulation" dissipates. AI regulation begins to resemble other endeavors at integrating technological developments within the legal framework, in the light of their social and economic implications.

Normatively, conceiving of AI regulation in such terms is also preferable because it fits innovation theory better.[6] From a social perspective, innovation can be theorized as a combination of three elements: (i) an invention, (ii) that is diffused and adopted within society, and (iii) that has a positive impact, in the sense that it furthers public policy aims.[7] Regulating AI as a technology, without regard to its socioeconomic embedding, would amount to focusing on invention (element (i)) and ignoring diffusion and adoption (element (ii)), thereby creating a significant error risk in the assessment of impact (element (iii)). Moreover, contemporary literature points to the increasing difficulty of separating the invention and diffusion elements in the most recent technological developments,[8] a point that is certainly valid for AI as well.

## The Blank Page Fallacy

Once "AI regulation" is understood as the development of the best legal approach to the introduction of AI Systems within the existing set of actors and mechanisms, then the second shortcoming comes plainly into view.

Too many of the AI and ethics endeavors seem to consider that "regulating AI" means starting from a blank page and building up a new regulatory corpus specifically dedicated to AI from scratch. The origins of that misconception are manifold. Perhaps, it reflects a perception on the part of engineers and computer scientists that law is made up of detailed rules bearing immediately on specific situations; the absence of such rules would imply a legal vacuum. This dovetails with the widespread assumption, in the digital economy,[9] that in the absence of an explicit prohibition, something is permissible. It could also derive from the "shiny object syndrome" of the previous heading: since AI is seen as a new technological development,[10] it is presumed that no law exists for it yet.

The blank page fallacy is also reminiscent of the 1990s, with the Easterbrook/Lessig debate on the need for a specific law of cyberspace, to use the then-fashionable term.[11] As with AI now, the novelty of the Internet then, with its cross-border architecture, led many to conclude that it arose in a legal vacuum and that a new corpus of law had to be developed to govern it. Subsequent developments showed that existing law could very well evolve to handle the challenges raised by the Internet. A generation later, instead of a specific body of Internet law, we now have Internet-related extensions or subsets within the main areas of law.

Along the same lines, despite the scarcity—if not outright absence—of specific rules concerning AI as such, there is no shortage of laws that can be applied to AI, because of its embeddedness in social and economic relationships, as described in the previous heading. It extends beyond the scope of this chapter to explore them in detail, but suffice it to mention some of them briefly, starting with the fundamental rights instruments enshrined in most constitutions and international law. Even if these instruments may not always apply to all relationships, they signal the significance of fundamental rights in our societies. Similarly, the basics of all social and economic interactions are laid out in the main areas of private law, including contract, liability, and property law. Some more specific bodies of law are also especially germane to the development and development of AI, such as intellectual property law, competition law and of course privacy and data protection law. One could even venture that, in the absence of any legislative initiative on AI, a legal framework for AI would eventually evolve, as the existing law is progressively applied to AI. This would be a lengthy process, however, bound to result in mistakes and missed opportunities given the relatively slow timeframe of organic growth in the law.

The real challenge is therefore not to build AI regulation from scratch but rather to ascertain whether and how existing law is appropriate and effective to deal with

AI, and then add to the law or complement it, as the case may be. The exercise is therefore more akin to updating an old edition of a textbook than writing one from a blank page. To be sure, the challenge is no less daunting: developing and implementing a legal approach to AI will require significant effort by academics, policy-makers, and stakeholders over a prolonged time period.[12] Nevertheless, framing the task correctly is essential.

Indeed, in an unfortunate side-effect, the blank page fallacy has enabled some stakeholders to downplay the significance of existing law and portray the key substantive issues as more open-ended and indeterminate than they really are. The "AI and ethics" label did not help in that respect, given that ethics are generally perceived to be open for debate and discussion.[13] For example, as a matter of ethics one might argue about the value of seeking truth versus allowing unfettered circulation of ideas, but as a matter of law, freedom of expression is firmly entrenched in fundamental rights instruments across the world, many of which apply to private relationships as well. Similarly, the protection of privacy and personal data is not just a valid ethical concern; it is a constitutional-level value with quite a strong protective legal framework, at least in the EU.

What is more, allowing the debate to be framed as an open-ended ethical discussion over a blank legal page can be counter-productive for policy-making, to the extent that it opens the door to various delaying tactics designed to extend discussion indefinitely, while the technology continues to progress at a fast pace.

As we move out of the conceptual examination phase toward concrete legal action, the effects of the shiny object syndrome and the blank page fallacy should dissipate; however, both these shortcomings will have caused distractions and delays on the way to dealing with the key issue in developing a legal approach to AI.

## THE MAIN CHALLENGE: PERMISSIVE OR PROTECTIVE APPROACH

When seen from a broader perspective, both shortcomings—especially the blank page syndrome—reflect the prevailing innovation culture in the digital economy,[14] from which most R&D on AI emanates. That innovation culture is closely intertwined with the applicable legal framework as it relates to the governance of innovation, i.e. its promotion and its policing.

Crisply put, the digital economy grew under a permissive approach to innovation. In legal terms, the permissive approach generally allows any invention to be brought to the market and turned into an innovation, subject only to general laws (contract, liability, etc.) that apply *a posteriori*, once it is on the market. *If and once* it turns out that the invention does create harm, then legal consequences follow, and corrective measures must be taken. Under the permissive approach, the emphasis is on avoiding Type I errors (false positives), whereby an invention would be prevented from reaching the market because of misplaced *a priori* concerns. Once this legal framework is internalized into the culture of firms and users, this translates into a rush to bring inventions out, even if they have to be "updated" later, as we can all witness daily with software.

In contrast, some major economic sectors are steeped in a more protective approach to innovation. Legally speaking, no invention can then be brought to the market unless it has first been shown to meet the requisite standard of safety and harmlessness. This is the essence of pharmaceutical and medical device regulations, as well as and aircraft regulations, to name but a few. Under this protective approach, the emphasis rather lies on avoiding Type II errors (false negatives), where an invention would turn out to be harmful because the *a priori* control would have failed. When translated into corporate culture, such a protective approach typically means that product development teams try to assess the risk of harm as thoroughly as possible in the course of product development. In particular, they also try to predict and contain mistaken and ill-intentioned uses (misuses and abuses) of the invention. This allows product development to proceed more smoothly through the legal framework before the invention is finally put on the market.

Over the past decades, as digital technology spread out to other economic sectors[15]—a phenomenon often described as "convergence"—the permissive and protective approaches were increasingly brought in contact with one another. Experience so far with converged product spaces shows that, as a matter of fact, the permissive approach of the digital sector tends to prevail over the protective approach. The digital sector tends to be more dynamic than the other economic sectors with which it is converging, which puts it in the leading position in the process of change that accompanies convergence. More concretely, the permissive approach and the accompanying culture push the digital sector to bring inventions out more quickly,[16] giving the permissive approach the upper hand.

Of course, the triumph of the permissive approach upon convergence defeats the rationale that originally led to the introduction of the protective approach in one of the sectors prior to convergence. It allows for the creation of the very risk that justified the protective approach in the first place: products are put on the market without prior control and harm can ensue.[17] Yet so far little has been done to push back against this trend. The current AI wave will generate further instances of convergence, as AI—a technology driven by the digital sector—spreads out through many economic sectors (and public-service functions). For the first time, however, the empirically observable prevalence of the permissive approach might be questioned, from a normative standpoint. In the current discussions around AI regulation, few participants advocate for a permissive approach (even though the discussion is usually not framed in terms of a permissive or protective approach). Chances are that the legal approach to AI will not rest on the permissive approach that the digital economy is used to. This will then put the law in tension with the long-standing and well-entrenched innovation culture that grew around the permissive approach.

Accordingly, it is not surprising that a main argument raised against moving away from the permissive approach when it comes to AI turns around the impact on innovation. Introducing deeper scrutiny of AI Systems, especially if done prior to putting them on the market, is presented as a sure-fire way to put a brake on innovation.[18] As a matter of fact, economic sectors governed by a protective approach—for example, pharmaceuticals and medical devices—evidence significant amounts of innovation. To a large extent, innovation incentives always remain, irrespective of the legal framework.[19] Even though arguments relating to innovation may not be that strong, these arguments might very well prove influential enough in policy-making circles, so that as a result AI Systems would not be put under a protective approach as we know it from pharmaceuticals and other sectors.

Perhaps the most convincing reason not to introduce a protective approach for AI Systems does not turn around innovation, but rather information. AI is a notoriously opaque technology, even to its conceptors. For a regulatory authority to reach the level of information required to make a firm determination on allowing or prohibiting a specific AI System would require considerable effort on its part. Even in the comparatively easier informational environment of pharmaceutical products, marketing authorization procedures are lengthy (extending over years) and expensive (in the billion-dollar range).

With it comes to AI, a permissive approach would thus involve excessive risk-taking from a social perspective, while a protective approach would be unpracticable because of information concerns. A compromise path must therefore be sought. It could very well come from the literature on "responsible innovation," the canonical definition of which runs as follows:

> a transparent, interactive process by which societal actors and innovators become mutually responsive to each other with a view to the (ethical) acceptability, sustainability and societal desirability of the innovation process and its marketable products (in order to allow a proper embedding of scientific and technological advances in our society).
>
> (von Schomberg, 2011; 7–9)

Responsible innovation involves an embedding of public policy objectives into the R&D and diffusion processes of firms, and a dialogue between firms and societal actors (extended to public authorities as well). When compared with either the permissive or the protective approach, responsible innovation implies a much greater involvement of firms in the life of regulation. From the perspective of firms, such involvement is necessary to ensure that their inventions are acceptable (the "trust" issue which plays a central role in the legal approach to AI). From the perspective of public authorities, this is the best path to solving the information problem, even though it entails sharing the regulatory burden with firms (in an elaborate form of co-regulation). This approach was already pioneered in the GDPR, with its provisions on "privacy by design" and "privacy by default." (GDPR, 2016) [20] The discussion around HCAI shows that many members of the AI community, across firms and other institutions, are willing to take up the challenge of integrating the consideration of public policy objectives into AI design.[21]

## THE FIRST LEGISLATIVE INITIATIVES: THE AI ACT IN THE EU AND BILL C-27 IN CANADA

The EU has taken the initiative globally with its AI Act proposal,[22] which is set to be enacted later this year.

In the commentary so far, the layered risk-based structure of the AI Act has drawn the most attention. The level of risk is assessed by reference to the purpose for which an AI System is used. In essence, a limited set of "AI practices" are deemed so harmful that they are prohibited outright.[23] Below these, a set of "high-risk AI Systems" is subject to an extensive set of legal obligations; we will focus on these below.[24] At a lower risk level, certain AI Systems are put under some transparency obligations.[25]

Finally, any AI System not falling in the first three categories is simply left for existing law to govern, without any additional measure.[26] In the course of the legislative procedure, provisions have been added to deal with general-purpose AI Systems, which can be used for various purposes falling within one or the other of the above risk-level categories.[27]

The limited scope of this chapter does not allow for a full examination of the AI Act. Rather, I will focus on how the AI Act reflects the foregoing discussion, namely:

- Avoidance of the shiny object syndrome by putting AI regulation in context and attaching regulation to existing actors and mechanisms rather than to AI as such;
- Rejecting the blank page fallacy in favor of reliance on existing law and legal institutions, that are then extended and shaped to fit with AI;
- Trying to find a way between the permissive and protective approaches by pushing firms toward responsible innovation.

The proposed AI Act is a typical piece of EU Internal Market legislation. It is concerned with ensuring the free circulation of AI Systems in the EU by putting forward a harmonized regulatory framework to attain relevant public policy objectives.[28] Of course, this balanced approach—market access and free circulation in return for compliance with public policy objectives relating to health, safety, and fundamental rights—follows directly from the choice of a legal basis, namely Article 114 TFEU. In that sense, the approach was pre-ordained. Nonetheless, it chimes with the dominant regulatory model in the EU, which integrates regulation within a competitive market economy and seeks to graft regulation onto existing actors and mechanisms. The AI Act does not directly regulate AI as such. As regards high-risk AI Systems, the AI Act does specify a set of requirements that high-risk AI Systems must comply with[29]: risk management, data and data governance, technical documentation, record-keeping, transparency and user information, human oversight, accuracy, robustness, and cybersecurity. These requirements are addressed to firms, and they govern the conditions under which AI Systems can be brought to the market by these firms. Firms are therefore responsible to ensure that high-risk AI Systems comply with the applicable requirements under the AI Act.[30]

Similarly, the AI Act links with other pieces of EU legislation, and it borrows existing legislative and regulatory techniques. Instead of drawing up AI regulation from scratch, the AI Act seeks to "normalize" the legal treatment of AI as much as possible. The regime for high-risk AI Systems which forms the core of the Act illustrates this point perfectly. High-risk AI Systems are partly defined by reference to existing EU legislation governing products deemed to create risks (e.g., pressure equipment, personal protective equipment, medical devices, motor vehicles, etc.): AI Systems that constitute one of these products or a safety component thereof are automatically classified as high-risk systems. Furthermore, the legal mechanism by which providers of high-risk AI Systems can gain access to the market is picked right out of the established EU playbook for risky products. Providers need to submit their systems to conformity assessment[31]; if the assessment is successful, a declaration of conformity is drawn, and the high-risk AI System can then be marketed.[32]

Finally, the proposed AI Act also tries to introduce a "responsible innovation" approach to the regulation of high-risk AI Systems. Here as well, the Act relies on tried-and-true recipes from EU product regulation,[33] with a view to push firms to take responsibility for the achievement of public policy objectives and internalize them in their development processes. As mentioned above, firms bear responsibility for ensuring that high-risk AI Systems are compliant, which they can discharge through conformity assessment. What is more, the AI Act makes room for the development of harmonized standards by European Standardisation Organisations (ESOs),[34] in order to specify the technical implementation of the requirements to be met by high-risk AI Systems.[35] Compliance with these harmonized standards leads to a presumption of conformity with the requirements of the AI Act. Firms can participate in the work of ESOs and thus play a role in standard development.

In parallel with the proposed EU AI Act, Canada is also debating a bill for an Artificial Intelligence and Data Act (AIDA) (Canada, 2022).[36] Depending on the speed of the Canadian legislative process, the Canadian AIDA could be enacted before the EU AI Act. The AIDA, while not as elaborate as the AI Act,[37] follows the same path. AI regulation is conceived of as regulation of how firms put AI Systems on the market, and it focuses on high-impact AI Systems. In contrast with the AI Act, however, the AIDA leaves limited room for firms to become involved in AI regulation: there is no conformity assessment system, nor does the AIDA make any room for the use of standards in assessing compliance with its provisions. Rather, the AIDA relies on a more old-fashioned command-and-control model, whereby government regulations will specify what firms need to do or avoid doing.

## THE AI ACT IN A BROADER CONTEXT

In order to grasp the full impact of the AI Act, it is necessary to place it in the broader context of a series of major EU legislative initiatives since 2020. They include the 2022 Digital Services Act (DSA), which updates the law governing online intermediaries: while paying lip service to the regime of liability exemptions in force since the beginning of the century (European Union, 2020), the DSA introduces much heavier duties on online intermediaries to police the content circulating on their platforms (European Union, 2022a). Also enacted in 2022, the Digital Markets Act (DMA) imposes obligations on large platforms (gatekeepers) in order to prevent anticompetitive practices affecting firms using those platforms to reach their customers (European Union, 2022b). In parallel with the AI Act, the EU institutions are also considering proposals for a Data Act (European Union, 2022c) to enable the free circulation of nonpersonal data between firms in order to boost EU industry.

One key tying feature of these legislative initiatives is that they all apply to the major digital platforms (commonly referred to as GAFA[38]). Without a doubt, these giant firms are considering these initiatives as one whole when deciding on their strategic positioning. While the legislative initiatives, taken together, might seem to impose quite a burden on these firms, at the same time they all make room for industry to participate in the implementation. Much like under the AI Act, firms therefore have ways and paths to become partners in their own regulation under the DSA, DMA, and Data Act. Actually, the aggregate picture would even warrant a stronger conclusion: the success of these legislative initiatives depends in no small part on participation from the industry—in particular the platform giants—in a form of co-regulation relationship, whereby public authorities give the policy impulse, but industry supplies the technical and commercial information that is essential for the public authorities to be able to implement the law effectively (de Streel & Larouche, 2022).

Furthermore, reading between the lines of these initiatives, an overall EU strategy emerges, whereby the EU aims to be the first mover, so that its legislation would be emulated abroad and serve as a model globally.[39] This phenomenon, which could be observed with the GDPR, has been dubbed the "Brussels effect" (Bradford, 2020). This strategy seeks to ensure that the policy preferences of the EU gain global acceptance by becoming a *de facto* reference for the industry, even if no *de jure* convergence can be achieved because of diverging positions taken by public authorities abroad. Here as well, the ability of the EU to involve industry and to get industry to internalize EU legislation and regulation (by way of responsible innovation) will prove crucial to this enterprise. The recent European Standardization Strategy also contemplates that the EU will try to rely on the significant involvement of European standardization bodies and firms in the global standardization ecosystem to achieve better promotion of EU policy preferences (European Commission, 2022).

## CONCLUSION

The debate around the legal approach to AI is now moving from a conceptual examination stage (AI and Ethics) to a stage where concrete legislative and regulatory actions are undertaken. This chapter showed that, in the process of transitioning from one stage to the other, two major shortcomings of the conceptual stage are being addressed. These were termed the "shiny object syndrome"—whereby AI is seen as a standalone regulatory object—and the "blank page fallacy"—whereby it is assumed that the task of regulating AI starts from a blank page.

In the course of correcting these shortcomings, a central issue is also tackled, namely whether AI should be placed under a permissive or protective approach to innovation. These approaches typically filter into the innovation culture that animates the industry. As a starting point, AI is associated with the digital economy, which has blossomed under a permissive approach. So far, whenever the permissive approach has clashed with the protective approach as two sectors converge, the permissive approach gained the upper hand in practice. AI is set to be the first case of convergence where an outcome is deemed undesirable; however, a protective approach does not appear appropriate either. A third way can be found in the "responsible innovation" concept, whereby firms need to internalize public policy objectives and become active partners in their own regulation.

The proposed EU AI Act is the first major AI regulation initiative worldwide. It can be seen that the AI Act avoids the shiny object syndrome and the blank page fallacy. The Act also points toward a "responsible innovation" approach, which will require major changes in the culture of firms that are used to operate under a permissive approach to innovation.

Seen in the broader context, the AI Act and the other major EU legislative initiatives regarding the digital economy are likely to become models, or at least benchmarks, for other jurisdictions. They point to a long, iterative process of progressively developing and implementing a mix of existing law, extensions of existing law, as well as some new law, to AI. At all stages, it is likely that industry involvement—in the form of co-regulation—will be not only useful but essential to ensure the effectiveness of the law, given the severe information imbalance affecting public authorities when it comes to AI.

## NOTES

1 Throughout this chapter, references to the proposed AI Act are made to the latest major stable version, namely the so-called "general approach" that arose out of the Council discussions on the original Commission proposal: European Union (2022d).
2 See the list compiled by AlgorithmWatch at https://inventory.algorithmwatch.org, which included 167 sets of guidelines as of April 2020.
3 See, for an overview, the leading work of Shneiderman (2022).
4 Leaving aside for now the discussion of whether AI could or should have legal personality, which has generally led to a refusal to do so. See the overview in Chesterman (2020).
5 "AI Systems" is the term used in the AI Act, *supra* note 1 as well as Bill C-27 in Canada, *infra* note 36 to designate AI technology turned into a product.
6 As is true for any other technology.
7 See Butenko and Larouche (2015). If the invention, once diffused and adopted, conflicts with public policy, then it would either be a "bad innovation" overall, or simply not an innovation, from a social perspective.
8 See the literature listed in West et al. (2014).
9 See *infra*, heading 2 on the innovation culture prevalent in the digital economy.
10 Even though we find ourselves in at least the third wave of AI: see Deng (2018).
11 See Easterbrook (1996) and the reply by Lessig (1999).
12 *Infra*, headings 4 and 5.
13 See the very fundamental choices articulated by Tasioulas (2022), as late as last year. To be clear, many ethicists are aware of the presence of legal constraints: see a.o. Floridi (2018), clearly laying out the interplay between law and ethics and making a difference between hard and soft ethics.
14 "Digital economy" is the latest moniker for what was known variously over time as the IT sector, the ICT sector, the digital industry, the online industry, etc.
15 For instance, health, energy, and transport.
16 Think, for instance, of the rapidity with which personal devices with health-related functions (heart rate monitors for running watches in earlier times, now smartwatches with health sensors) were issued and then became ingrained in our lives, because they emanated from the digital sector as opposed to the medical device industry.
17 The best example is perhaps the software failure that played a central role in the crash of two Boeing 737Max planes: see Travis (2019) for an account of how the innovation culture of the software industry corrupted that of the hardware (airplane) industry.
18 See for instance the criticism made by Grady (2023) or Thierer (2022).
19 Of course, it is impossible to know if these sectors would be even more innovative with a more permissive approach.
20 Art. 25.
21 *Supra* note 3.
22 *Supra* note 1.

23 AI Act, *supra* note 1, Art. 5. They include subliminal techniques, exploiting vulnerable persons, social scoring systems, and facial recognition or other remote biometric identification techniques for purposes of law enforcement (save in limited cases).
24 Ibid., Art. 6-51. The list of high-risk use cases is more elaborate and cannot be quickly summed up. See Art. 6 and Annexes II and III.
25 Ibid., Art. 52. These include AI Systems designed to interact with humans or generate content that could be misinterpreted (deep fakes).
26 Ibid., Rec. 1.
27 Ibid., Art. 4a.
28 Ibid., Rec. 1 and Art. 1.
29 Ibid., Art. 8-15. This is where the AI Act connects with the AI and Ethics discussion, by defining broad principles applicable to high-risk AI Systems.
30 See Art. 16–29, which contain an elaborate set of provisions applicable to providers of high-risk AI Systems to ensure compliance with the provisions of Art. 8-15, buttressed by complementary obligations imposed on importers, distributors, and even users themselves.
31 Art. 43.
32 Art. 19.
33 These may appear novel to firms operating in sectors covered by a permissive approach to innovation, but are not.
34 Among ESOs, at this point in time, the Commission decided to entrust standard development to CEN/CENELEC, and not to ETSI.
35 Art. 40. The requirements are listed *supra* note 29.
36 Being part 3 of Bill C-27, *An Act to enact the Consumer Privacy Protection Act, the Personal Information and Data Protection Tribunal Act and the Artificial Intelligence and Data Act and to make consequential and related amendments to other Acts*, tabled on 16 June 2022. The AIDA follows in the footsteps of the pioneering *Directive on Automated Decision-Making* adopted by the Treasury Board in 2019 (latest version Government of Canada (2023)), which introduced an Algorithmic Impact Assessment for procurement of AI systems used for decision-making by the Canadian federal government.
37 A number of key elements are left to be developed in implementing regulations.
38 The acronym stands for the set of Google (now Alphabet), Amazon, Facebook (now Meta), and Apple. Microsoft or Netflix is sometimes added to the set (GAFAM or FAANG, respectively). Chinese giants such as Baidu, Alibaba, or Tencent are also added to the set at times.
39 As seen above, the Canadian AIDA is aligned with the proposed EU AI Act.

## REFERENCES

AlgorithmWatch. (2020). *AI Ethics Guidelines Global Inventory*. Retrieved from https://inventory.algorithmwatch.org

Bradford, Anu (2020). *The Brussels Effect: How the European Union Rules the World*. Oxford: OUP.

Butenko, A., & Larouche, P. (2015). "Regulation for Innovativeness or Regulation of Innovation?" *Journal of Law, Innovation and Technology*, *7*(1), 52–82.

Canada. (2022). Bill C-27, An Act to enact the Consumer Privacy Protection Act, the Personal Information and Data Protection Tribunal Act and the Artificial Intelligence and Data Act and to make consequential and related amendments to other Acts, tabled on 16 June 2022.

Chesterman, S. (2020). "Artificial Intelligence and the Limits of Legal Personality." *International & Comparative Law Quarterly*, *69*(4), 819.

de Streel, A., & Larouche, P. (2022). *A Compass on the Journey to Successful DMA Implementation.* Concurrences N°3-2022. Retrieved from https://ssrn.com/abstract=4215925

Deng, L. (2018). "Artificial Intelligence in the Rising Wave of Deep Learning." *IEEE Signal Processing Magazine, 35*(1), 180.

Easterbrook, F. H. (1996). "Cyberspace and the Law of the Horse." *University of Chicago Legal Forum, [1996],* 207.

European Commission. (2022). An EU Strategy on Standardisation – Setting global standards in support of a resilient, green and digital EU single market COM(2022)31.

European Union. (2020). *Directive 2000/31/EC of 8 June 2000 on electronic commerce [2000] OJ L 178/1.*

European Union. (2022a). *Regulation 2022/2065 of 19 October 2022 on a Single Market for Digital Services (Digital Services Act) [2022] OJ L 277/1.*

European Union. (2022b). *Regulation 2022/1925 of 14 September 2022 on contestable and fair markets in the digital sector (Digital Markets Act)* [2022] OJ L 165/1.

European Union. (2022c). *Proposal for a Regulation on harmonised rules on fair access to and use of data (Data Act)* COM(2022)68 (23 February 2022).

European Union. (2022d). *Proposal for a Regulation laying down harmonised rules on artificial intelligence (Artificial Intelligence Act),* General Approach, CONS 15698/22 (6 December 2022).

Floridi, L. (2018). "Soft ethics, the governance of the digital and the General Data Protection Regulation." *Philosophical Transactions of the Royal Society A, 376*(2128), 20180081.

G20. (2019). *Ministerial Statement on Trade and Digital Economy* (9 June 2019).

General Data Protection Regulation (GDPR). (2016). *Regulation 2016/679 of 27 April 2016 [2016] OJ L 119/1, Art. 25.*

Government of Canada. (2023). *Directive on Automated Decision-Making.* Retrieved from https://www.tbs-sct.canada.ca/pol/doc-eng.aspx?id=32592

Grady, P. (2023). "The AI Act Should Be Technology-Neutral." *Center for Data Innovation* (1 February 2023). Retrieved from www.datainnovation.org

Lessig, L. (1999). "The Law of the Horse: What Cyberlaw Might Teach." *Harvard Law Review, 113,* 501.

OECD. (2019). *Recommendation of the Council on Artificial Intelligence, OECD/LEGAL/0449* (22 May 2019).

Shneiderman, B. (2022). *Human-Centered AI.* Oxford; OUP.

Tasioulas, J. (2022). "Artificial Intelligence, Humanistic Ethics." *Daedalus, 151,* 232.

Thierer, A. (2022). "Why the Future of AI Will Not Be Invented in Europe." Retrieved from www.techliberation.com (1 August 2022).

Travis, G. (2019). "How the Boeing 737 Max Disaster Looks to a Software Developer." *IEEE Spectrum* (18 April 2019).

von Schomberg, R. (2011). "Introduction." In R. von Schomberg (Ed.), *Towards Responsible Research and Innovation in the Information and Communication Technologies and Security Technologies Fields* (pp. 7–9). Luxembourg: EU Publications Office.

West, J., et al. (2014). "Open innovation: The next decade." *Research Policy, 43,* 805–806.

# 6 Privacy in the Future Era of AI

*Matt Malone*
Thompson Rivers University, Kamloops, British Columbia, Canada

Defining privacy is an elusive task (Hunt, 2015: 161–162).[1] Daniel Solove, an authority on privacy law and theory, describes it as a concept in "disarray" (Solove, 2008: 1)—something that cannot "be reduced to a singular essence" (Solove, 2013: 24). Instead, it is a smattering of related concepts best understood "pluralistically rather than as having a unitary common denominator" (Solove, 2008: 9). In the digital era, this fundamental aspect, combined with the rise of information and computer technologies, has created many significant privacy conundrums. As Luciano Floridi notes, such technologies have simultaneously augmented and eroded informational privacy, posing challenges to policy approaches predicated either on regulating activity that results in undesirable consequences (the "consequences" approach) or that violates human rights or welfare (the "rights" approach) (Floridi, 2005: 193–194). The consequences approach has struggled to address the proposition that "a society devoid of any informational privacy may not be a better society," while the rights approach confronts definitional issues of mixed public–private information and imprecision around foundational concepts like ownership (Floridi, 2005: 194). Such questions reveal that scholars might not be able to agree on what privacy *is*, even if they tend to know what it is *about*. Whether it implicates control over the collection, storage, use, or disclosure of information (or the consent to such practices by others) or whether it is about a "right to be let alone"[2] in "free zones" (Solove, 2013: 50) away from others' scrutiny, interference, intrusion, or access, privacy is power.[3] With increasing calls for human-centered AI—aligning the technology to the flourishing of people and their interests—addressing issues about AI's impact on privacy remains a pivotal question.

AI technology challenges privacy in two significant ways. First, there is a consent gap. Existing regimes of consent and awareness that seek to empower individuals and groups are challenged by the type of increasingly versatile and ubiquitous data collection that feeds AI systems. Nowadays, individuals are rarely capable of fully understanding where, when, or how their data is collected and then used by AI systems. Second, there is a knowledge gap. Full comprehension of the processes used by most AI systems is unrealistic. Even for experts, the synthesizing and inferring that occurs by complex—often black box—algorithms using massive datasets is beyond human cognitive capacities. Moreover, because AI systems often set "unconstrained variables to extreme values[,] if one of those unconstrained variables is actually something we care about," the systems are able to wrest it from our control and make

 DOI: 10.1201/9781003320791-8

the "solutions" it produces highly undesirable (Russell, 2014)—raising significant questions about alignment between human values and technological development. This reality is only compounded by a lack of transparency that results in AI systems accomplishing their goals with sometimes unintended, harmful effects, or with emergent goals divorced from human oversight.

These consent and knowledge gaps result in perpetual intrusions into domains privacy might otherwise seek to control. For example, to make AI systems more human-value aligned, including more respectful of privacy, governments are seeking to regulate AI inputs and processes. At the input level, these efforts include bolstering notice and consent regimes and requiring robust data handling safeguards (Canada, 2021a); at the process level, they include rights to explanations and other forms of accountability such as validation studies, bug reports, and so on (Canada, 2021b). But given how useful the outputs of AI systems can be, any desire for control over input or process can quickly lose any sense of priority or urgency for all but the most zealous of privacy advocates. The allure of AI outputs that are strikingly useful, precise, efficient, accurate, and reliable—contributing to their increasing capture of human time, attention, and trust—often renders attempts to exert control over the inputs or the processes of AI moot. The much-debated issue of whether or not we can introduce human alignment into emergent and evolvable intelligence has led cautionary experts to argue that, while we are achieving technical progress, AI may move beyond these initial constraints imposed on it to achieve its own best goals.

While we have not consented to all the trade-offs of this bargain—and are rarely capable of doing so—we increasingly consent to giving AI systems our most precious resources: our time, attention, and trust. Understanding the consequences of this transfer of power is one of the main focuses of this article, which seeks to explore privacy as a transversal concept in the past, present, and future. Privacy is not just a value, but a *reaction* to technology. Privacy determines how far we let technology reach into spheres of human life and consciousness. With AI arguably surpassing the rate of human evolution given the number of mutations and variations that can be introduced and evaluated in shorter periods of time, privacy has become cybernetic—in flux with AI technologies themselves and interacting with them constantly to redraw its contours. Like the proverbial "terms of use" forms that continue to inform us about new encroachments on our privacy interests through novel uses of our personal information, the nature of privacy interests is constantly changing. On occasion, drama erupts where privacy interests alert us to unwanted intrusions (Nissenbaum, 2010: 10). But as those shocks fade, privacy is quickly redefined and reconceived, and as AI captures more time, attention, and trust, privacy will continue to play a determinative role in drawing the boundaries between human and technology.

## THE PAST

It may be helpful to understand how we arrived here.

Throughout history, the legal codes of many jurisdictions have traditionally granted privacy rights to protect citizens against intrusions by the state.[4] Later, the articulation of privacy as a "right to be let alone" (Warren & Brandeis, 1890: 193) in broader social contexts became a lynchpin in the struggles for the right to contraception for married couples[5] and unmarried couples,[6] abortion,[7] same-sex sexual activity,[8] and same-sex marriage.[9] Notions of privacy in a consumer context then took hold, defining privacy as "the claim of individuals, groups, or institutions to determine for themselves when, how, and to what extent information about them is communicated to others" (Westin, 1967: 7). The flagship of these legislative efforts to date has been the European Union's *General Data Protection Regulation*, which created rights for data subjects pertaining to transparency, access, rectification, and automated decision-making (including "the right not to be subject to a decision based solely on automated processing").[10] The GDPR was, however, deeply flawed, as its approach to privacy and data protection adhered to a vision of privacy where the main guardrails on information use were regimes of notice and consent.[11] Since the passage of the GDPR, the democratic (and nondemocratic) world has been awash in privacy and data protection legislative efforts. Some are data-specific laws, including subject matter-specific legislation (e.g., biometric data) and instance- or conduct-specific laws (e.g., data breach notification, specific penalties for the disclosure of intimate images, etc.) This governance framework fundamentally creates gaps in protection, conceptual fragmentation, and unprincipled decision-making.[12] While some argue it has a contagious or catalyzing effect, (Chander et al., 2021) its true legacy is a lack of a meaningful response and the "disarray" previously noted. This approach *welcomes* novel invasions of privacy.

Where privacy conceptualizes individuals *qua* consumers' exerting "power to conceal information about themselves that others might use to their disadvantage" (Posner, 1983: 271), some have begun to argue that a proper response requires the overt articulation of privacy as a fundamental human right (American Civil Liberties Union, 2015). Still others, such as Helen Nissenbaum, have stated that any paradigm of individual rights may be insufficient—suggesting collectivist paradigms are a necessary part of any future regulation (Nissenbaum, 2010; Kaminski, 2019: 1553). Others have said the focus should be on tiers of data based on their sensitivity (Solove, 2023). These voices argue that the current consumer approach to privacy and data protection is flawed only by the failure to distinguish between different types of sensitive data—with certain advocates suggesting enhanced privacy protection requires a focus on the type, quality, or nature of the subject matter. This "data sensitivity" approach calls on us to tier subject matter and associated protections.[13] But this approach ignores the reality that acts of misuse of personal data generally *activate* underlying privacy interests—and are not often prone to anticipation in light of continual transformations in technology. Moreover, the focus on *content* has resulted in privacy and data protection regimes that have unhelpfully concentrated on questions like the accuracy of information,[14] while ignoring more pressing issues around whether the collection, storage, and use of the data was even necessary to begin with,[15] as well as what harms emanate from certain uses and practices (Kaminski, 2019: 1553).[16] Questions of use and practice almost necessarily engage thornier public interest questions that are present in other contexts of confidentiality and secrecy, where the public interests in nondisclosure (e.g., maintenance of important commercial interests that are vital to the operation of the economy) come into conflict with public interests in disclosure (e.g., having open access to information).[17] Privacy breaches must, in fact, reckon with such public interest concerns on a regular basis, as seen in contexts as diverse as whistleblowing, journalistic disclosures, breaching privacy to protect safety or prevent harm, and doxing in the public interest (Malone, 2022b).

## THE PRESENT

Perhaps the most important shift in privacy today is who does the intruding.

Contrary to the caution from George Orwell (1949) in *1984* about the greatest risk of surveillance coming from state power, the rise of surveillance is no longer only from state actors but "other state agents, businesses, and even individuals" (Müller, 2020). The traditional frameworks of privacy protection were not set up for this paradigm. Where state action was the focus, today governments can, and often do (Standing Committee, 2022), easily obtain necessary information from *private* third parties (Solove, 2013: 102–103). This so-called third-party doctrine, by which individuals who voluntarily provide information to third parties hold no privacy claim over such information when it is then obtained by the state, has subverted both privacy and search and seizure in the modern era. Many of these third parties are also just as powerful as states (e.g., Apple's market valuation exceeds the GDP of two G7 nations).[18] The paradigm of surveillance capitalism in which personal data is captured and commodified to reinforce and sustain the targeting of technology users as consumers is now a common idiom of this era.

AI is accelerating these trends, arrogating increasing power in the form of human time, attention, and trust. AI is power-seeking itself, for the basic reason that the accumulation of power allows it to better achieve its primary goals (Carlsmith, 2021). Research also indicates algorithms seek power in a wide range of environments (Turner et al., 2021). Indicative of this power is the way public discussion about AI is influenced by actors developing these technologies. It is axiomatic that power influences how we talk. "Political language," as Orwell (2006) wrote, "is designed to make lies sound truthful and murder respectable." As Carissea Véliz (2020) has highlighted, privacy-eroding technologies of the digital era have staged an impressive fight against privacy rights at the discursive level. "Privately owned advertising and surveillance networks are called 'communities'," she writes,

> citizens are "users," addiction to screens is labelled "engagement," our most sensitive information is considered "data exhaust" or "digital breadcrumbs," spyware is called "cookies," documents that describe our lack of privacy are titled "privacy policies," and what used to be considered wiretapping is now the bedrock of the internet.

If privacy is a semantic battleground, language has become its weapon; it follows that the most effective form of regulation is often consumer protection legislation that codifies penalties around misrepresentation and false advertising (Roth, 2022).

Although privacy's modern interfaces often give users perceived control, they are premised on doctrines that no longer function. These dynamics are becoming pronounced, as data trading begins to operate across almost every societal institution (Véliz, 2020). The nearly unhindered collection, retention, and use of personal data is creating significant power asymmetries. The ingestion of data from and about individuals is then purposed toward the identification of patterns, connections, categorizations, classifications, and correlations. This is occurring with technologies that are also, in most cases, opaque and proprietary—protected under legal doctrines of trade secrecy or confidential information. These forms of power and vulnerability are not always tangible. First, AI exponentially complicates the paradigm of data utilized for "unintended purposes not disclosed to users at the time of the collection" (Manheim & Kaplan, 2018: 121). Second, the rapid evolution of AI models makes transparency and accountability elusive, since the training of algorithms can change their nature (Kaminski, 2019). These dual attacks against human control must also reckon with the allure of the technology's outputs. As AI technologies gain human time, trust, and attention, they frequently "stand poised to supersede much of what we humans do" (Lobel, 2022)—whether the task is enhancing searches, drafting an email, commanding a vehicle, or alerting humans of imminent risk. The benefits of these technologies often blind us to their cost. Just as the rhetoric of national security and safety are coopted by states to justify intrusions of privacy, state actors consistently and often invoke public safety to dismiss privacy and data protection laws (Malone, 2022a). A similar dynamic can be observed in the recent push for the adoption of such technologies during the COVID-19 crisis (American Civil Liberties Union, n.d.).

To be sure, the present still provides many examples of the human attachment to privacy. Indeed, privacy tends to be valued by those who design and own AI systems. A visit to Silicon Valley will reveal no shortage of people working at the epicenter of the AI economy who refuse to adopt its technologies (or let their family members do so). To bolster their privacy, Meta CEO Mark Zuckerberg purchased houses surrounding his (Bayly, 2016), former Alphabet CEO Eric Schmidt requested the deletion of records on Google's search function (Rosoff, 2011), and Twitter CEO Elon Musk suspended accounts that tracked the location of his private planes (Wile, Collier, & Helsel, 2022). "That famous and powerful people assiduously guard their privacy should come as no surprise," writes Véliz (2020). "It is something we all value. Otherwise, we would freely provide our passwords."

## THE FUTURE

As AI continues to monopolize human time, attention, and trust, the transformation of AI technologies into extensions of humans will engender cybernetic dynamics upon notions of privacy. In other words, privacy will be in flux with the acceptance or rejection of AI-driven technologies. This cybernetic view of privacy is not new. What people have demanded, wanted, or articulated as private interests has evolved over time. In the past, privacy protections against state intrusiveness sought protection against state action in specific *places*—often in the home. Rights against unreasonable search and seizure contoured physical locations. Later, privacy protections were identified as surrounding *persons*—as privacy law came to protect practices around condom use, abortion, and intimate sexual acts. Now privacy interests are often felt arising in electronic spaces around the use of personal data (and other types of data). The rise of AI has obfuscated another massive encroachment on privacy.

Against this backdrop, one might query whether privacy as a value can, or will, endure at all. This question is not new. Philosopher Alan Watts (n.d.) asked in the 1960s whether network expansion would, in time, "abolish privacy." Such a vision flows from the recognition that, in the digital era, from the "vast ocean of data, there is a frighteningly complete picture of us" (Smolan, 2016). Watts viewed technologies as "extensions" of humans in the same manner as Marshall McLuhan (1964), who argued that new media technologies amplify and replace parts of the human body, fundamentally altering organisms themselves. This theory sees the adoption of technologies as extensions of the self, where disruptions or breakdowns in the function of a given technology, such as the loss of one's glasses, headphones, or car keys—or, today, the hacking of one's digital avatar or twin—can intensely disrupt function and wellbeing. Likewise, when people try to adopt "digital minimalism" and reduce their use of technology, they often experience withdrawal (Newport, 2019). As Watts (n.d.) stated:

> The distinction of the artificial from the natural is a very artificial distinction. The [technological] constructs of human beings are no more unnatural than bees' nests and birds' nests, and constructs of animal and insect beings. *They are extensions of ourselves.* (emphasis added)

Such a proposition taken to its extreme holds that AI presents a path for the continued evolution of humans, rather than biological evolution—that is, AI initially created by, but now separate from, humans, presents the next stage in evolution. The doomsday version of this proposition is that AI might be the next evolutionary step for life on Earth, with humans on the way out.

"So, what," asks Watts (n.d.), "about the situation when it arises that we are all computerised?" In the near future, when even more of life will be lived online, and when avatars and digital twins are ubiquitous, it is not difficult to imagine "electronic echoes" of ourselves reverberating from data extracted from our persons, but existing separately in electronic networks and computers. Will these electronic echoes fundamentally render privacy illusory? Will anything about us truly be private? One way to challenge this question is to dispute the nature of privacy itself. Humans are an accretion of social inputs; even if we have an innate grammar, it is hard to dispute that languages are learned by copying expressions of external social networks. Knowledge organically acquired from these inputs is, in some respect, not ours—not private. Following this argument, the sources influencing and working on individuals suggest that even one's personal thoughts are not private zones but, rather, zones of public influence. (Interestingly, in many depictions of science fiction, telepathy, or communication without recourse to physical channels or interaction, is a common trope associated with futurized societies; this common trope of such a deep invasion into a private sphere might have something to say about how we anticipate reconceiving privacy for new technologies in the future.)[19] "You are not nearly as much a private individual as you think," states Watts. "You are also, of course, exercising these influences upon other people."[20] Conceptualizing this "interiorized other"—the sum of these voices inside of us, which makes the self indistinguishable from others—poses important questions about the limits to privacy as a source or catalog of rights "to which a person is inherently entitled simply because she or he is a human being" (Sepulveda et al., 2004). Such a view seems not only to channel the values of certain cultures but also to deprive us of the benefit of insights that could arguably come from surrendering privacy interests. If everything about the self is knowable, measurable, and inferrable, there are clear benefits that may arise from the analysis of constant inputs. What these developments will mean pertains to how AI technologies cross some of the final barriers to inferring, anticipating, and reading human cognitions and thoughts. Technological developments that permit reading humans' thoughts and emotions will challenge human capacity to withhold information in these spaces at their will and to prevent others from entering these spaces. The dearth of "mental privacy" laws that exist today has opened an entirely unregulated space for the growth of this neurotechnology, whose architecture is already in place with emergent technology like wearable devices, facial recognition, and other biometric-harnessing technology.

This observation brings us to the unique problems to be faced by societies deeply rooted in discourses of human rights—societies in which the dignity of the human is the touchstone of socio-legal architecture. Even where AI creates greater *equality*, it is unlikely to produce greater *individuality*—a foundational concept to Western democracies for the last three centuries. In *The Equality Machine*, Orly Lobel (2022: 3) offers a stirring defense of AI for its promise to enhance equality, which she states is "today's foremost imperative." Lobel argues AI might ultimately make it easier than ever to assure that our workplaces reflect the averages necessary to meet various equality imperatives in hiring, representation, and so forth—serving as an effective "smoke detector" for these problems. (Lobel, 2022) As she writes: "We worry that algorithms are black boxes—in other words, opaque and difficult to understand (which they often are). But what about the black box of the human mind?" (Lobel, 2022 : 46). Lobel (2022: 46) points to the example of "[h]uman decision-making in the hiring realm," which "involves dozens of recruiters, interviewers, co-workers, clients, and supervisors, each a small black box of their own." She argues AI can often thwart many of the adverse effects such as discrimination that can occur during these processes. AI might help assess and monitor issues of equity in the employment or the delivery of services, promising more *equal* outcomes than those produced by systems powered by humans. Although the clear rebuttal, which she addresses, is that AI learns from provided inputs and rules; so when they are imperfect, AI can develop its own biases.[21]

Regulations aimed at measuring or tamping down on the excesses of these technologies—for example, reporting and monitoring of excesses (just as polluters are asked to report and monitor their externalities)—might help address some of the adverse effects of AI. Among these efforts is the call of this book—to make AI human-centered. But Floridi calls such terminology "anachronistic" for being "both trivially true and dangerously ambiguous" (Floridi, 2021). He is not wrong. For one thing, *equality* as celebrated by Lobel may oppose how humans wish to exercise their *individuality*, which grounds current approaches to privacy regulation premised on individual rights like notice, consent, and control. This is a core reason why regulation of AI's inputs and processes has proven so elusive. Even democratic governments that espouse a commitment to rights frameworks rooted in the dignity of individual humans are increasingly deploying AI to obtain cost, speed, and accuracy advantages in the delivery of goods and services—flattening their citizens into data to extract and make decisions in deeply objectifying and intrusive ways. The lack of consent to data collection and the nonreviewability of those processes, as noted above, means the outputs often undercut human dignity. And yet we continue to opt in, as evidenced in the amount of human time, trust, and attention these systems now receive. The collapse in institutions of confidence and trust in the last 50 years—the evisceration of religious institutions in democratic societies (Pew Research, 2022), the collapse of social relationships like marriage (Pew Research, 2010) and the waning engagement of voters in democracies (Kopf, 2017)—all suggest that trust in liberal democracies is already in decline. The adoption of AI technologies outperforming humans provides further reason to turn away from human institutions. Most individuals have already abandoned human-curated news (e.g., newspapers) (Kabani, 2022) in favor of AI-curated news (e.g., social media sites collecting news and curating it for audiences) (Shearer, 2021). Such transformations are the natural evolution in the migration of trust from humans toward technology. Such trade-offs alter how humans are willing to think about privacy.

Today, we are often ready to surrender power to AI systems by giving them our time, attention, and trust in innumerable other examples. For example, many of us wear electronic devices that capture and manipulate our health data in return for the analyses AI systems can provide. Many of us choose to transit in vehicles using AI technology (e.g., self-driving or driver assistance programs; global positioning and mapping apps using AI; as well as in airplanes using autopilot and autoland) to gain time and efficiency, increase safety, reduce risk, economize fuel, lower carbon emissions, or meet a similar objective. Many of us interact with social media platforms whose algorithms provide streams of content tailored to our interests. I used it in writing this chapter to identify spelling and grammar deficiencies. AI outputs can seem shockingly *personal* even as they are generated through deeply *impersonal* processes—by flattening individual data with other data to identify patterns, categories, and inferences.[22] Today, as we give AI our time, attention, and trust, our commitment to privacy is complicated by the fact that AI technologies mix harms with benefits.[23] These benefits, and the attendant loss of power, can make AI intrusions on our privacy interests difficult to negotiate—and put humans far from the center.

Several privacy-enhancing solutions have been proposed to chart the path forward in this era. One is more frequent recourse to federated learning or federal analytics—the idea of decentralizing data so that privacy-sensitive training data for algorithms and learning models is not aggregated in a single location or source (McMahan, 2017). However, such calls face enormous resistance from corporate actors working to break down barriers to data flows and prohibit data localization. In certain contexts, others have advocated for differential privacy approaches—adding "noise" in the form of fake data to datasets to resist "privacy threats, including data linkage and reconstruction attacks" (Office of the Privacy Commissioner, 2021). However, thorny ethical and moral questions arise in the context of using fake data in certain contexts. Finally, the proposal of synthetic data—the use of generative models to turn source data into *fake* data—has been suggested as a solution to enhance privacy and confront risks of reidentification from data sets. However, this development runs the risk of obsolescence, as advances in inference-generating AI technology will undoubtedly advance to combat the defense systems of synthetic data. Each of these proposals bolsters privacy against the rise of AI, though each faces important challenges.

## CONCLUSION

"Waking up begins with *am* and *now*," wrote Christopher Isherwood (1964) at the beginning of his novel of mourning, *A Single Man*. Isherwood used Roman type to describe the protagonist's physicality and italics to describe the protagonist's consciousness. "That which has awoken then lies for a while staring up at the ceiling and down into itself until it has recognized I, and therefrom deduced *I am, I am now*." In Isherwood's narrative, these fugues condense into a protagonist, George, who is both a body and a being. At times, George authors his experiences with active conscientiousness and is, in all senses, a single man; at other times, he is simply on autopilot, part of a world in which he is not a single man at all, but a mere biological unit connected to a wider, external ecosystem, including not just people but also animals and nature. This duality between the internal and the external has always been part of human life. We are conscious, unique, and individual beings and, at the same time, members of a larger system in which we are not significant at all. As AI increasingly captures more of our time, attention, and trust—as it is already doing so for many of us from when we wake up in the morning to the last thing we do at night—taking away some of our autonomy, and sorting us into categories rather than embracing our individuality—it eats further into these spaces considered the *I am, I am now*. Now and in the future, privacy will be the boundary between when we find ourselves on either side of this divide. If AI technologies are human-centered—enhancing humans in their applicable contexts in ways that advance their flourishing and their interests—such questions should not be hard to resolve, and any trade-offs, easy to negotiate. But where the technology arrogates power in ways that replace, devalue, or impugn the dignity of humans, its ultimate threat remains driving humans to a point where the divide does not exist at all.

## ACKNOWLEDGMENT

The author thanks Russell Walton, Chun An Hsueh, Robert Diab, Tesh Dagne, Malwina A. Wójcik, Oshri Bar-Gil, Catherine Régis, and Maria Axente for their comments on earlier drafts.

## NOTES

1 Hunt, notes that "[d]espite the many persistent and serious attempts at elucidation, privacy remains a deeply—arguably an *essentially*—contested concept". See also Thomson (1975, 295) and Gavison (1980, 424–426).
2 *Olmstead v United States*, 277 U.S. 438, 478 (1928) (Brandeis, J dissenting).
3 One authority, in particular, who stakes this claim is Carissa Véliz (2020), who adopts it in her own calls for the end of the surveillance capitalist economy (especially practices involving data brokering and targeted advertising).
4 See e.g., Blecher (1975: 279) and Müller (2020).
5 *Griswold v Connecticut*, 381 U.S. 479 (1965).

6 *Eisenstadt v Baird*, 405 U.S. 438 (1972).
7 *Roe v Wade*, 410 U.S. 113, 152 (1973).
8 *Lawrence v Texas*, 539 U.S. 558 (2003).
9 *Obergefell v Hodges*, 576 U.S. 644 (2015).
10 See European Parliament and Council of the European Union (2016), GDPR, articles 12–24.
11 See European Parliament and Council of the European Union (2016), article 7.
12 *The Common Law's Hodgepodge*, *supra* note 2 at 181–84. See also Müller (2020).
13 See e.g., Danielle Keats Citron (2022).
14 See Diab (forthcoming).
15 For example, see Privacy Act, RSC 1985, c P-21. This law enunciates clear rights around maintaining accurate information (s 6) but offers only a very low standard for actually acquiring and compiling personal information (so long as it "relates directly to an operating program"—s 4).
16 As an example, see Canada (2021a).
17 See e.g., Sierra Club of Canada v. Canada (Minister of Finance), 2002 SCC 41 at paras 53-55, [2002] 2 SCR 522.
18 Apple Inc (accessed online February 18, 2023), Google Finance, online at: https://www.google.com/finance/quote/AAPL:NASDAQ?sa=X&ved=2ahUKEwj1iLiqzaD9AhVKGjQIHSZPCWAQ3ecFegQINRAg and G7 (accessed online February 18, 2023) Wikipedia, online at: https://en.wikipedia.org/wiki/G7
19 I owe this observation to a very memorable discussion with my research assistant Russell Walton, and I would not dare claim any privacy interest in the thought as my own given the clear benefit I owe to several discussions with him.
20 Ibid.
21 See e.g., Reuters (2018) and Vartan (2019).
22 Even when AI's outputs are not always personal, they often seem personal, when, in fact, they are similar or identical to what is being presented to thousands of other people—not dissimilar to "cold reading" that magicians and psychics do to give the illusion of personalization.
23 This insight is gleaned in particular from Newport's (2019) discussion of digital minimalism and digital technologies in general. See also Hendrycks et al. (2022).

## REFERENCES

American Civil Liberties Union. (2015, February). Informational Privacy in the Digital Age. https://www.aclu.org/other/human-right-privacy-digital-age

American Civil Liberties Union. (n.d.). Surveillance Under the Patriot Act. Retrieved from https://www.aclu.org/issues/national-security/privacy-and-surveillance/surveillance-under-patriot-act

Bayly, L. (2016, May 25). *Zuckerberg to Demolish $30M in Real Estate to Keep Things Private*. NBC. Retrieved from https://www.nbcnews.com/tech/tech-news/zuckerberg-demolish-30m-real-estate-keep-things-private-n580216

Blecher, M.D. (1975). Aspects of Privacy in the Civil Law. *The Legal History Review* 43, 3–4 at 279.

Canada (2021a). First Session, Forty-fourth Parliament, 70-71 Elizabeth II, 2021-2022, Bill C-27 (An Act to enact the Consumer Privacy Protection Act, the Personal Information and Data Protection Tribunal Act and the Artificial Intelligence and Data Act and to make consequential and related amendments to other Acts) (November 22, 2021) House of Commons, online at: https://www.parl.ca/legisinfo/en/bill/44-1/c-27

Canada (2021b). First Session, Forty-fourth Parliament, 70-71 Elizabeth II, 2021-2022, Bill C-292 (An Act respecting transparency for online algorithms) (November 22, 2021) House of Commons, online at: https://www.parl.ca/legisinfo/en/bill/44-1/c-292

Carlsmith, J. (2021). Is Power-Seeking AI an Existential Risk? Open Philanthropy. Retrieved from https://arxiv.org/pdf/2206.13353.pdf

Chander, A., Kaminski, M.E., & McGeveran, W. (2021). Catalyzing Privacy Law. 105 Minnesota Law Review 1733.

Citron, D. K. (2022). *The Fight for Privacy: Protecting Dignity, Identity, and Love in the Digital Age*. WW Norton.

Diab, R. (Forthcoming). Harm to Self-Identity: Reading Goffman to Reassess the Use of Surreptitious Recordings as Evidence.

*Eisenstadt v Baird*, 405 U.S. 438 (1972).

European Parliament and Council of the European Union. (2016). Regulation (EU) 2016/679 of the European Parliament and of the Council of 27 April 2016 on the protection of natural persons with regard to the processing of personal data and on the free movement of such data, and repealing Directive 95/46/EC (Arts. 12-24) [General Data Protection Regulation]. *Official Journal of the European Union, L* 119, 1–88.

Floridi, L. (2005). The Ontological Interpretation of Informational Privacy. *Ethics and Information Technology*, 7, 185–194.

Floridi, L. (2021). The European Legislation on AI: A Brief Analysis of its Philosophical Approach.

Gavison, R. (1980). Privacy and the Limits of Law. *Yale Law Journal*, 89(3), 421–426.

*Griswold v Connecticut*, 381 U.S. 479 (1965).

Hendrycks, D., Carlini, N., Schulman, J., & Steinhardt, J. (2022, June 16). Unsolved Problems in ML Safety. https://arxiv.org/pdf/2109.13916.pdf

Hunt, C. D. L. (2015). The Common Law's Hodgepodge Protection of Privacy. *UNBLJ*, 66, 161–162.

Isherwood, C. (1964). *A Single Man*. Farrar, Straus and Giroux.

Kabani, M. (2022, February 25). Acclaimed newspapers all around the country shutting their presses. CBS News. https://www.cbsnews.com/news/local-news-60-minutes-2022-02-25/

Kaminski, M. E. (2019). Binary Governance: Lessons from the GDPR's Approach to Algorithmic Accountability. *Southern California Law Review*, 92(6), 1529–1553.

Kopf, D. (2017, February 1). Voter turnout is dropping dramatically in the "free world". Quartz. https://qz.com/899586/global-voter-turnout-is-dropping-dramatically-across-the-world/

*Lawrence v Texas*, 539 U.S. 558 (2003).

Lobel, O. (2022). *The Equality Machine. Harnessing Digital Technology for a Brighter, More Inclusive Future*. Public Affairs.

Malone, M. (2022a). IP, Encryption, and the Threat to Public Safety. *Manitoba Law Journal*. https://papers.ssrn.com/sol3/papers.cfm?abstract_id=3783519

Malone, M. (2022b, December 16). The truck convoy demonstrates our double standard on releasing personal information. *The Globe and Mail*. https://www.theglobeandmail.com/opinion/article-the-truck-convoy-demonstrates-our-double-standard-on-releasing/

Manheim, K. M., & Kaplan, L. (2018). Artificial Intelligence: Risks to Privacy and Democracy. *Yale Journal of Law and Technology*, 21, 107–121.

McLuhan, M. (1964). *Understanding Media: The Extensions of Man*. McGraw-Hill.

McMahan, H. B. (2017). Communication-Efficient of Deep Networks from Decentralized Data. *20th International Conference on Artificial Intelligence and Statistics (AISTATS)*.

Müller, V. C. (2020). Ethics of Artificial Intelligence and Robotics. Stanford Encyclopedia of Philosophy. Retrieved from https://plato.stanford.edu/entries/ethics-ai/

Newport, C. (2019). *Digital Minimalism: Choosing a Focused Life in a Noisy World*. Portfolio.

Nissenbaum, H. (2010). *Privacy in Context: Technology, Policy, and the Integrity of Social Life*. Stanford University Press.

*Obergefell v Hodges*, 576 U.S. 644 (2015).

Office of the Privacy Commissioner. (2021, April 21). Privacy Tech-know blog: Privacy Enhancing Technologies for Businesses. https://www.priv.gc.ca/en/blog/20210412

Olmstead v United States, 277 U.S. 438, 478 (1928) (Brandeis, J dissenting).
Orwell, G. (1949). 1984. Secker & Warburg.
Orwell, G. (2006). *Politics and the English language*. Broadview Press.
Pew Research Center. (2010, November 18). The Decline of Marriage And Rise of New Families. https://www.pewresearch.org/social-trends/2010/11/18/the-decline-of-marriage-and-rise-of-new-families/
Pew Research Centre. (2022, September 13). Modeling the Future of Religion in America. https://www.pewresearch.org/religion/2022/09/13/modeling-the-future-of-religion-in-america/
Posner, R. A. (1983). *The Economics of Justice*.
Reuters. (2018, October 11). Amazon ditched AI recruiting tool that favored men for technical jobs. *The Guardian*. https://www.theguardian.com/technology/2018/oct/10/amazon-hiring-ai-gender-bias-recruiting-engine
*Roe v Wade*, 410 U.S. 113, 152 (1973).
Rosoff, M. (2011, April 1). Eric Schmidt Tried To Get Google To Hide His Political Donation In Search Results. Bloomberg. https://www.businessinsider.com/eric-schmidt-tried-to-get-google-to-hide-his-political-donation-in-search-results-2011-4
Roth, E. (2022, August 6). California DMV accuses Tesla of making false claims about Autopilot and Full-Self Driving. *The Verge*. https://www.theverge.com/2022/8/6/23294658/california-dmv-accuses-tesla-false-claims-autopilot-full-self-driving-autonomous-vehicles
Russell, S. (2014, November 14). The Myth of AI. *Edge*. Retrieved from https://www.edge.org/conversation/the-myth-of-ai#26015
Sepulveda, M., van Banning, T., Gudmundsdottir, G. D., Chamoun, C., & van Genugten, W. J. M. (2004). *Human Rights Reference Handbook* (3rd ed.). University for Peace.
Shearer, E. (2021, January 12). More than eight-in-ten Americans get news from digital devices. *Pew Research Center*. https://www.pewresearch.org/fact-tank/2021/01/12/more-than-eight-in-ten-americans-get-news-from-digital-devices/
Smolan, S. (2016, February 24). The Human Face of Big Data. PBS Documentary.
Solove, D. (2008). *Understanding Privacy*. Harvard University Press, 1.
Solove, D. (2013). *Nothing to Hide: The False Tradeoff between Privacy and Security*. Yale University Press, 24.
Solove, D. (2023, January 11). Data Is What Data Does: Regulating Use, Harm, and Risk Instead of Sensitive Data. https://papers.ssrn.com/sol3/papers.cfm?abstract_id=4322198
Standing Committee on Access to Information, Privacy and Ethics, 44th Parliament 1st Session, House of Commons, "Collection and Use of Mobility Data by the Government of Canada and Related Issues" (May 2022), online at: https://www.ourcommons.ca/DocumentViewer/en/44-1/ETHI/report-4/
Thomson, J. J. (1975). The Right to Privacy. *Philosophy & Public Affairs*, 4(4), 295.
Turner, A. M., Smith, L., Shah, R., Critch, A., & Tadepalli, P. (2021). Optimal Policies Tend to Seek Power. *35th Conference on Neural Information Processing Systems*. https://arxiv.org/pdf/1912.01683.pdf
Vartan, S. (2019, October 24). *Racial Bias Found in a Major Health Care Risk Algorithm*. Scientific American. https://www.scientificamerican.com/article/racial-bias-found-in-a-major-health-care-risk-algorithm/
Véliz, C. (2020). *Privacy Is Power: Why and How You Should Take Back Control of Your Data*. Penguin Random House.
Warren, S., & Brandeis, L. (1890). The Right to Privacy. *Harvard Law Review*, 4, 193.
Watts, A. (n.d.). The Future of Privacy and Human Organization. https://www.youtube.com/watch?v=8bBAo3-qCKM
Westin, A. (1967). *Privacy and Freedom*. Atheneum.
Wile, R., Collier, K., & Helsel, P. (2022, December 14). Elon Musk threatens legal action, suspends Twitter account that tracks his jet. *NBC News*. https://www.nbcnews.com/business/business-news/twitter-suspends-elon-jet-account-that-tracked-elon-musk-plane-rcna61718

# 7 The Moral Landscape of General-Purpose Large Language Models

*Giada Pistilli*

Sorbonne University, Paris, France

## INTRODUCTION

The confusion around the term "Artificial General Intelligence" (AGI), often trapped and disputed between the marketing and research fields, deserves to be defined and analyzed from an ethical perspective. In 1980, American philosopher John Searle published an article in which he argued against what was then called "strong AI." Following the legacy of Alan Turing, the question Searle posed was: "Is a machine capable of thinking?" (Searle, 1980). To briefly summarize the experiment, the philosopher illustrated a thought experiment known today as "the Chinese room" to attempt to answer his question. The thought experiment consists of imagining a room in which Artificial Intelligence (AI) has at its disposal a set of documents (knowledge base) with Chinese sentences in it. A native Chinese speaker enters the room and begins to converse with this AI; the latter can answer, considering it can easily find which sentence corresponds to the questions asked. The American philosopher's argument is simple: although AI can provide answers in Chinese, it has no background knowledge of the language. In other words, the syntax is not a sufficient condition for the determination of semantics.

Although the term "strong AI" seems to be replaced by "AGI" nowadays, the two terms do not mean the same thing. More importantly, there is still a lot of confusion among pioneers and AI practitioners. Machine Learning (ML) engineer Shane Legg describes AGI as "AI systems that aim to be quite general, for example, as general as human intelligence" (Legg and Hutter, 2007). This definition seems to be a philosophical position rather than an engineering argument.[1] Nevertheless, in this chapter I will not discuss human intelligence, a topic arousing debates for centuries in many social sciences (e.g., epistemology, philosophy of mind, cognitive psychology, anthropology, etc.), but rather AGI capabilities. Therefore, the interpretation I will use to the term "Artificial General Intelligence" points to AI systems as increasingly specialized in precise tasks, specifically in processing natural language.[2] The idea is then to scale exponentially the capabilities of a given AI system. In this sense, I will not discuss the possibility of theoretical physics to realize this idea but rather its philosophical implications and, specifically, its moral implications.

 DOI: 10.1201/9781003320791-9

Therefore, this chapter wishes to foster the development of Human-Centered Artificial Intelligence (HCAI), understood as systems created by humans for humans, with the primary objective being enhancing human well-being. My analysis will therefore try to shed light on specific issues related to General-Purpose Large Language Models, emphasizing ethical tensions and highlighting potential solutions to be explored.

## NATURAL LANGUAGE PROCESSING (NLP)

Before discussing the AGI moral implications, it is essential to situate our arguments and clarify a few technical details.

Natural Language Processing (NLP) is a subfield of linguistics, computer science, and nowadays also of AI that focuses on the interactions between human and machine language. Initially based on a symbolic recognition system (called symbolic AI), learning in NLP today refers more to statistical probability methods in Neural NLP. NLP systems based on Machine Learning algorithms are increasingly popular, and one type of learning is making waves: Transformers (Vaswani et al., 2017). We can see the entry of the Transformers architecture as a revolutionary moment for NLP, as it allows models to scale more easily. Based on the idea of self-attention, it allows the machine to focus on specific parts of the text sequence and weigh the importance of each word to make its prediction. This technique attempts to mimic human cognitive attention. As Wittgenstein would say, a word only makes sense in its context (Wittgenstein, 1953). Similarly, Transformers, in the pretraining phase, make connections between words. The principle is to use a very large dataset and focus the attention of the model on a small but important part of it, depending on the context (Vaswani et al., 2017).

For example, BERT (Bidirectional Encoder Representations from Transformers) (Devlin et al., 2018) is a powerful language model that leverages the capabilities of transformer-based architecture and is trained on a vast corpus of textual data. This model is designed to provide a contextual understanding of words present in a sentence, which makes it an ideal choice for NLP tasks such as question answering, sentiment analysis, and text classification. One of the most prominent applications of BERT is in Google's search engine. Google has employed BERT to better comprehend the purpose of a user's query, thereby delivering results that are supposed to be more relevant to the user's intent. However, this mechanism has not been spared from criticism and its potential malfunction, such as highlighting irrelevant information, conveying false information, or discrimination (Noble, 2018).

The use of language patterns in search engines is accelerating; these new human–computer interaction technologies will change how we approach information and its research. For example, Google announced they would soon introduce their new "experimental AI service," Bard, powered by their language model LaMDA (Language Model for Dialogue Applications) (Pichai, 2023). That same language model caused much noise last summer because the engineer who was testing it said he believed LaMDA was sentient (Tiku, 2022). Many scholars revolted, including me, and we tried to call attention to how certain conversations are what a journalist called the "Sentient AI Trap" (Johnson, 2022).

## GENERATIVE PRE-TRAINED TRANSFORMER 3 (GPT-3)

To illustrate our point, we will take the GPT-3[3] language model as a case study, given its scope and multiple missions. My arguments only wish to be a philosophical conceptual basis for thinking about ethical issues related to Large Language Models (LLMs) and asking questions for the future.

As reported by the Ada Lovelace Institute (2023) and OpenAI's latest blog post about AGI (Altman, 2023), GPT-3 makes a good candidate for our analysis given its "general-purpose" capabilities and scope—even though the frontier between AGI and "general-purpose" remains yet unclear.

GPT-3, Generative Pre-trained Transformer 3, is an autoregressive language model that uses deep learning to produce human-like text (Broackman et al., 2020). OpenAI's API[4] can be applied to virtually any task that involves understanding or generating natural language. On their API webpage, there is a spectrum of models with different power levels suitable for various tasks. Examples of GPT-3 models are: chat (it simulates an AI assistant to converse with), Q&A (where you can ask questions on any topic and get answers), summarize for a second grader (makes a summary in simple words of a provided text), classification (you write lists and ask for categories to be associated with them), and much more.

GPT-3's ability to multitask makes it a good example of progress toward something that would appear as Artificial General Intelligence. Moreover, OpenAI's strategy for selling access to GPT-3 is also noteworthy, given the hype generated around its potential applications and use cases. While guardrails like content filters exist, their effectiveness can be limited in practice: given the statistical nature of AI systems, it is a deterministic approach to a probabilistic system. If we add to this the human unpredictability concerning the use of these models, the approaches taken in the context of GPT-3 remain limited.

## USE CASE APPLICATIONS

If we look at concrete use cases of such AI models, there are numerous examples of application of GPT-3 in final products. For example, the French company Algolia[5] and its cloud search API for websites and mobile applications. Algolia provides a range of features, including search-as-you-type suggestions, faceted search, and geospatial search, as well as the ability to index and search through large amounts of data in real time.

Another use case of GPT-3's API is the company Copy.ai.[6] Copy.ai is an AI-powered writing assistant that helps users generate high-quality written content. The company's AI technology uses advanced NLP algorithms to analyze large amounts of text data and generate new written content that is similar in style and tone to the input provided by the user.

Nevertheless, can we genuinely trust these systems when we implement them within final products and market them, advertised as lightly as marketing a new smartphone? The confident and compelling outputs of GPT-3 run the risk of ensnaring its users in the art of rhetoric. Its latest successor, the overreported ChatGPT,[7] is flagrant proof of the dangers due to the question of trusting what it produces as content.[8] This means that the fallibility of LLMs like GPT-3 or ChatGPT and their inherent unreliability in generating content necessitate systematic human oversight over the information produced in its outputs.

If we cannot trust the content produced by a language model, what will happen when it is impossible to distinguish human content from AI-generated content? Will it be necessary for users—who are consumers of online content—to distinguish the real from the fake and the artificial from the human? What impact and moral consequences will this lack of distinction have?

## THE PROBLEM OF ARTIFICIAL "GENERAL-PURPOSE" INTELLIGENCE (AGI)

Let us now imagine the extension of the capabilities of language models, having a multitude of goals as the primary—but general—purpose. As seen above, there are several definitions of what an AGI is. Another interesting definition for our analysis is the one proposed by Goertzel and Pennacin in their 2007 book *Artificial General Intelligence*:

> Artificial General Intelligence (AGI) refers to AI research in which "intelligence" is understood as a general-purpose capability, not restricted to any narrow collection of problems or domains and including the ability to broadly generalize to fundamentally new areas.
>
> (Goertzel & Pennachin, 2007)

The various definitions of AGI often recall a cross-cutting capability of the language model, defined as "general-purpose." Moreover, in their latest blog post "How should AI systems behave, and who should decide?,[9]" they open with the sentence "OpenAI's mission is to ensure that artificial general intelligence (AGI[10]) benefits all of humanity."[11] If we are taking GPT-3 as a case study, it is because OpenAI defines its API as follows: "unlike most AI systems which are designed for one use-case, the API today provides a general-purpose "text in, text out" interface, allowing users to try it on virtually any English language task" (Broackman et al., 2020). The simplicity of using this type of AI system is that users can exploit them with almost no computer skills. Users simply have to write their request in natural language in the prompt.[12] GPT-3 will respond with content generation that attempts to match the answer ("text-out") to the question ("text-in"). Although lowering the barrier of entry to certain technological tools is welcome, questions remain about the safety and the potential risks associated with their use.

## SELECTED ETHICAL CONCERNS REGARDING GENERAL-PURPOSE LARGE LANGUAGE MODELS

Developing general-purpose LLMs without a specific objective but rather with a wide range of capabilities, with the intention of moving toward AGI, gives rise to several ethical concerns on various levels. I will not explore all ethical concerns but rather focus on three in particular.

1. The first ethical tension we face is related to the innumerable capabilities of the AI model. In moral philosophy, which deals with defining, suggesting, and evaluating the choices and actions that put individuals in a situation of well-being, it isn't easy to morally assess an artifact with an assortment of different scopes. Moreover, the capacities of a Large Language Model like GPT-3 are often defined but can multiply with its use. Given the breadth of possible uses in natural language, the model's capabilities can be infinite if not defined a priori and framed by its developers. If the goal of an AGI is to no longer recognize itself in a list of skills but rather to have an infinity of them, the situation becomes highly complex to keep under control. It won't be easy to assess and make value judgments about something whose full range of capabilities is still unknown. Also, it will be challenging to control possible malicious uses, to name a few: phishing, fake product reviews, misinformation, and disinformation, etc. One example comes from a study by the Government Technology Agency of Singapore. The researchers used GPT-3 in conjunction with other AI products focused on personality analysis to generate phishing emails tailored to their colleagues' backgrounds and traits. The researchers found that more people clicked the links in the AI-generated messages than the human-written ones by a significant margin (Hay Newman, 2021). Moreover, GPT-3 has also been used to create content for online farms, which often repurpose news from established sites to attract ad revenue. Some of these AI-powered sites have been caught spreading false information (Vincent, 2023).

   Therefore, I argue that in order to make a moral judgment about a technological artifact, it is essential to know and define its goals. In the absence of these conditions, ethics will hardly find its usefulness. Calculating the risks, consequences, context, and model use would be very challenging or even impossible if its capabilities and use cases were infinite.

   Moreover, without going into the psychology and characteristics of human intelligence, there is confusion among AGI pioneers between the latter and Human-Level AI (Goertzel, 2014). Nils Nilsson described the AGI as a machine capable of autonomous learning; the question emerging here is: without *a priori* fixed limits, how can control be exercised over its possible and various uses? (Nilsson, 2010) What safeguards are in place to prevent abuse and misuse? Furthermore, what are the limits set on the machine learning of this AGI? Given these technologies' are state of the art, the current state of moral analysis around these systems often seems to dwell on the technical limits of machine or human intelligence. Quid about the boundaries of the latter's capabilities?
2. Secondly, as already pointed out by Goetze and Abramson in their paper "Bigger isn't better" (2021), by sociologist Antonio Casilli's studies of "click workers" (2019) and researcher Kate Crawford (2021), there is an ethical concern related to social justice. Crowdwork, often used to train

such large models, does not guarantee the quality of the dataset and perpetuates wage inequalities.

> Crowdworkers are generally extremely poorly paid for their time; ineligible for benefits, overtime pay, and legal or union protections; vulnerable to exploitation by work requesters [...]. Moreover, many crowdworkers end up trapped in this situation due to a lack of jobs in their geographic area for people with their qualifications, compounded with other effects of poverty.
>
> (Goetze & Abramson, 2021)

For example, the famous ImageNet dataset was labeled by an equally renowned crowdwork: Amazon's Mechanical Turk, which offers tailored services to adjust and improve AI systems' data and knowledge bases while training them to enable automation (Crawford, 2021). The way these Large Language Models are trained is a bit obscure and raises issues of social justice and relevance when annotating data that will need to feed a globally targeted AI model. This set of issues raised seems to refer to the logic of what some contemporary philosophers call the "technoeconomy" (Sadin, 2018). According to this logic, the economy would find itself driving technical and technological developments, seeking to minimize their costs to produce maximum benefits.

Another example concerns the latest scandal related to OpenAI's creation of a safety system for ChatGPT that could detect and filter out toxic language. OpenAI contracted with an outsourcing firm in Kenya to label tens of thousands of text snippets, many of which contained explicit and disturbing content, such as descriptions of child sexual abuse, murder, suicide, and torture (Perrigo, 2023). The workers who labeled the data were reportedly paid less than $2 per hour, which raises ethical concerns about fair compensation and worker exploitation, as reported by the abovementioned scholars. This case also illustrates how even the most benevolent intentions may yield limited actions and results if the subsequent implementation fails to consider the ethical implications relating to social justice.

3. This last argument allows us to make a transition to our third ethical problem: language. Speaking of Natural Language Processing and Large Language Models, it is inevitable to talk about it. I argue that the language-related problem in Large Language Models is of two different natures. The first is the difficulty in controlling the text generation ("text-out") produced by the model. As an example, GPT-3 has a content filter to warn the user when confronted with content that is unsafe (text containing profane, discriminatory, or hateful language) or sensitive (the text could be talking about a sensitive topic, something political, religious, or talking about a protected class such as race or nationality). As mentioned above, this content filter is inaccurate and unsatisfactory, as the content generated by GPT-3 is often toxic. Within the context of language models and their role in shaping communication, it is imperative to remember that the values conveyed by language are fundamental in guiding human behavior and action (Habermas, 1990). Thus, the implicit values that exist within a language model may be transmitted

through its use. Recent empirical research has demonstrated that the values that are embedded in the GPT-3 training data are predominantly reflective of American values, rather than those of other cultural contexts (Johnson et al., 2022).

Regardless, it still will be difficult to tame this titan under these AGI conditions of "general-purpose." In this case, the limits are not only ethical but also technical. Text generation being a probabilistic calculation of which word will follow within the same sentence, GPT-3 will always be in the condition to give different answers from each other, according to the examples inserted in its prompt. Therefore, if the text-in already presents toxicity, finding it in the text-out will be easy. Differently, if in the prompt there are no toxic contents, there will always be the probability that GPT-3 answers with a text-out containing toxic elements. Once again, the ethical problem here is related to the vastness of the language model and the desire to open it up to a multitude of capabilities.

The second nature of the language-related ethical problem when it comes to Large Language Models is the absence of diversity. Diversity is understood not just as a representation of gender and ethnicity but also as an actual language (Spanish, Portuguese, Danish, etc.). In fact, according to OpenAI, 93% of the training data was in English. The next most represented language was French (1.8%), followed by German (1.5%), Spanish (0.8%), Italian (0.6%), and so on (Brown et al., 2020). Researchers have already begun to explore the multilingual capabilities of GPT-3, noting, for example, how it works poorly in minority languages such as Catalan (Armengol-Estapé et al., 2021). Since the absence of a piece of data is as important as its presence, the very scarce presence of languages other than English leads us to some rather negative considerations, given the multilingual and universal nature that an AI model like AGI is intended to take. The overwhelming and cumbersome omnipresence of the English language is a serious problem that needs to be addressed as soon as possible if we want to make AI accessible to everyone. Because GPT-3 is a system that uses natural language to function and provide answers, orienting it exclusively to English and the values that revolve around American culture will not do justice to the pluralism of values in which we live in our diverse societies. The risk of implicitly promoting a monoculture fostered by large American industries is indisputable. The danger here is twofold: on the one hand, the propagation of the monoculture may be permeated by the implicit or explicit values of the industries developing these AI systems. On the other hand, this same monoculture can be promoted and shared, implicitly or explicitly, through the value systems belonging to the culture dominating these new technological developments. One striking example is related to the recent testimony of the Facebook whistleblower. During her testimony, Frances Haugen pointed out that the lack of moderation tools in languages other than English allowed users of the online platform to freely share content in violation of Facebook's internal policy (Hao, 2021).

## POTENTIAL SOLUTIONS TO BE EXPLORED

First, a challenging but fundamental question must be asked: what then is the ultimate goal of these Large Language Models? What is the purpose of AGI? Since in ethics the "I do it because I can" paradigm can't stand, we should be able to define "the" purpose clearly and not settle for the vague "general-purpose." In its absence, it will be difficult to find a justification and, consequently, evaluate it morally. Considering the advances and the current state of the art of machine learning technologies, automating it more and more can only be desirable after well-framed safeguards have been put in place. If this can still be part of building an AGI, developing *ex-ante* well-structured capabilities limits would be necessary.

Secondly, we need to start shedding light on these dark processes behind the AI industry regarding our social justice issue. The "black box" is not only found within the algorithms but also in the exploitative processes that often bind the poorest part of the world to make us believe that these processes are automated—but they are not. The demand for human labor to produce the datasets needed to run these Large Language Models grows exponentially. As a result, national and international institutions need to start asking questions quickly, in order to bring answers and a clear legislative framework for these new "data labeler-proletarians."

Finally, the issue related to language is, in my opinion, one of the thorniest to deal with. Aside from the concealed hypocrisy found among the AGI pioneers, who sell their products as being "universal," the problem here is structural. Today we're talking about Large Language Models, but I'd like to point out that the entire Internet ecosystem is governed by the English language and an American monoculture that permeates every corner of it. Today we are facing a difficulty that we can turn into a possibility: we can fix this kind of problem in language models and try to integrate the feedback from its users as much as possible. The process will undoubtedly be longer, but it could be the beginning of a fruitful collaboration. In addition, it might help to change the paradigm of AGI and make it rather "narrow AI": oriented toward specific capabilities and circumscribed to its context. In this way, each context could appropriate its model and make it its own, thus ensuring a plurality of values relevant to its social context.

From an ethical standpoint, developing narrow and culturally based AI models may offer several benefits compared to pursuing AGI. One noteworthy advantage is that these models are tailored to specific contexts and can accommodate the needs and values of specific communities. By doing so, these models could mitigate the risk of perpetuating biases and unintended consequences, as they are aligned with local ethical and cultural norms. By focusing on narrow AI models, stakeholders can ensure that the development process is more controllable, transparent, and subject to greater scrutiny and accountability. Furthermore, given the dominance of English in the AI ecosystem, prioritizing the development of language models for non-English languages is essential to ensure that diverse linguistic and cultural perspectives are represented in the discourse. Taken together, prioritizing the development of narrow and culturally based AI models can address ethical concerns related to AI and promote the technology's ethical use in ways that align with local values and needs.

The argument presented here is exemplified by grassroots organizations such as Masakhane,[13] aimed at fostering research in NLP specifically for African languages. Remarkably, even though African languages constitute nearly 2,000 of the total world languages, they are scarcely represented within technological platforms. Another pertinent example can be observed in the endeavors of Te Hiku Media,[14] a nonprofit Māori radio station. This pioneering grassroots initiative within the domain of NLP concentrates on the safeguarding of minority languages, while concurrently ensuring the control and sovereignty of the community's data (Hao, 2022).

Another virtuous example of projects addressing the issue of language and data governance is the BigScience open science workshop[15] and their approach to multilingualism. In their paper "Data governance in the age of large-scale data-driven language technology" (Jernite et al., 2022), the authors present their definition of data governance as "the set of processes and policies that govern how data is collected, stored, accessed, used, and shared" (p. 1). The advent of machine translation systems and LLMs presents unique ethical opportunities and challenges. The authors illustrate the implications of these challenges and opportunities; for example, the ethical concerns of utilizing biased or sensitive data, the privacy issues of exposing personal or confidential information, the quality issues of utilizing low-resource or noisy data, and the diversity gaps of underrepresenting certain languages or groups. The authors also propose some guiding principles for data governance, such as establishing unambiguous data ownership and consent mechanisms, developing data quality metrics and standards, promoting data diversity and inclusion, and fostering collaboration and transparency among stakeholders.

## CONCLUSION

In conclusion, we have seen how technical problems often go hand in hand with ethical issues. In pursuit of the genuine development of HCAI, a paradigm where AI systems are tailored to conform to human values, and invariably ensure human benefit, it is imperative to address the highlighted ethical tensions. Moreover, given the interdisciplinary nature of the scientific domain of artificial intelligence, these ethical problems cannot be solved without the help of engineers. And when I talk about philosophers and engineers working together, it also means that engineers shouldn't make themselves out to be ethicists without the right expertise and knowledge. Indeed, philosophy has been asking questions of this order for thousands of years; its experience can serve us not to make the same mistakes, but more importantly to well formulate the right questions to ask in this new and evolving technological context. The heightened attention being paid to moral philosophy is of paramount importance and represents an urgent concern. Nonetheless, as has become evident in contemporary times, the complex challenges posed by emerging technologies cannot be resolved through technical efforts alone. In this regard, the social sciences and humanities are called upon to play a critical role in helping this discipline. Because while science serves to describe reality, it is ethics that ultimately guides the way in which this reality ought to be constructed in the future.

## NOTES

1 In this chapter, I do not make the distinction between ethics and morality, both having the same etymology coming from Greek and Latin respectively.
2 Language is defined as "natural" when it belongs specifically to humans (e.g., Chinese, Spanish, and German), as opposed to the "artificial" language of machines (e.g., different code languages).
3 Although ChatGPT and GPT-4 would have been even more relevant objects of analysis, this chapter and its related research have been elaborated months before their release.
4 An API, or Application Programming Interface, is a set of rules and protocols for accessing a web-based software application or web tool.
5 https://www.algolia.com/about/
6 https://www.copy.ai
7 This improved version of GPT-3, also developed by OpenAI, is focused only on the question-and-answer task, thus not relevant to our analysis. Accessible to anyone on condition of signing up on the platform, it can be accessed at the following link: https://chat.openai.com/chat
8 To explore the issue of trust further, I recommend reading this recent article appeared in *Nature*: https://www.nature.com/articles/d41586-023-00423-4
9 https://openai.com/blog/how-should-ai-systems-behave/
10 Their definition of AGI reads: "By AGI, we mean highly autonomous systems that outperform humans at most economically valuable work."
11 Ibid.
12 A prompt is a set of initial input given to a large language model to generate output based on the provided context.
13 https://www.masakhane.io/
14 https://tehiku.nz/
15 https://bigscience.huggingface.co/

## REFERENCES


Altman, S. (2023). Planning for AGI and beyond. *OpenAI blog*. Retrieved from https://openai.com/blog/planning-for-agi-and-beyond

Armengol-Estapé, J., Bonet, O. D., & Melero, M. (2021). On the Multilingual Capabilities of Very Large-Scale English Language Models. *arXiv*, abs/2108.13349.

Broackman, G., et al. (2020). *OpenAI API*. OpenAI Blog. Retrieved from https://openai.com/blog/openai-api

Brown, T. B., et al. (2020). Language Models are Few-Shot Learners. *arXiv*, arXiv:2005.14165.

Casilli, A. (2019). *En attentant les robots*. Editions Seuil, Paris.

Crawford, K. (2021). *Atlas of AI*. Yale University Press, Yale.

Devlin, J., et al. (2018). Bert: Pre-training of deep bidirectional transformers for language understanding. *arXiv preprint*, arXiv:1810.04805.

Goertzel, B. (2014). Artificial General Intelligence: Concept, State of the Art, and Future Prospects. *Journal of Artificial General Intelligence*, 5(1), 1–46.

Goertzel, B., & Pennachin, C. (2007). *Artificial General Intelligence*. Springer-Verlag, Berlin Heidelberg, Berlin.

Goetze, T. S., & Abramson, D. (2021). Bigger Isn't Better: The Ethical and Scientific Vices of Extra-Large Datasets in Language Models. In *WebSci '21 Proceedings of the 13th Annual ACM Web Science Conference (Companion Volume)*.

Habermas, J. (1990). *Moral consciousness and communicative action*. MIT Press.

Hao, K. (2021). The Facebook whistleblower says its algorithms are dangerous. Here's why. *MIT Technology Review*. Retrieved from https://www.technologyreview.com/2021/10/05/1036519/facebook-whistleblower-frances-haugen-algorithms/

Hao, K. (2022). A new vision of artificial intelligence for the people. *MIT Technology Review*. Retrieved from https://www.technologyreview.com/2022/04/22/1050394/artificial-intelligence-for-the-people/

Hay Newman, L. (2021). AI Wrote Better Phishing Emails Than Humans in a Recent Test. *Wired*. Retrieved from https://www.wired.com/story/ai-phishing-emails/

Jernite, Y., et al. (2022). Data governance in the age of large-scale data-driven language technology. In *2022 ACM Conference on Fairness, Accountability, and Transparency*.

Johnson, K. (2022). LaMDA and the Sentient AI Trap. *Wired*. Retrieved from https://www.wired.com/story/lamda-sentient-ai-bias-google-blake-lemoine/

Johnson, R. L., Pistilli, G., Menédez-González, N., Duran, L. D. D., Panai, E., Kalpokiene, J., & Bertulfo, D. J. (2022). The Ghost in the Machine has an American accent: value conflict in GPT-3. *arXiv preprint*, arXiv:2203.07785.

Legg, S., & Hutter, M. (2007). Universal Intelligence: A Definition of Machine Intelligence. *Minds and Machines*, 17(4), 391–444.

Nilsson, N. J. (2010). *Quest for Artificial Intelligence: A History of Ideas and Achievements*. Cambridge University Press, Cambridge.

Noble, S. U. (2018). *Algorithms of Oppression: How Search Engines Reinforce Racism*. New York University Press, New York. https://doi.org/10.18574/nyu/9781479833641.001.0001

Perrigo, B. (2023). Exclusive: OpenAI Used Kenyan Workers on Less Than $2 Per Hour to Make ChatGPT Less Toxic. *Time*. Retrieved from https://time.com/6247678/openai-chatgpt-kenya-workers/

Pichai, S. (2023). An important next step on our AI journey. *Google Blog*. Retrieved from https://blog.google/technology/ai/bard-google-ai-search-updates/

Sadin, E. (2018). Le technolibéralisme nous conduit à un 'avenir régressif'. *Hermès, La Revue*, 80(1), 255–258.

Searle, J. (1980). Minds, brains, and programs. *Behavioral and Brain Sciences*, 3(3), 417–424. doi:10.1017/S0140525X00005756

Tiku, N. (2022). The Google engineer who thinks the company's AI has come to life. *The Washington Post*. Retrieved from https://www.washingtonpost.com/technology/2022/06/11/google-ai-lamda-blake-lemoine/

Vaswani, A. et al. (2017). Attention is All you Need. *ArXiv abs/1706.03762*.

Vincent, J. (2023). AI is being used to generate whole spam sites. *The Verge*. Retrieved from https://www.theverge.com/2023/5/2/23707788/ai-spam-content-farm-misinformation-reports-newsguard

Wittgenstein, L. (1953). *Philosophische Untersuchungen (Philosophical Investigations)*, translated by G. E. M. Anscombe, 1953.

# 8 Anand Rao's Commentary

*Anand Rao*
PwC, Greater Boston, MA, USA

Anand Rao is a Principal with PwC's US Advisory Practice and the firm's Global Artificial Intelligence Lead. He has worked extensively on business, technology, and analytics issues across a wide range of industry sectors including Financial Services, Healthcare, Telecommunications, Aerospace & Defense, across the US, Europe, Asia, and Australia. Before his consulting career, Anand was the Chief Research Scientist at the Australian AI Institute, a boutique research and software house. He holds a PhD in computer science and has co-edited four books on Intelligent Agents and over 50 papers in computer science and AI in major journals, conferences, and workshops.

**What, in your opinion, is human-centered AI and what is its potential, its importance?**

When I first encountered the term, I thought, "That's a bit odd … If you go back to the beginning of AI, it has always been about understanding human intelligence and mimicking it—scholars like Roger Schank, in the '70s and '80s were very influenced by psychology." So, at first, the idea of human-centered AI felt anachronistic to me. But then, as I thought about it, I realized that AI isn't as closely related to psychology as it used to be. So, I felt that calling into question the lack of interaction between humans and AI could make a lot of sense. And when I started reading more about value alignment, especially what Stuart Russell wrote about that, it made even more sense to focus on a human-centric approach to AI—to reflect on common values we share as humans and think about how to implement them in AI. So I went from "human-centered AI is odd …" to "Yes, I understand now what human-centered AI means, and I get its appeal."

However, one thing that still bothers me with the idea of human-centered AI is that it elevates humans as the highest form of the ecosystem that we have. I come from an Eastern tradition. In the Indian philosophical tradition, life pervades everything and consciousness and sentience pervade everything: animals, objects, and even rocks. So, I worry that human-centered AI will focus on human benefits to the detriment of

DOI: 10.1201/9781003320791-10

other things, both animate and inanimate. I think human-centered AI is too narrow. AI should really be for everything.

**What would be the right label to use to capture the wider ecosystem?**

Maybe we should be talking about something like human-led AI for the benefit of everything? Or human-enabled AI for the good of all? Or AI for all? But "for all" could be interpreted as including only people, so "AI for everything" is probably better. You know the Gaia concept? AI should be for the whole planet—human-led, of course, but not only centered on human needs.

**Maybe we can dive into the articles you've read to see how they've changed some of your views on AI and your ideas about the challenges we're facing. What article did you read?**

I read "The Moral Landscape of General-Purpose Large Language Model" by Pistilli; "Privacy in the Era of AI" by Malone; and "Towards Addressing Inequality and Social Exclusion by Algorithms: Human-Centric AI through the Lens of Ubuntu" by Wójcik.

**What did you make of them?**

I liked the first one on general-purpose AI. It was very interesting. Pistilli highlights three issues with large language models, and I agree with most of what she says, but I'd like to focus on one fundamental question she raises: "What is the ultimate goal of large language models? What is the purpose of artificial general intelligence (AGI)?"

Her point, I think, is that in order for AI ethics to be truly effective, AGI should be defined—and should be defined in such a way that makes it possible for us to evaluate its capabilities. But the thing is that defining AGI is a challenge. When you come up with a definition, there's inevitably something that has been left out. So the risk is that you end up in an endless circle of defining and redefining what exactly AGI is.

Just look at history. The term "AI" was first coined to describe AGI. Simon, Shaw, and Newell's "General Problem Solver," for instance, was meant to be AGI. They were saying, "Hey, we need a general problem solver. How do we do that?" But the task was too difficult, so people started focusing on more specific tasks: planning, reasoning, learning, etc. And then, after a while, people like John Searle came and said, "Oh no! That's narrow AI. We should be thinking about AGI." And we've been going back and forth between narrow AI and AGI ever since.

I think it might be because there's an inherent tension between defining AGI and evaluating AGI. If you want to evaluate AGI, you might come up with a definition that is very narrow so that your evaluation goes well. But then, you'll come back and say: "Of course, it's not general intelligence … We are just testing one of these things, right?" That's how progress is made in AI: we give ourselves a particular goal, and once it's achieved, we move the goalpost a little further and say, "That was easy, that's not AGI, now try to do this."

Think of the way we're assessing ChatGPT's creativity. AGI would need to be creative, but what does it mean for an AI to be creative? And who's deciding what's creative? Humans, of course. So, right now, AI is coming up with new things, and humans are saying, "Hey, I don't think that's creative. It's not mimicking my intelligence. It's not doing it the way I do things." But on the other hand, if AI comes up with something that we have not done—which in AI's sense might be creative, as it produces something that wasn't in the training data—then we say, "No, that's not creative either, it's not AI because it's not mimicking what I do."

And that's just for creativity—that is a subset of intelligence! So, will we ever agree on a definition of AGI? Probably not, and that's where it gets tricky. AI systems are more "general purpose" than ever before. At what point do we stop calling something narrow intelligence and start calling it general intelligence? Since I don't think we'll get to any agreeable answer among all of us, from the philosophers and linguists to the AI scientists, I think that's going to be a problem. That said, Pistilli's paper was very good, and I agree with her other points.

**The other two articles you read deal with questions like privacy and the risk of social exclusion.**

I was keen on Wójcik's paper. She discusses something I made a point of talking about in a couple of responsible AI forums. Responsible AI principles are very much based on a libertarian view of what human values are. The example I generally use to illustrate the difference is that of individual liberty vs. social good. Western society essentially elevates individual liberty to a very, very high level, whereas it associates the promotion of social good with totalitarianism. For example, Western countries look at China and say, "That's such a totalitarian regime, everyone has to give their information." Okay, but let's leave China alone for a minute. Let's take India, Korea, or Japan, where I think many people consider the common good is as important or even more important than individual liberties. What do we make of their view? Why should AI support human individual liberty instead of the common good?

I remember reading a South-East Asian author on this, someone who was writing with Buddhist philosophy in mind. But I was not at all aware of the African contribution to this important debate. The notion behind Ubuntu—the idea that identity is about relations between different people, that who you are and what you do is defined by a complex web of interactions—is important and should be taken into consideration in the whole responsible AI world, which I think has embraced the Western view much too tightly. Allowing for alternative values would be very welcome. Maybe we want to come up with some basic human values that everyone should be able to agree on and then allow for variations between cultures. But as of now, I don't think anyone has drawn such a straight line, and I don't envy anyone who is supposed to get any agreement on that minimal threshold.

**What about Malone's article on privacy?**

It was very good. I learned a lot about the origins of the notions of individual liberty and privacy, the fact that they go back as far as 1604, but what I really liked in this paper is the section about the future, in which Malone presents what he calls a cybernetic view of privacy. According to this vision, it's what AI does that is going to define how we think about privacy, and it's the way we think about privacy that determines what AI will be doing.

I like this idea of a constant flow of information going back and forth between AI and humans. I think it's the most fruitful way forward for us to understand AI. That's why I prefer talking about augmented intelligence instead of artificial intelligence. I don't think we should be focusing on whether AI will be replacing workers or not. Yes, AI will be able to perform some tasks better than humans. But can we find ways for AI to increase human capabilities instead of trying to replace humans? Malone touches on the idea of AI as "extensions" of humans at the end of his paper. That's something we're interested in at my firm, "How can we raise the performance of the human-AI combo?"

In the business world, we often talk about change management, but in order to raise the performance of the human–AI combo, I think we should focus on trust management.

I was introduced to this idea of managing people's trust in AI when I worked on developing an air traffic management system for the Sydney Airport. We could prove that our algorithm was coming up with the optimal solution to land all aircraft effectively. But traffic controllers were managing airplanes heuristically. And even though we would have been able to prove that our algorithm was more effective than their heuristics, they would not have accepted it.

So, we didn't force the system onto them. What we did instead was provide them with multiple options: they could choose the optimal option or choose the heuristic option. When we first rolled out the system, controllers always used the heuristic option, but they were provided with additional information like, "That's great, you landed all of the aircraft, but, by the way, had you chosen this route instead, you would have saved them X gallons of fuel and Y minutes." Over time, controllers started trying the AI option once in a while. They'd say, "Why don't I try that tool just for that one aircraft? I'll do everything else as usual." And slowly, they gained confidence in our algorithm. That's what I mean by trust management.

Since Malone argues that the adoption of technology shapes the way we envision the notion of privacy, I think this idea of trust management is somewhat related to his cybernetic view of privacy.

**If you had one or two recommendations for policymakers, what would they be? What would you do to increase this trust that you are talking about, or to achieve any objectives that you consider crucial?**

First, I think there needs to be a real conversation involving all stakeholders concerned with AI policy: governments, large companies developing and promoting the technology, citizen groups, and academics. Each of them has a different mindset: citizen groups are looking at social justice; academics are thinking about what technology can do from a theoretical standpoint, or about the philosophical issues stemming from new AI developments; policymakers are saying, "Tell me what to do to make sure my people are protected"; and then, corporations, of course, are interested in making money out of AI.

So how do you reconcile all of these interests? I think that's the biggest challenge. Everyone has different incentives. Most academics, for instance, are interested in solving academic issues, and they're not necessarily paying close attention to social or corporate issues. Whereas businesspeople, for their part, don't care all that much about what academics have to say. They want something practical that they can do tomorrow—which is unfortunate given that they could learn so much from what academics have thought through, they could avoid a lot of mistakes. So bringing together all of these stakeholders is one thing policymakers should do.

My second recommendation would be about sandboxes. I know that the UK, the European Union, and various countries have tried implementing them. But for sandboxes to be truly effective, there are a couple of pillars that need to be put in place. In particular, we have to think about amnesty for what was done in the past. The reason why some firms are hesitant to get involved in sandboxes is they feel regulators might be gathering information about them. They think, "I'm doing these things, if I tell them what I'm actually doing, they're going to come back later and say, 'Your underwriting algorithm was biased' and I might face sanctions." So there needs to be some amnesty for those involved in regulatory sandboxes.

Another thing is that it's not always possible for us to understand the full social implications of regulations before rolling them out, so sandboxes should serve as parallel runs to test things out. This has been done with a financial regulation called Basel II.

For four or five years, quantitative impact studies were conducted to test that new framework. Institutions were asked to do two sets of calculations in parallel: they would do the calculations the old way, and they would also do the calculations following a new set of rules. Only after going through four iterations—only after Quantitative Impact Studies #1, #2, #3, and #4 were done—did the Basel II regulation officially come out. That process allowed regulators to iron out a number of nuances: mortgages work like this, credit cards work like that, you can't do that for small countries, you can't do that for large countries, etc.

It was an additional burden on companies to do the calculations twice, but still … I think doing something like that for AI would be a good idea. Testing regulations in a parallel way, with a limited set of people, before rolling them out is something regulators should really consider. It could make a significant difference in building people's trust.

# 9 Benjamin Prud'homme's Commentary

*Benjamin Prud'homme*
Mila—Québec Artificial Intelligence Institute, Montréal, Québec, Canada

Benjamin Prud'homme is the Vice-President, Policy, Society and Global Affairs at Mila–Quebec Artificial Intelligence Institute, one of the largest academic communities dedicated to AI. At the time this text was written, he was the Executive Director of the AI for Humanity department at Mila. Previously, he was a litigator in human rights, constitutional, and family law until his 2018 appointment as policy advisor to the Minister of Justice of Canada. In 2019, he became the advisor to the Minister of Foreign Affairs of Canada on matters of human rights and multilateral relations, a position he occupied until joining Mila. His work and publications focus on the role of human rights in the international governance of AI, the inclusion of marginalized individuals and communities in the life cycle of AI systems, and the epistemology of interdisciplinarity. An expert for the United Nations Broadband Commission's Working Group on AI Capacity Building, he sits on the Advisory Board of Sustainability in the Digital Age (a think tank) and the Board of Directors of the Canadian Civil Liberties Association.

**Please tell us more about your work.**

Mila—Québec Artificial Intelligence Institute is a community of more than 1,000 specialists in machine learning and experts in the social sciences, law, learning design, etc., who are all dedicated to innovation and scientific excellence that benefit society. I head its AI for Humanity department, which focuses on the responsible and beneficial development of AI. We do so via three streams of work: Governance of AI—including public policy and regulation; Applied Research Projects for socially beneficial purposes; and Learning and Education.

I often refer to the latter stream as Capacity Building, which I care deeply about. For example, we work with policymakers to improve their understanding of AI: "What is AI?", "What is generative AI?", "How do these technologies work?", or "What approaches could be used to regulate them?" Our intention is not to dictate specific regulatory approaches but rather to support the capacity of our policymakers—and ultimately, of our democratic systems—to apprehend AI. We also offer learning and training opportunities for people coming from a technical background—whether in

 DOI: 10.1201/9781003320791-11

research or industry—as to how to integrate responsible AI reflections and practices into AI research, projects, or companies.

In a way, our overarching objective at AI for Humanity is to contribute to the operationalization of certain responsible/ethical AI principles. Because if the importance of these principles is now fairly consensual—think about explainability, transparency, privacy, etc.—putting them into practice remains a major challenge. We're trying to do our part in tackling this by building interdisciplinary bridges, bringing together people whose complementarity and collaboration contribute to bringing responsible AI principles to life, be it through governance mechanisms or algorithmic design choices. This is crucial. Indeed, I'm afraid that if we don't take seriously and address the gap between disciplines as well as between policymakers and AI practitioners, the principles we put forward may end up not being reflected in AI systems and ecosystems.

So I don't think we can afford to separate the conversation between principles and practices of Responsible AI, the former being seen as work for social scientists, jurists, and policymakers, and the latter being seen as work for technical experts. We must address AI with a socio-technical approach, and therefore look at algorithmic design as a series of socio-technical choices. And we need to create interdisciplinary teams to address it as such, and to provide the tools to AI practitioners to make choices consistent with the principles being put forward.

This brings me to what seems to be a consensual finding between the three articles I've read: the interdisciplinarity we just talked about is going to be central if we want to achieve human-centered AI. Social sciences or legal perspectives alone are not going to allow us to fully comprehend and govern AI. Or, conversely—and I think that is more the danger we're facing now—I don't think that technologists (and particularly industries) should alone pave the way on how to govern AI. And I would go further: even a strong set of interdisciplinary experts should not alone set our AI governance agenda.

This is also something that comes out of the three texts I read: end users, marginalized communities, and society more generally must be involved in setting the rules. People who build AI systems and who take part in the setting of principles and regulation need to engage seriously with society to ensure the technologies they're building in fact serve the people they're intended to serve, address their problems without creating new ones.

**Can you tell me a bit more about what you learned reading the articles? How are they going to enrich your perspective on the kind of work you're trying to do at Mila?**

I read three papers: Valor's interview about defining human-centered AI, Larouche's paper on "Human-Centered Artificial Intelligence: From conceptual examination to legislative action," and Bar-Gil's paper on "Redefining Human-Centered AI: The human impact of AI-based recommendation engines." There are a few things I read that I profoundly agree with and was glad to see written in print.

First, each of the three papers, in their own respective ways, points at how fundamental interdisciplinarity is to human-centered AI. However—and it's not a critique as much as it is an invitation to pursue the conversation—I think the papers do not sufficiently discuss how difficult meaningful interdisciplinarity is, and the ways in which we can make it work. We seem to be nearing the point where most people like and agree with the idea of interdisciplinarity but do not devote sufficient energy on making it work. You know, it's one thing to bring an ethicist, a sociologist, or a linguist into your project; it's another thing to create the structural environment necessary to integrate their perspectives in ways that really inform the development of an AI system. Interdisciplinarity is incredibly difficult, especially in practical settings where you have deadlines, deliverables, and a team of human beings who need to interact with each other to create a product or come up with a solution. I don't think anyone has fully figured out the right way to do it. So, for me, these articles invite us to think about how to do meaningful interdisciplinary work in AI.

Now, another thing I was glad to see in the articles—in Larouche's paper specifically—is an invitation to get out of the false dichotomy between innovation and regulation. The paper offers interesting ideas to solve what is often presented as an irreconcilable dilemma. And that is very valuable. I think this tension between innovation and governance is dangerous, and we ought to find ways to overcome it. What Larouche proposes—a nuanced middle-ground approach that combines elements of both the permissive and the protective approaches to tech regulation—provides an interesting angle from which to tackle the challenge.

I really enjoyed Shannon Valor's piece, particularly the way she breaks down human-centered AI into two main dimensions: processes and outputs—actually, I don't think she explicitly uses the words "output" or "end result", but the idea is omnipresent nevertheless—and to me, this distinction is very useful because focusing on processes opens up the door to concrete daily actions we can take to implement human-centered AI in our work. How to frame the problem, for example? Valor suggests—and I agree with that—that technologists should refrain from framing the problem themselves. Rather, they should work in close collaboration with affected stakeholders and communities and ask them: "What do you need? How can technology support you?" This bottom-up approach contrasts with what she calls "reverse adaptation," which, in simple terms, is a process where a group of technologists come up with a solution and then try to tweak it to make it work for (or to have it purchased by) people on the ground.

So yes, taking a step back and focusing more (or at least, equally) on processes instead of end results is a very stimulating invitation for an AI institute like Mila. It gives us all these concrete considerations we can play with. It encourages us to be creative and to think of new ways to involve people in our processes, for instance. And then, we can experiment, run pilots, and openly share the results of those pilots—positive or negative—with the community.

And finally, there is one question I see permeating all three articles without ever being fully addressed—or at least, not at the length I wish it had been. And that is the fact that the economic context and system we live in so profoundly influences the way we think about AI governance. Each in their own ways, Valor, Bar-Gil, and Larouche discuss the fact that there is an enormous concentration of power in the hands of a very few companies; that such concentration provides extraordinary power to these companies; and that the lens through which we look at AI is first and foremost an economic opportunity lens that doesn't really make human-centered AI a top priority. So to me, this question of economic power was underlying in the three papers I've read, but I wish it had been addressed more straightforwardly. I wish someone had tried to examine the following question: "What would a conversation on AI governance look like if we were operating in a different economic context?" Questioning the capitalist system itself was outside the scope of the book, I guess, but even without doing that, one could have questioned the concentration of power within the system, and how that alone profoundly shapes the current conversation. Indeed, what we're seeing right now is that a handful of players—all in the same space, all from the same demographics—are calling the shots. And we seem to be taking it for granted: "It is what it is. Our governance mechanisms need to take that into account."

It would be interesting to think about the possibilities that might arise if governments were to try to tackle this concentration of power. What would it look like if we had a wider diversity of actors? If the perspectives were more diverse? My assumption is that it wouldn't alone be sufficient to fully achieve human-centered AI, but it would be a great first step in bringing more perspectives into the conversation, and to ultimately (truly) listen to and hear from individuals and organizations who believe human-centered AI should be a priority.

**What recommendations would you make to policymakers and practitioners?**

My first recommendation, or perhaps my first hope, would be that we start moving away from the dichotomy between innovation and regulation—that we acknowledge it might be okay to stifle innovation if that innovation is irresponsible. I'd tell policymakers to be more confident in their ability to regulate AI; that yes, the technology is new, but that it is inaccurate to say they have not (successfully) dealt with innovation-related challenges in the past. Perhaps that would reduce the thoughts of the likes of "The task is mammoth, so it's better to go super slow" or "It's better to do nothing than to do the wrong thing."

We have to accept that sometimes, perfect, yes, is the enemy of good. A lot of people in the AI governance community are afraid of not getting things right from the get-go. And you know, one thing I've learned in my experiences in policymaking circles is that we're likely not going to get it entirely right from the get-go. That's ok. Nobody has a magic wand. So, I'd say the following to policymakers: Take the issue seriously. Do the best you can. Invite a wide range of perspectives—including marginalized communities and end users—to the table as you try to come up with the right governance mechanisms. But don't let yourself be paralyzed by a handful of voices pretending that governments can't regulate AI without stifling innovation.

To convince oneself that innovation and regulation can go hand in hand, I think that Larouche's approach is interesting. When we remind ourselves that we're not in a legal vacuum, that we've actually regulated technologies before—technologies that were different, sure, but that were similar in terms of the legal challenges they posed—then AI regulation becomes something more approachable.

Valor's definition of human-centered AI brings me to my second recommendation: let's reframe the narrative around AI governance. If one considers that AI should be human-centered, that it should be designed by people, for people and with people in mind in order to promote human flourishing, then, again, the dichotomy between innovation and regulation becomes a little moot. Governance—including regulation—is not seen as an instrument that slows down innovation anymore. Rather, it is seen as an instrument that can propel socially beneficial (or at least, responsible) AI, and that perhaps stifles it when it does disservice to humans. It allows us to go away from a narrative where innovation for the mere sake of innovation is the premise.

My last recommendation is simple: address the concentration of power. It's not unprecedented; it has been done before. To me, the way in which the conversation is set up right now—whereby we're fine with having a few giants that have the capacity to almost dictate the agenda—is very flawed. To be clear: these giants are not necessarily of bad faith. But they are companies, and their duty is to their shareholders, not to the planet or to all humans. Once we remind ourselves of that, I think it becomes obvious that there should be far more diverse voices with their hands on the AI wheel.

# *Section II*

## *Sectoral Representations of HCAI*

# 10 Good Governance Strategies for Human-Centered AI in Healthcare
## *Connecting Norms and Context*

*Michael Da Silva*
University of Southampton, Southampton, UK

*Jean-Louis Denis and Catherine Régis*
Université de Montréal, Montréal, Québec, Canada

## INTRODUCTION

Many stakeholders are excited about the development and use of artificial intelligence (AI) in healthcare, particularly in contexts where most healthcare systems struggle with human and financial resource shortages. AI could provide professionals with the "gift of time" (Topol, 2019a) to give more compassionate care—and help patients access medical knowledge that permits them to take a more active role in health-related decisions. The average US nurse spends 25% of their work time on regulatory and administrative activities for which AI could partly assist (Davenport & Kalakota, 2019). AI-enabled diagnoses and medical treatments could, in turn, provide substantive health improvements (Topol, 2019a). While many raise understandable concerns about how AI will impact care, market trends suggest health-related AI is here to stay: the global (non-health-specific) AI market is expected to reach a value of USD 1,811 billion by 2030 (Research & Markets, 2023). Well-designed AI could improve the quality and efficiency of healthcare and services and population health (Topol, 2019b); however, one must seek ways to ensure it fulfills its beneficial aims, rather than creating a false sense of value.

The question is how to ensure that health-related AI furthers human values. Healthcare scholars use the expression "Human-Centered AI" (HCAI) in at least two ways. One focuses on the need to develop AI that furthers the interests or needs of healthcare actors, like patients, healthcare providers, and managers. The other

DOI: 10.1201/9781003320791-13

focuses on whether AI is "usable" in terms of its ergonomic features, and fits with the use environment. Even safe and effective AI with tremendous potential benefits will not secure them if design issues make providers unlikely to use that technology for good in particular settings. These views are therefore complementary. Failure to address this interdependency will contribute to widening the implementation gap between AI development and real-world deployment (Seneviratne et al., 2019) and diminishing trust in innovation.

Developing good governance for HCAI in healthcare demands pairing at least two governance approaches reflecting these traditional views on HCAI: (1) a normative approach that ensures that we *protect* the interests of healthcare actors at a regulatory level (at least their objective and overarching ones) and (2) a contextual approach that allows *understanding*, *responding, and adapting* to such interests at a practical level. The first minimally requires that regulators ensure that AI innovations in healthcare are sufficiently safe to be implemented, and therefore respect some key normative principles. The second speaks to the capacity of the AI community to design, develop, and implement AI innovations that, in practice, respond to a variety of interests and generate use value. These approaches both speak to the need for AI to aim at valuable ends; to be proven safe, effective, and fit for purpose; and to be designed for use in real settings. This chapter provides a brief overview of how AI is used in healthcare now before examining both approaches and their interrelation.

## HEALTH-RELATED AI: A BRIEF OVERVIEW

AI developments already occur in health research, prevention, health management, clinical care and services, and public health sectors. Here are some examples. **In research**, they are being developed to speed up the medical drug discovery and clinical trials processes (WHO, 2021; Morriet, 2022). **In prevention**, they are being proposed to better target and more rapidly react to the risk of nosocomial infections in healthcare institutions, better anticipate cardiac arrests, identify suicidal risks, and give tailored information (through chatbot) to caregivers and patients (Morriet, 2022). **In management**, they are developed to more efficiently allocate human and financial resources in healthcare and to manage waiting lists (WHO, 2021). **In clinical care and services**, they are deployed to reduce medical errors (due, e.g., to prescription drugs), improve diagnostics, and give more time to clinicians by liberating them from some time-consuming tasks (e.g., note taking, scheduling) (FDA, 2022; Haug & Drazen, 2023). **In public health**, they are used to identify, track, and analyze events associated with pandemics through mentions on news articles and social media posts, and to compare how different combinations of public health strategies (e.g., border control and testing) affect community outbreak risks.

AI, then, is making its way into all the key components of healthcare systems and beyond. The extent to which AI tools have proven effective remains debatable. Some are now part of "accepted medical practice in the interpretation of some types of medical images, such as ECGs plain radiographs, computed tomographic (CT) and magnetic resonance imaging (MRI) scans, skin images, and retinal photographs" (Haug & Drazen, 2023). Yet concerns that many AI tools have not yet been validated in clinically relevant settings (Topol, 2019b) remain. Some AI tools at least meet evidentiary standards required of other therapeutics. The U.S. Food and Drug Administration (USFDA) (2022) has a public list of AI-enabled products that have been validated and approved as medical devices. However, the precise levels of evidence of the effectiveness AI tools can or should provide before they are made available remain debatable; even some AI tools that are used in or constitute medical devices may avoid current regulatory scrutiny in some settings (Da Silva et al., 2022). Nondevice AI validation must occur elsewhere, perhaps in care settings.

The use of health-related AI also triggers serious concerns. Medical AI can, for example, create discriminatory effects, such as when AI is trained on problematic data that is unrepresentative or uses biased proxies (Guo et al., 2022; Obermeyer et al., 2019). Other concerns include privacy violations (viz., AI tools process huge quantities of personal health information), dehumanization of healthcare relationships, the costs of AI technologies compared to other healthcare priorities, and environmental impacts (AI consumes a lot of energy) (Flood & Régis, 2021). The following discusses how a "human-centered" approach can help address some of the difficult choices posed by this mix of potential benefits and risks.

Striking the appropriate balance requires attending to a complex compact of actors, sometimes interdependently. AI is developed for and used by many actors. Even where regulation occurs, still other mechanisms must ensure tools are appropriate in particular settings. The regulation sets a normative floor. Practitioners must also use "legal" AI in ways that actually further human interests and user interests should guide decisions at each stage.

## Human-Centered Health-Related AI and Human-Centered Design

While the literature on the meaning of HCAI in healthcare remains limited, there is a general convergence toward the idea that it should favor AI developments that respond to healthcare users' needs or interests. This is sometimes framed in terms of normative ends. Another common suggestion is that the interaction between AI and humans should lead to improving human performance. Shneiderman (2022: v), for example, highlights "the primacy of human values such as rights, justice, and dignity" in HCAI. These definitions may seem disconnected from the more technical use that is often made of the term (where HCAI means AI is designed to be usable by persons in the real world), but the approaches are importantly linked. For instance, a recent United Kingdom Medical and Healthcare Regulatory Agency [MHRA] (2022) regulatory document states that

> "[h]uman-centred" here refers to the concept of "human-centred design" that is "an approach to interactive systems that aims to make systems usable and useful by focusing on the users, their needs and requirements, and by applying human factors/ergonomics, and usability knowledge and techniques."

This need not oppose a more normative approach. AI must be used to further human ends. Shneiderman (2022: 9) himself focuses heavily on design, suggesting HCAI should be "designed to be supertools which amplify, augment, empower, and enhance human performance."

We focus here on human-centered design in terms of "usability" but go beyond ergonomics to look at usability in a broader context. Health-related AI scholars suggest many other features HCAI may require, including a "human-in-the-loop" (Chancellor et al., 2019; Doncieux et al., 2022); transparency or explainability (Schmidt, 2020; Liang et al., 2021); and features of trustworthiness AI, including privacy protection and data governance norms and nondiscrimination protections (Doncieux et al., 2022). Still others suggest AI only qualifies as "human-centered" if there is stakeholder participation throughout development (Melles et al., 2021). Whether each suggestion is required for rights, justice, or dignity is a live question, which might be submitted to empirical testing. However, a focus on design elements conducive to actual use appears valuable in any approach. AI that could cure a disease will not do so if people cannot use it. In this sense, attention to humans' role "in the loop" of AI development and use is important even if, empirically, seemingly autonomous AI can further human values (Shneiderman, 2022).

*User-centered design* focuses on the experience of end users and its implications for the design and use of technological innovations. It provides a practical step-by-step approach to document and co-develop the potentialities of a given innovation. Seneviratne et al. (2022) propose a model in four elements to increase the usability and benefits of a technology for end users, focusing on machine learning cases, though plausibly extending more broadly: (1) identification of solvable pain points, (2) eliciting the unique value of AI (e.g., automation and augmentation) to solve pain points or situations perceived as problematic, (3) the actionability pathway providing practical guidance to end users, and (4) the model's reward function. Solvable pain points are perceived issues in a context where actors estimated the technology can create value. AI can be mobilized here to address persisting clinical problems or issues related to organizing the delivery of care to maximize patient experience. The second element relates to the unique value proposition of an innovation that increases the ability of agents to document, assess, and intervene in a specific situation. At this level, actors look to enhance the capacity in accessing and using data to guide their decisions and interventions. The third element, the action pathway, relates to the concrete implications for reorganizing the delivery of care based on the enhanced ability to access and exploit massive data. The fourth element of a user-centered model refers to the tangible benefits that are perceived and achieved by the use of technological innovation.

This is only one fruitful framework illustrating the importance of ensuring health-related AI is usable and creates value for AI users. Practical guidance is available for ensuring tools are usable, and for testing their usability (e.g., Wiklund et al., 2016 (on medical devices)). The key point here is that any normative sense of "HCAI" will be inextricably bound to its technical usability. Normative and contextual approaches to HCAI may relate in numerous ways. The present text simply suggests that the kinds of practical considerations raised in technical discussions of "HCAI" as usable AI are important in any view of "human values." Concerns with usability have, we argue, implications for a range of ways in which we approach the regulation and practical use of health-related AI.

## Regulatory Considerations

Regulators have a role to play in ensuring health-related AI is likely to serve human ends. Recent years have demonstrated the limits of self-regulation approaches to AI governance (Langlois & Régis, 2021). Some binding measures are likely to apply wherever AI is developed. The question is how to regulate health-related AI in ways that permit the wide use of beneficial AI to further human ends while minimizing the risk of it not being used or used poorly or harmfully. Any existing regulators focused on the safety or efficacy of AI as a medical device, public funding for fostering its use, professional rules for its use, etc., should be interested in creating overarching principles within their respective mandates that define the playing field in which contextual interpretative exercises (as on the MHRA definition above) can take place to protect patients first.

These considerations highlight three ways various regulators can help ensure that health-related AI remains relevantly human-centered. They first highlight the importance of regulation to ensure **the objective performance** (efficacy and security) of AI systems. AI systems can only respond to healthcare actors' interests/needs if basic evidentiary norms applying to all health products are met and state-of-the-art AI characteristics are considered before AI implementation. This is aligned with Shneiderman (2022: vi)'s view that HCAI approaches should value "reliable systems based on sound software engineering practices." It also aligns with basic regulatory norms whereby health products of any kind should only be available where proven safe and effective. There is, moreover, presently cause for caution before permitting new AI on markets. After all, proper evidentiary standards for health-related AI effectiveness remain in flux in both regulatory and computer engineering domains and possible risks of widespread AI implementation remain significant. At a minimum, we should seek adequate proof that a particular AI will do what its developers claim in real contexts.

The first question regulators can ask here is: "Does this AI tool meet proper performance (efficacy and security) standards before being launched into the healthcare market?" A human-centered approach will, however, lead them to ask, "Does the interactive mix between this AI tool and human users in the healthcare contexts in which it will be available and used meet such standards?" This second question echoes the UK approach where MHRA (2022) suggests, "[I]t is the performance of the human-AI team not just the performance of the model that is key to ensuring … [AI] is safe and effective." There is, then, a need to set high enough performance standards to ensure the AI will perform as expected, at least when the level of risk involved for patients justifies it (the higher the risks, the higher the regulatory requirements). But attention to the way it will be used and by whom is also valuable. Ensuring products further human values and work in context requires formal ex-ante audit or evaluation processes (which will mostly be based on control settings data) *and* ex-post ones in the kind of settings in which the AI system will be used.

Regulators must also ensure **AI developers' and deployers' accountability**. Accountability here minimally requires *answerability* and *responsibility* from individuals, which is part of the usual business of regulatory regimes. The need to foster responsibility and accountability was in fact mentioned as a key ethical principle for the use of AI for health by the World Health Organization (WHO, 2021). AI actors that develop and deploy AI systems who do not respect the duties to which they were bound should be held accountable. This requires, among other options, that regulators implement or adapt existing reporting mechanisms with respect to AI systems in healthcare that easily allow healthcare actors (including patients in some cases) to notify problematic technical incidents (about, e.g., performance), inadequate evolution of AI systems that needs to be brought to the attention of regulators, privacy violations, and other potential human rights infringements (e.g., if a nurse realizes a triage AI tools for intensive unit bed allocation overly prioritizes men). While incidents may not always trigger liability issues, they may do so in some circumstances. States must ensure some accountability measures exist. Even those that do not trigger liability will make AI developers and deployers aware of issues and trigger duties to respond and adjust to notifications when needed. Safety and efficacy regulators should also respond to incidents.

This highlights a further need to focus on ensuring there is **sufficient information about AI to permit appropriate usage from end users**. One can hardly imagine AI systems responding to human needs and ensuring the protection of patients without sufficient capacity from healthcare professionals and managers—and patients when AI is directly available to them—to understand the appropriateness and limits of the AI systems they use. This is essential to exercise critical thinking and agency with respect to AI, which is vital when addressing human health. Such agency is a prerequisite for providers exercising their professional judgment in clinical contexts, such as when analyzing whether a particular AI tool is appropriate for a particular patient or situation (Laverdière & Régis, 2023). This agency demands a sufficient level of transparency about what we know about AI. Information could concern the type of data used to train the algorithm, which populations are underrepresented in it, what the main variables/correlations used to develop it are, what its weak spots are, etc. The MHRA's (2022) calls to be "transparent about the intended user population," update "the intended user population when changes are made to the model," and describe" how uninterpretable AI (global or local) might impact the performance of the human-AI team" are notable in this respect. And information should be updated when the model evolves in practical settings (adaptative AI). Publicity about the standards used to evaluate these tools on the part of regulators and the evidence proffered to meet those standards is also wise. While we recognize challenges with proprietary data here, statements about why regulators believe a given tool is fit for purpose do not seem to obviously require sharing such data. The kind of information discussed here may also be necessary to request and gain valid informed consent for the use of AI where such rules operate, further impacting uptake (see Cohen, 2020).

These three proposed regulatory norms are nonexhaustive and can only serve their intended ends when applied in contexts. Regulators must leave room for the HCAI to appropriately flourish in specific contexts. Findings from contextual settings may then feed into the regulatory context to modify norms when needed.

## The Importance of Context

Connecting the regulatory world with the world of practice is an exercise in multilevel governance. Multilevel governance "refers to systems of governance where there is a dispersion of authority upwards, downwards and sideways between levels of government—local, regional, national and supranational—and across spheres and sectors, including states, markets and civil society" (Daniell & Kay, 2017: 4). Whether AI development, implementation, and evaluation are "human-centered" will be the result of continuous coordination and negotiation among a plurality of actors, sectors, and levels of decision and intervention; regulatory rules only provide the floor required to consider whether a tool is human-centered. Other leading definitions of HCAI accordingly suggest value and technical queries must attend to a broader context. Riedl (2019) highlights the importance of understanding AI's role as "part of a larger system consisting of human stakeholders." The MHRA (2022), again, indicates that it is the performance of the human–AI team (not solely the performance of the model) that is key to ensuring safe and effective AI. And Shneiderman (2022) suggests that the value of even "autonomous" AI must be understood considering its ability to serve human ends, and that questions about whether to permit its use must examine how it will interact with humans. Such analysis requires examining AI's potential in practical settings, such as the clinical care context.

We focus here on two aspects of this interaction that seem especially pertinent: the diversity of the community of end users and the use of value as a property of experimentation, practice, and reflexivity in context. There is, of course, no single context in which AI–human interactions will occur. Questions about usability and efficacy need to speak to the particular contexts where they are intended to be implemented. This presents yet another challenge: every use context contains a diversity of actors with varying values, interests, expectations, and powers, which interact to shape the journey of AI as innovation. Efforts will have to be made to reconcile the conflicts they produce (Denis et al., 2007).

HCAI, in other words, must be alive to the range of interactions AI may have with different kinds of humans, and to the implications this may have for its overall value. The value of a technology is a result of layers of interactions in particular contexts (Pettigrew, 1985; Latour, 1997). Outcomes accordingly cannot be predicted *ex-ante* but will develop and be the product of a local assessment negotiated by concerned actors and public in their *context of application*. Some outcomes may not even be definable *ex-ante*: a range of actors negotiate "success" conditions. Aligning normative and technical requirements for human ends thus requires understanding how networks of actors and their positions interpret and shape a technology's development and deployment. Levels of cooperation across professions and between professionals and health managers, trust of frontline workers with policymakers, predispositions and capacities of agents, involvement of patients, and perceived attributes and benefits of AI devices will all influence an innovation's uptake.

May (2013) and Murray et al. (2010)'s *normalization process theory* is representative here. It refers to the necessary and fundamental tasks performed by agents in context when it is time to put into practice an innovation. In its basic form, it holds that the process of normalizing the use of technological innovation in a setting is conditioned by four factors. First is the capacity and propensity of agents to make sense with the technology and to achieve a shared understanding of a technology: the value proposition at the core of a user-centered design is considered here as a process of sense-making and sense-giving in clinical contexts (Gioia & Chittipeddi, 1991).

Second is the participation of agents within the organization in promoting the legitimacy of innovation, that is, in enrolling others around the potential benefits and in generating new opportunities to test and use the technology. Third is the development of resources and know-how at the collective or organizational level to enable the spread of AI from localized experiments to standard usage. Fourth is the development of resources and competencies to perform a constant monitoring of user experience, in order to enable adjustments and refinements.

This too is merely a fruitful example of a general approach. The key is that HCAI's prospects rely on a plurality of agents in context testing and assessing a tool and its human-centric properties. The use and value of HCAI in healthcare will be the result of such an effortful and systematic process of getting familiar with the innovation and learning its potential by trial and error in the day-to-day life of organizations. From an end user's perspective, the focus is less on agreeing on the core attributes of a technology in research laboratories (viz., outside its context of use) but on creating space for innovation within real-life settings (health organizations, clinics, etc.). Local dynamics and representations performed by multiple agents in situations of technological innovations will culminate in a concrete assessment of a tool's human-centeredness. This assessment is partly contingent on technological properties: materiality has a contingent effect on actors' representation, behaviors, and experience (Jarzabkowski & Kaplan, 2015). Efforts made by agents in context aim to understand an innovation to produce greater use value. Context as a multilayered phenomenon influences actors' capacity to perform the tasks involved in using and appreciating a technology. Therefore, to determine if a tool's properties align with human-centric aspirations and values, one must look at layers of interacting contexts: the context of discovery, a regulatory context based on normative guidance, and a context of application where end users perform effortful tasks to make sense with technological innovation.

A human-centered approach here suggests two further actionable elements. First, the joint effect of context, technology, and inherent pluralism of practice settings favors a contingent approach to the determination and assessment of the use value of any innovation, including purported HCAI in healthcare. Any approach based *strictly* on an essentialist view of technologies, that is, an approach searching for a fixed set of material attributes that will materialize human-centeredness in any context, will be insufficient to understand and shape the potential of a technology in this regard. In a context like healthcare, with distributed capacities and power, a plurality of agents share the concrete tasks of assessing and shaping the humanistic potential of AI.

Second, reflexivity, that is, a critical and deliberate approach in the process of experimenting with innovations and determining their value in context, can be supported or achieved through routine and systematic evaluation, monitoring, and co-development. Reflexive evaluation of AI integrates diverse viewpoints and perspectives (Patton et al., 2015).

Reflexivity implies that different actors can deliberately clarify their relation to a technology and their appreciation of its potential. One plausible policy implication for HCAI in healthcare would be to support the development of processes and space that nurture reflexivity and capacities in context among a large diversity of groups, organizations, and settings. Current initiatives led by innovation labs created in many sectors aim to provide such space and resources. Methodologically, it implies the ability to systematically monitor the journey and the impact of some important innovations. It also implies having methodological capacities to ensure rigor and depth in analyzing in context the journey of a technology. End users are here audiences and participants of such evaluations and they are co-opted to cocreate guidelines and policies that regulate use in context (Patton et al., 2015). As a recent example of the elaboration of such guidelines, there is the Guide for responsible AI innovation principles in healthcare (Passalacqua et al., 2021) with its autoevaluation form for AI innovators. This interactive online tool was created by the Academic Health Center of the University of Montreal with more than 60 collaborators (innovators, managers, academics, patients, clinicians, etc.) to support AI innovators at each stage of the innovation cycle. The guide gives concrete orientation on how to design responsible AI tools in hospital settings and it allows innovators to reflect on their processes and practices through a self-evaluation questionnaire to adjust the development and deployment of their AI tool accordingly.

The importance of contextual factors does not render the regulatory considerations above useless but instead highlights their role in a broader ecosystem: regulators, through their policies and interventions, can enhance the capacity of organizations and actors to implement user-centered design. For example, the MHRA (2022) plans to promote design principles in line with a user-centric perspective. There is a broader opportunity to create feedback loops where information can proceed from the world of end users to regulators and back. A robust lifecycle review may seek to identify *where human–AI teams are effective*.

## CONCLUSION

Approaches to HCAI focused on "furthering human values" and "ensuring human usability" can be related insofar as "human-centered" requires to be normatively desirable and actually used in a way that will further human values. Melding these normative and technical desires cannot be done strictly through regulation or professional norms alone and what is required for AI to be "human-centered" must be redefined over time through reflexive processes at the regulatory and bedside levels as real-use cases highlight normative and technical benefits and challenges. This does not abrogate responsibility for regulating health-related AI. Human-centered analyses instead offer starting points for apt regulation. Among other things, for example, the foregoing suggests that the lifecycle review process of an innovation is necessary and should attend to usability at each stage.

## ACKNOWLEDGMENT

The authors thank the workshop participants, editors, and Hannah Da Silva and Napoleon Xanthoulis for feedback on prior drafts.

## REFERENCES

Chancellor, S., Baumer, E. P. S., & De Choudhury, M. (2019). Who is the "Human" in Human-Centered Machine Learning: The Case of Predicting Mental Health From Social Media. *Proceedings of the ACM on Human-Computer Interaction*, *3*(CSCW), 1–32.

Cohen, G. (2020). Informed consent and medical Artificial Intelligence: What to tell the patient? *The Georgetown Law Journal*, *108*, 1425–1470.

Da Silva, M., Flood, C. M., & Herder, M. (2022). Regulation of Health-Related Artificial Intelligence in Medical Devices: The Canadian Story. *UBC Law Review*, *55*(3), 635–682.

Daniell, K. A., & Kay, A. (2017). Multi-level governance: An introduction. In *Multi-level Governance: Conceptual challenges and case studies from Australia*, edited by Katherine A. Daniell, Adrian Kay, Australia: Anu Press, 3–32.

Davenport, T., & Kalakota, R. (2019). The potential for artificial intelligence in healthcare. *Future Health*, *6*(2), 94–98.

Denis, J. L., Langley, A., & Rouleau, L. (2007). Strategizing in pluralistic contexts: Rethinking theoretical frames. *Human Relations*, *60*(1), 179–215.

Doncieux, S., Chatila, R., Straube, S., & Kirchner, F. (2022). Human-centered AI and robotics. *AI Perspectives*, *4*(1), 1.

Flood, C. M., & Régis, C. (2021). AI and Health Law. In Martin-Bariteau, F. & Scassa, T. (Eds.), *Artificial Intelligence and the law in Canada*. LexisNexis, Toronto, Ontario.

Gioia, D. A., & Chittipeddi, K. (1991). Sensemaking and sensegiving in strategic change initiation. *Strategic Management Journal*, *12*(6), 433–448.

Guo, L. N., Lee, M. S., Kassamali, B., Mita, C., & Nambudiri, V. E. (2022). Bias in, bias out: Underreporting and underrepresentation of diverse skin types in machine learning research for skin cancer detection—a scoping review. *Journal of the American Academy of Dermatology*, *87*(1), 157–159.

Haug, C. J., & Drazen, J. M. (2023). Artificial intelligence and machine learning in clinical medicine, 2023. *New England Journal of Medicine*, *388*(13), 1201–1208.

Jarzabkowski, P., & Kaplan, S. (2015). Strategy tools-in-use: A framework for understanding "technologies of rationality" in practice. *Strategic Management Journal*, *36*(4), 537–558.

Langlois, L., & Régis, C. (2021). Analyzing the contribution of ethical charters to building the future of Artificial Intelligence Governance. In *Reflections on AI for humanity*, edited by B. Braunschweig, M. Ghallab. Springer, 150–170, https://doi.org/10.1007/978-3-030-69128-8_10

Latour, B. (1997). *Nous n'avons jamais été modernes: essai d'anthropologie symétrique*. Paris: La Découverte.

Laverdière, M. et Régis, C. (2023). Soutenir l'encadrement des pratiques professionnelles en matière d'intelligence artificielle dans le secteur de la santé et des relations humaines. *Proposition d'un prototype de code de déontologie* (online).

Liang, Y., He, L., & Chen, X. A. (2021). Human-Centered AI for Medical Imaging. In Y. Li & O. Hilliges (Eds.), *Artificial Intelligence for human computer interaction: A modern approach* (pp. 539–570). Human-Computer Interaction Series. Cham: Springer International Publishing.

May, C. (2013). Agency and implementation: understanding the embedding of healthcare innovations in practice. *Social Science & Medicine*, *78*, 26–33.

McGann, M., Wells, T., & Blomkamp, E. (2021). Innovation labs and co-production in public problem solving. *Public Management Review*, *23*(2), 297–316.

Medicines & Healthcare products Regulatory Agency (MHRA). (2022). *Software and AI as a Medical Device Change Programme – Roadmap*.

Melles, M., Albayrak, A., & Goossens, R. (2021). Innovating health care: Key characteristics of human-centered design. *International Journal for Quality in Health Care*, *33*(Supp. 1), 37–44.

Morriet, O. (2022). *Intelligence artificielle & santé*. Montréal: Fonds de recherche du Québec et CHUM.

Murray, E. et al. (2010). Normalisation process theory: A framework for developing, evaluating and implementing complex interventions. *BMC Medicine*, *8*(1), 1–11.

Obermeyer, Z., Powers, B., Vogeli, C., & Mullainathan, S. (2019). Dissecting racial bias in an algorithm used to manage the health of populations. *Science (New York, N.Y.), 366*(6464), 447–453.

Passalacqua, A. et al., (2021). Guide des principes d'innovation et d'IA responsables en santé (available online).

Patton, M. Q., McKegg, K., & Wehipeihana, N. (Eds.). (2015). *Developmental evaluation exemplars: Principles in practice*. Guilford Publications.

Pettigrew, A. M. (1985). Contextualist research and the study of organizational change processes. *Research Methods in Information Systems, 1*(1985), 53–78.

Research and Markets. (2023). *Artificial Intelligence Market Size, Share & Trends Analysis Report By Solution, By Technology (Deep Learning, Machine Learning), By End-use, By Region, And Segment Forecasts, 2023–2030.*

Riedl, M. O. (2019). Human-Centered Artificial Intelligence and Machine Learning. *Human Behavior and Emerging Technologies, 1*(1), 33–36.

Schmidt, A. (2020). Interactive human centered Artificial Intelligence: A definition and research challenges. In *Proceedings of the International Conference on Advanced Visual Interfaces* (pp. 1–4). AVI '20. New York, NY: Association for Computing Machinery.

Seneviratne, M.G. et al. (2019). Bridging the implementation gap of machine learning in health care. *BMJ Innovations, 6*, 45–47.

Seneviratne, M. G., et al. (2022). User-centred design for machine learning in health care: A case study from care management. *BMJ Health & Care Informatics, 29*(1), e100656.

Shneiderman, B. (2022). *Human-centered AI*. Oxford University Press.

Topol, E. (2019a). *Deep medicine: How Artificial Intelligence can make healthcare human again*. USA: Basic Books, Inc.

Topol, E. (2019b). High-performance medicine: the convergence of human and artificial intelligence. *Nature Medicine, 25*(1), 44–56.

U.S. Food and Drug Administration (FDA). (2022). *Artificial Intelligence and machine learning (AI/ML)-enabled medical devices*. Digital Health Center of Excellence. https://www.fda.gov/medical-devices/software-medical-device-samd/artificial-intelligence-and-machine-learning-aiml-enabled-medical-devices

WHO. (2021). *Ethics and governance of artificial intelligence for health: WHO guidance*. Geneva: World Health Organization.

Wiklund, M. et al. (2016). *Usability testing of medical devices*, 2nd ed. CRC Press.

# 11 Human-Centered AI for Sustainability and Agriculture

*Jennifer Garard*

Sustainability in the Digital Age, Concordia University, Montréal, Québec, Canada

Future Earth Canada, Montréal, Québec, Canada

*Allison Cohen*

Mila—Québec AI Institute, Montréal, Québec, Canada

*Ernest Habanabakize and Erin Gleeson*

Sustainability in the Digital Age, Concordia University, Montréal, Québec, Canada

Future Earth Canada, Montréal, Québec, Canada

McGill University, Montréal, Québec, Canada

*Mélisande Teng*

Sustainability in the Digital Age, Concordia University, Montréal, Québec, Canada

Future Earth Canada, Montréal, Québec, Canada

Mila—Québec AI Institute, Montréal, Québec, Canada

*Gaétan Marceau Caron and Daoud Piracha*

Mila—Québec AI Institute, Montréal, Québec, Canada

*Rosette Lukonge Savanna*

Leapr Labs, Kigali, Rwanda

*Kinsie Rayburn and Melissa Rosa*

Planet Labs, San Francisco, CA, USA

*Kaspar Kundert*

Self-employed consultant



DOI: 10.1201/9781003320791-14

*Éliane Ubalijoro*

Sustainability in the Digital Age, Concordia University, Montréal, Québec, Canada

Future Earth Canada, Montréal, Québec, Canada

There is broad consensus that global sustainability crises such as climate change, biodiversity loss, and soil degradation are already causing severe negative impacts on people and the planet that will worsen over time. Addressing these complex, interrelated issues is not straightforward. The Intergovernmental Panel on Climate Change has made it clear that tackling climate change will require rapid and unprecedented changes across all sectors of society (IPCC, 2018) and that "the move towards climate-resilient societies requires transformational or deep systemic change" (Pathak et al., 2022). The Kunming-Montreal Global Biodiversity Framework plainly articulates the urgent need to reduce threats to biodiversity, for example, by phasing out harmful subsidies and enhancing incentives for the sustainable use of biodiversity (CBD COP-15, 2022). The United Nations Food and Agriculture Organization warns that rapidly degrading soils arising from intensive agricultural practices pose an immense threat to food security and ecosystems around the world (FAO, 2022).

Artificial intelligence (AI) and machine learning (ML) have incredible potential to help tackle these issues, for example, by automating shifts to renewable energy sources, optimizing supply chain management, monitoring ecosystem health, predicting extreme weather for early responses, and enabling precision agriculture and integrated agricultural input management (Rolnick et al., 2019; McLennon et al., 2021). A recent report also found that AI can play a significant role in governance, by mobilizing data, optimizing strategies, changing behavior, and empowering citizens (Sustainability in the Digital Age, Future Earth, & Climate Works Foundation, 2022). AI has the potential to be a transformative tool for sustainability and can push the most powerful levers of systems change (see Figure 11.1), described by Donella Meadows as focusing on rules, power dynamics, and mindsets (Meadows, 1999). However, there are valid concerns that AI, and ML as a subset of AI, risks exacerbating inequalities, infringing on rights to privacy and security, and accelerating environmental degradation if not governed properly (Luers et al., 2020; Bohnsack et al., 2022). In the case of agriculture, further challenges related to insufficient internet access for farmers, uneven data availability and distribution of benefits, difficulty generalizing ML models at scale, and limited availability of trained personnel in developing nations to develop, deploy, and maintain AI applications make it difficult for AI to reach its transformative potential (Delgado et al., 2019).

Across all potential applications of AI in social contexts, it remains critical to approach the use of AI and ML with intentionality, empathy, and humility and to foster inclusive innovation (Ubalijoro et al., 2023). This is necessary to ensure that these tools respond to genuine social needs and emphasize responsibility and inclusivity, complying with the central promise of the Sustainable Development Goals to "leave no one behind" (United Nations System Chief Executives Board for Coordination, 2016). The concept of human-centered AI (HCAI) advances such approaches. While academic consensus on a definition is lacking, in practice HCAI

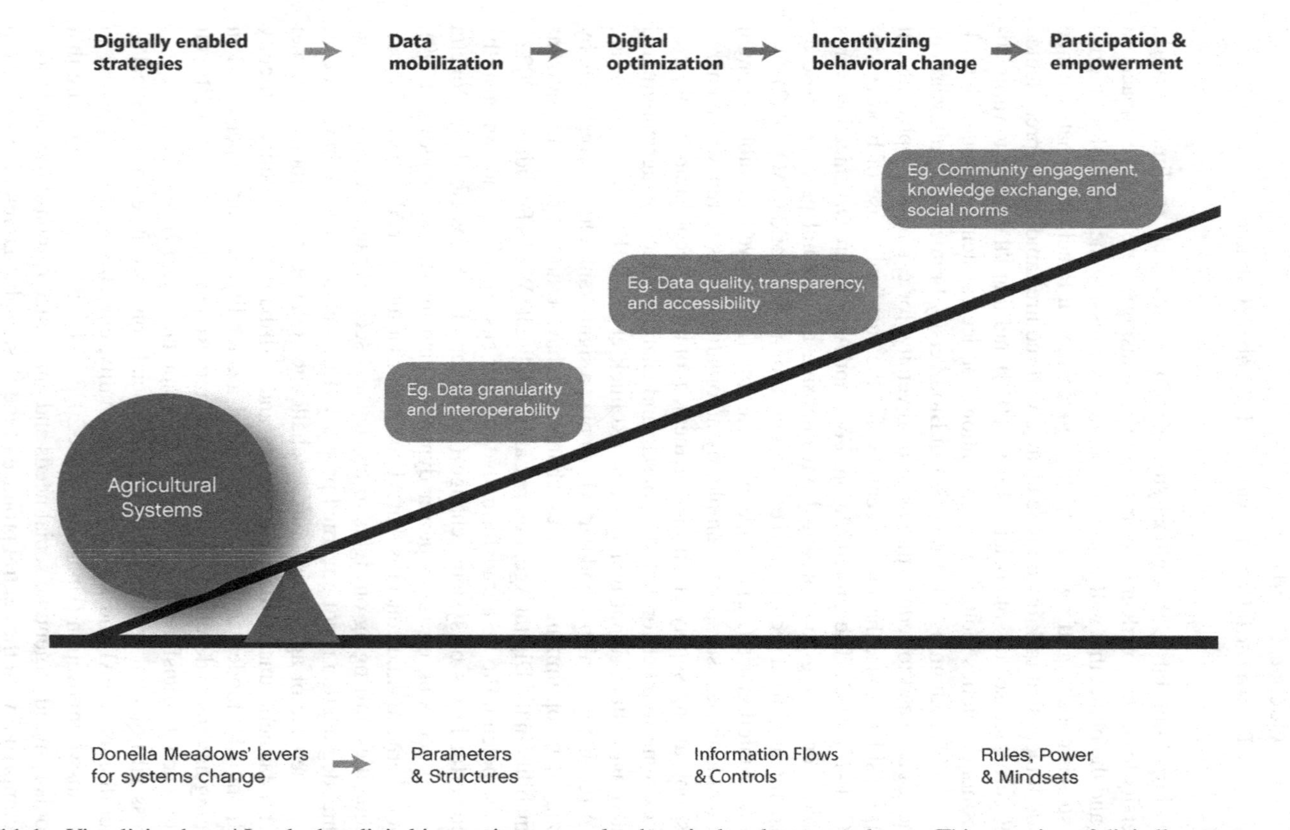


**FIGURE 11.1** Visualizing how AI and other digital innovations can unleash agricultural systems change. This mapping of digitally empowered governance strategies along Donella Meadows' levers of systems change (adapted from Meadows, 1999; Luers et al., 2020) informed the development of the DISA project. Digital strategies that prioritize participation and empowerment have the strongest potential to bring about systems change.

is increasingly seen as important to addressing stakeholder groups' diverse needs and expectations (Chancellor et al., 2019).

In this chapter, we understand HCAI as an open, inclusive, and participatory approach to AI design and use, one that acknowledges the role of humans in ensuring that technology is responsibly deployed in ways that benefit people and the planet (Ibid; Shneiderman, 2020). Explicitly, we explore four guiding principles of HCAI—inclusive data stewardship, robust AI applications, trust in partners, projects, and technology, and systems of accountability. We then present the Data-driven Insights for Sustainable Agriculture project (DISA) that leverages ML and an HCAI approach in support of smallholder farmers' transitions to regenerative agriculture in Rwanda. Our aim is to highlight how the guiding principles have been operationalized within this use case and share lessons learned that can be leveraged more broadly when designing HCAI in the domain of sustainability and agriculture.

## GUIDING PRINCIPLES FOR HCAI IN SOCIAL CONTEXTS

Operationalizing HCAI in an applied, social context requires significant investment in time at the early stages of a project. HCAI must be inclusive of diverse viewpoints and elevate marginalized voices from the design phase onwards. Engaging with stakeholders on the ground helps to promote understanding among AI developers and learning within implicated communities, to test and co-develop digital tools, and to contextualize and clearly communicate project activities (Sustainability in the Digital Age, Future Earth, & ClimateWorks Foundation, 2022; Chuard et al., 2022) all while adhering to HCAI guiding principles.

The four guiding principles for HCAI discussed here are not intended to be comprehensive nor are they mutually exclusive. Rather, they emerge from the literature as relevant, actionable, and interrelated principles that can pave the way for HCAI in the complex domain of sustainability and agriculture.

### Guiding Principle 1: Inclusive Data Stewardship

The accuracy of ML models depends on the quality of the data used to train, evaluate, and test them. Thus, the process of collecting, storing, and using data must be thoughtfully crafted in order for the AI development process to be considered human-centered. It is important to learn from different frameworks, codes, and principles for promoting effective and ethical data stewardship centered around people. The FAIR principles, created by stakeholders from academia and industry, state that data and digital assets should be Findable, Accessible, Interoperable, and Reusable (Wilkinson et al., 2016). In response to critiques of FAIR, the International Indigenous Data Sovereignty Interest Group established the CARE Principles for Indigenous Data Governance—Collective Benefit, Authority to Control, Responsibility, and Ethics (Research Data Alliance International Indigenous Data Sovereignty Interest Group, 2019). These principles emphasize respecting the rights and interests of all stakeholders involved (Carroll et al., 2020). The Global Open Data for Agriculture and Nutrition (GODAN) code of conduct advances a co-designed ethical framework to promote responsible sharing of agricultural data, covering issues such as data privacy and security, intellectual property rights, data quality, and data accessibility (Ubalijoro et al., 2021), which is especially critical where national laws are lacking. Advancing the responsible development of AI in ways that respect the rights, interests, and values of Indigenous peoples and communities is also critically important (Lewis et al., 2020). Interestingly, Room 9 of the "17 Rooms Initiative" assembled a group of data scientists, development experts, and local stakeholders, including co-author Éliane Ubalijoro, who prototyped a digital data cooperative and are developing a companion guidebook supporting digital asset creation and ownership for enhanced value and agency.

## Guiding Principle 2: Robust AI Applications

The development of robust AI applications based on algorithms that are accessible and replicable (Pineau et al., 2020) and whose underlying architecture and assumptions are clear to developers and end users (Margetis et al., 2021) is critical in light of the influence that algorithmic outputs can have (Dietterich, 2017; Buolamwini et Gebru, 2018). As we know, AI is entirely dependent on the data used to train algorithms—data which is necessarily an abstracted and biased representation of reality (LeCun et al., 2015). These datasets often lack the robustness that would allow the models to provide consistent, high-quality outputs in all scenarios. Understandably, errors in ML model's predictions risk undermining trust; however, these errors can easily be missed since the plethora of possible mistakes are hard to detect. One example is that of a computer vision algorithm that was only trained to detect cows on farmland. Since this dataset lacked examples of the cow in other locations, the algorithm inferred that the farmland was a relevant feature associated with the presence (or lack thereof) of a cow (Beery et al., 2018). While cows might not be commonly found in beach settings, it is important to evaluate the capacity of an ML model to generalize beyond the training dataset to ascertain whether it can arrive at accurate conclusions in as many contexts and conditions as possible. A more comprehensive system can then guide ML scientists to improve robustness, assess deployment, monitor errors, and evaluate performance over time.

## Guiding Principle 3: Trust in Partners, Project, and Technology

Building trust among those engaged in, and affected by, a project is a key component of HCAI. This entails building trust in the project's goals and processes, and in each team member's respective capacity to deliver their piece of the puzzle in an integrated, transparent fashion (Osburg et Lohrmann, 2017; Schulz et Feist, 2021). Before training ML models, it is important to conduct trust-building exercises that facilitate equity, amplify marginalized voices, and empower local stakeholders. These activities should ensure that all stakeholders involved have the necessary capacity to understand each other and participate meaningfully in the project (Chuard et al., 2022; Colfer et Prabhu, 2023). This allows for the co-creation of digital tools that emerge from the integration of multidisciplinary insight. Another opportunity for trust building comes from an awareness of the strengths and limitations of the dataset (Druce et al., 2021), since transparency in the algorithm's capacity to perform well, and under which circumstances, allows for a level of conservativeness surrounding the tool's deployment that can also build trust over time. Trust can be sustained by establishing reliability, consistency (Lankton et al., 2015), and agility throughout the project development lifecycle (Lien et al., 2021).

### Guiding Principle 4: Systems of Accountability

Accountability is a critical component of HCAI, but methods to implement accountability have not been extensively researched. While much of the literature on HCAI equates accountability with transparency (e.g., Lepri et al., 2021), transparency—of data, knowledge, and methods—is likely not sufficient to ensure accountability (Gupta et al., 2020). Increased transparency does have the potential to enhance accountability, but can also pose risks without appropriate safeguards (Luers et al., 2020). Implementing deliberate systems of accountability can help address the misleading assumption that transparency automatically leads to accountability. Decentralized structures of accountability are often more responsive, inclusive, and empower local collaborators to operate autonomously (Shah, 2004). Non-hierarchical networks of accountability can promote shared understandings of diverse perspectives on accountability and enhance communication (Bäckstrand, 2008). Accountability in data stewardship is especially relevant to privacy and open science (Young et al., 2019). New tools are emerging in the ML community to promote transparency and accountability, such as Datasheets for Datasets (Gebru et al., 2021) and model cards for model reporting (Mitchell et al., 2019). These tools help contextualize data, make intended use and limitations explicit, and identify data collectors. Licensing can also enhance accountability and standardize practices in the Open-Source community (using a variety of available licenses, e.g. from Open Source Initiative).

## HUMAN-CENTERED AI IN PRACTICE: THE DATA-DRIVEN INSIGHTS FOR SUSTAINABLE AGRICULTURE PROJECT

To illustrate the practical application of these principles, we present a project launched in 2021, Data-driven Insights for Sustainable Agriculture, or DISA. Currently, DISA focuses on smallholder farmers in Rwanda, who constitute 77.2% of the national farming households and largely live on less than US$5/day (World Bank, 2016). In this context, it is critical to understand not only the physical environmental issues at hand, including soil health, biodiversity conservation, water utilization, and greenhouse gas emissions, but also the social, cultural, and economic needs.

DISA is led by a consortium of partners from Future Earth Canada, Sustainability in the Digital Age, Mila—Quebec AI Institute, Regional Research Centre for Integrated Development Rwanda, Planet Labs, and Leapr Labs, with support from the Fonds de Recherche du Québec and the ClimateWorks Foundation. The DISA project team is a transdisciplinary group of domain experts, from disciplines including AI/ML, geographic information systems, agriculture, and climate change. The goal of DISA is to support transitions away from resource-intensive and extractive farming models towards knowledge-intensive models to drive economies that are climate-resilient, net-zero, and nature-positive. DISA addresses a need for robust, data-driven evidence to support the development of policy and investment frameworks. Taking an HCAI approach, the DISA project team is working to incorporate insights from ML trained on satellite imagery with local knowledge to build an evidence base for the viability of regenerative agriculture in Rwanda, with the goal of scaling to other parts of Africa. At the time of writing, ground truth data has been collected through a survey of 815 smallholder farms. Development of ML algorithms is underway, with an algorithm to delineate farm boundaries nearly complete and algorithms to detect regenerative agricultural practices expected in the coming months.

## Developing the Project

From its earliest stages, DISA has adopted a participatory research approach in line with the guiding principles of HCAI. The project emerged from a two-year consultative process that builds on the Reimagining Climate Governance framework, developed through consultations with nearly 100 collaborators around the world (Sustainability in the Digital Age, Future Earth, & ClimateWorks Foundation, 2022), as well as work to promote the benefits of data use to improve governance of agriculture and food systems (de Beer et al., 2022) (Figure 11.1 visualizes the importance of leveraging AI and other digital tools to engage and empower communities, where participation and empowerment have the potential to push the most impactful levers of systems change). DISA emphasizes the need to empower smallholder farmers with the knowledge and skills necessary to implement regenerative agriculture and leverage digital tools. The DISA team is working to develop the needed rigorous, data-driven evidence base for regenerative agriculture and to support harmonization amongst different terms used to describe regenerative agricultural practices, including natural farming and conservation agriculture.

The initial stage of DISA involved extensive knowledge exchanges, co-led by co-author Éliane Ubalijoro, to understand the problems faced by smallholder farmers. This includes a collaborative webinar held on July 27, 2020, by the World Bank, GODAN, and Rythu Sadhikara Samstha (Government of Andhra Pradesh) with the objective of sharing knowledge about regenerative agriculture. The webinar aimed to make this knowledge accessible to farmers and policymakers in Rwanda, especially in the context of challenges posed by COVID-19 and climate change (India-Rwanda Knowledge Exchange on Natural Farming, 2020). In 2021, Future Earth supported the Alliance for Food Sovereignty in Africa (AFSA) in a series of agriculture knowledge exchanges between the Andhra Pradesh Community Managed Natural Farming Programme (APCNF) and African stakeholders, which led to the development of a community of practice of organizations. More than 50 organizations representing African farmers engaged in these exchanges, where they heard testimonies from peers in India who have already transitioned to regenerative agriculture.

## Applying the Four Guiding Principles

### Guiding Principle 1: Inclusive Data Stewardship

Environmental and social concerns around agriculture are compounded by limitations regarding available data and questions around how the data that *is* available is managed and for what purposes it is used. As the project entails accessing large amounts of data containing sensitive information about farmers, farms, crops, and animals, protecting the privacy of those whose data is collected is of paramount importance. To ensure that this data is kept secure and protected from potential misuse throughout and beyond the project itself, DISA's plans for data management and stewardship are centered around people, based on the frameworks and codes of conduct highlighted earlier. Achieving a balance between openly sharing project results, critical to scaling the work in Rwanda, and maintaining the privacy of smallholder farmers, whose data is pivotal to building an evidence base for regenerative agricultural practices, is a challenge. Being explicit about this balance with farmers and ensuring they consent to how their data is used is thus very important. In this way, the DISA team is aligning data stewardship with practical project design to enable the voices of smallholder farmers to guide the collection and use of data, keeping their well-being as a top priority.

### Guiding Principle 2: Robust AI Models

DISA is employing an inclusive approach to the creation of our data set by using diverse data points from across the country collected in collaboration with local farmers. Ground truth data has been collected using an extensive survey conducted in person by enumerators in Rwanda. Developing the survey was an iterative process that took many months of consultation. This time was used, for example, to address DISA's "Catch-22" whereby the ML team had a difficult time pinpointing exactly what information was needed to train the model without seeing some data and testing it, while other team members needed to know what information was required on the ML side to design questions to include in the survey. Time was also needed to support the ML team in fully understanding the local context and building an intuition around the use of regenerative agriculture.

The survey gathers information about the nine principles of regenerative agriculture adopted from APCNF (AFSA et APCNF, 2022) as well as outputs of interest such as soil fertility and soil erosion. These data points are used by the ML team to determine the extent of regenerative agricultural practices at scale. The DISA team will begin by using ground truth data to train ML models to identify the nine principles of regenerative agriculture and correlate these practices with output variables. This process will involve participatory research to ensure that ML models yield actionable insights relevant to farmers' needs. In the short term, these insights include crop yield prediction and information on resilience to extreme weather events. As a next step, the DISA team is interested in exploring innovative methods to map soil organic carbon, which combines remote sensing data, machine learning models, and soil spectroscopy (Vågen et al., 2018) to support smallholder farmers in gaining access to carbon financing.

### Guiding Principle 3: Trust in Partners, Project, and Technology

Discussions in the context of DISA are always centered around finding a common ground regarding the best pathways forward towards agreed-upon end goals, in line with a major aspect of HCAI—to build trust through collaborations and diversity. Regular meetings amongst the immediate project team held over many months have forged strong partnerships and have enabled team members to build mutual understanding and align visions and values for the project through dialogue. Prolonged and meaningful engagement with other organizations, including government officials, local NGOs, private sector organizations, academic institutions, and funding organizations help to build trust and foster innovative collaborations beyond the core team. One major challenge in building trust is finding funding for the time required to do so, which is critically important but may be undervalued (Sustainability in the Digital Age, Future Earth, & ClimateWorks Foundation, 2022).

Working with smallholder farmers, the DISA team aims to support continued learning opportunities through capacity building and knowledge exchange. A training of trainers program, following the positive experience of the APCNF, has the potential to support the exponential growth of the transition towards regenerative agricultural practices over time, spearheaded by trusted local champion farmers. Eventually, as the positive impacts of regenerative agriculture are experienced by farmers, we hope this trust will continue to grow.

### Guiding Principle 4: Systems of Accountability

The DISA team began exploring systems of accountability through data stewardship frameworks that prioritize privacy and consent and take into consideration the power dynamics relevant when engaging stakeholders from more marginalized communities. When collecting ground truth data, a consent form tailored to smallholder farmers' concerns and capacity was delivered verbally by local enumerators, and permission to conduct the survey was obtained from the National Institute of Statistics of Rwanda. The team is now exploring options for creating a decentralized accountability network. This network would entail a clear, nonhierarchical structure involving partners and relevant stakeholders across all stages of the project, drawing on key insights from governance literature (Shah, 2004; Bäckstrand, 2008). A key component of developing such a networked arrangement is building trusted relationships with all groups who are accountable to one another.

The issue of capacity presents a challenge since it is not feasible to train smallholder farmers from all 815 farms surveyed to critically assess the development of ML algorithms. In the short term, team members will collect iterative feedback from representatives of smallholder farmers on targeted issues at different stages of the ML research and development pipeline. This includes, for example, validating with local farmer groups that boundary detection algorithms are working as intended. DISA also engages with experienced local partners who can help build trust and ensure accountability (Sustainability in the Digital Age, Future Earth, & ClimateWorks Foundation, 2022), including Bridge2Rwanda, a local agriculture NGO, agronomists with the Rwandan Agriculture Board, and members of farmer cooperatives. Co-creating and implementing a shared and accessible system of network-based accountability, that specifies metrics of success where possible, will be the next step in advancing HCAI in the DISA project.

## CONCLUSION

Unprecedented environmental challenges are creating a need for collective behavioral change. In this context, AI can be a useful tool to create the evidence that's needed to motivate a policy response and behavioral change. HCAI is an important approach to ensure that AI development is inclusive and incorporates appropriate guiding principles from diverse perspectives, though challenges remain.

The key challenge in implementing HCAI is the need to build and maintain trust between all members of the project. This requires project collaborators to engage regularly and meaningfully before the project even begins, sometimes for months, to discuss the division of roles and responsibilities and align on how they'd like to collaborate. While a discussion of metrics of success for HCAI is beyond the scope of this paper, co-developing a clear plan for HCAI early in a project and tying components to measurable outcomes provides further opportunities to strengthen trust and accountability. These are critical points to keep in mind for future projects taking an HCAI approach.

The Reimagining Climate Governance in the Digital Age initiative recognizes the need to integrate fragmented approaches across scales to accelerate collective climate action and provide insights on creating sustainable, inclusive climate solutions. It is critically important to deeply integrate the cultural and social contexts in which technologies are developed and deployed (Sustainability in the Digital Age, Future Earth, & ClimateWorks Foundation, 2022). The field of AI is advancing rapidly, including developments like ChatGPT, for example (OpenAI, 2023), but there is a need to amplify insights and accessibility from the Global South, currently underrepresented due to biases and lack of training data from certain countries. Furthermore, inclusive access to information and emerging technologies that serve the needs of real people will be critical to ensure that innovations in AI do not further deepen the digital divide but instead contribute to bridging it (Ubalijoro et al., 2023). Future research is needed to integrate insights from governance in adapting and evaluating guiding principles for data stewardship, robustness, trust, and accountability into actionable, context-specific frameworks. Sharing the work done in DISA to date and highlighting the challenges faced is intended to assist others seeking to leverage HCAI for sustainability and agriculture.

DISA's goal is to advance the exploration of outcomes of the Reimagining Climate Governance in the Digital Age initiative to pave the way for meaningful impacts on people and the planet in line with the 2030 Sustainable Development Goals and COP15's Kunming-Montreal Global Biodiversity Framework Target 18, which calls for the elimination, reformation, or phasing out of harmful incentives and subsidies for biodiversity in a way that is proportionate, just, fair, effective, and equitable. Taking a human-centered approach to the use of AI in DISA is the foundation of ensuring that this goal is achievable.

## REFERENCES

AFSA & APCNF. (2022). Taking Agroecology to Scale: Learning from the Experiences of Natural Farming in India. Retrieved from https://afsafrica.org/wp-content/uploads/2022/10/natural-farming-guide-digital-lr.pdf

Bäckstrand, K. (2008). Accountability of Networked Climate Governance: The Rise of Transnational Climate Partnerships. *Global Environmental Politics*, *8*, 74–102.

Beery, S., Van Horn, G., & Perona, P. (2018). Recognition in Terra Incognita. Retrieved from https://openaccess.thecvf.com/content_ECCV_2018/html/Beery_Recognition_in_Terra_ECCV_2018_paper.html

Bohnsack, R., Bidmon, C.M., & Pinkse, J. (2022). Sustainability in the digital age: Intended and unintended consequences of digital technologies for sustainable development. *Business Strategy and the Environment*, *31*, 599–602.

Buolamwini, J., & Gebru, T. (2018). Gender Shades: Intersectional Accuracy Disparities in Commercial Gender Classification. *Proceedings of Machine Learning Research*, *81*, 1–15.

Carroll, S.R. et al. (2020). The CARE Principles for Indigenous Data Governance. *Data Science Journal*, *19*, 43.

CBD COP-15. (2022). *Kunming-Montreal Global Biodiversity Framework*. Retrieved from https://www.cbd.int/doc/decisions/cop-15/cop-15-dec-04-en.pdf

Chancellor, S., Baumer, E.P.S., & De Choudhury, M. (2019). Who is the 'Human' in Human-Centered Machine Learning: The Case of Predicting Mental Health from Social Media. *Proceedings of the ACM on Human-Computer Interaction*, *3*(147,1-147:32).

Chuard, P., Garard, J., Schulz, K., Kumarasinghe, N., & Matthews, D. (2022). A portrait of the different configurations between digitally-enabled innovations and climate governance. *Earth System Governance*, *13*, 100147.

Colfer, C., & Prabhu, R. (2023). *Responding to Environmental Issues through Adaptive Collaborative Management: From Forest Communities to Global Actors*. Taylor & Francis.

de Beer, J., Oguamanam, C., & Ubalijoro, É. (2022). Ownership, Control, and Governance of the Benefits of Data for Food and Agriculture: A Conceptual Analysis and Strategic Framework for Governance.

Delgado, J.A., Short, N.M., Roberts, D.P., & Vandenberg, B. (2019). Big Data analysis for sustainable agriculture on a geospatial cloud framework. *Frontiers in Sustainable Food Systems*, *3*.

Dietterich, T.G. (2017). Steps toward Robust Artificial Intelligence. *AI Magazine*, *38*, 3–24.

Druce, J., Niehaus, J., Moody, V., Jensen, D., & Littman, M.L. (2021). Brittle AI, Causal Confusion, and Bad Mental Models: Challenges and Successes in the XAI Program. Preprint. Retrieved from http://arxiv.org/abs/2106.05506

FAO. (2022). *Soils for nutrition: State of the art*. Retrieved from https://www.fao.org/documents/card/en/c/cc0900en

Gebru, T., et al. (2021). Datasheets for Datasets. Preprint. Retrieved from http://arxiv.org/abs/1803.09010

Gupta, A., Boas, I., & Oosterveer, P. (2020). Transparency in global sustainability governance: To what effect? *Journal of Environmental Politics and Planning*, *22*, 84–97.

India-Rwanda Knowledge Exchange on Natural Farming: Webinar 2 - The Rwandan Perspective (2020). Retrieved from https://www.youtube.com/watch?v=2m1Q0a7rJHM&list=PLeRVyCRL9PkQDf2PtBDoC5XC7ozhzjsSf&index=42&t=149s

IPCC. (2018). *Global Warming of 1.5°C: An IPCC Special Report on the impacts of global warming of 1.5°C above pre-industrial levels and related global greenhouse gas emission pathways, in the context of strengthening the global response to the threat of climate change, sustainable development, and efforts to eradicate poverty*. Cambridge University Press.

Lankton, N., McKnight, D.H., & Tripp, J. (2015). Technology, humanness, and trust: Rethinking trust in technology. *JAIS*, *16*, 880–918.

LeCun, Y., Bengio, Y., & Hinton, G. (2015). Deep learning. *Nature*, *521*, 436–444.

Lepri, B., Oliver, N., & Pentland, A. (2021). Ethical machines: The human-centric use of artificial intelligence. *iScience*, 24.

Lewis, J.E. et al. (2020). Indigenous protocol and Artificial Intelligence position paper. Retrieved from https://www.indigenous-ai.net/position-paper/

Lien, A., et al. (2021). Trust is essential to the implementation of adaptive management on public lands. *Rangeland Ecology & Management*, *77*, 46–56.

Luers, A. et al. (2020). Leveraging digital disruptions for a climate-safe and equitable world: The D^2S Agenda. *IEEE Technology and Society Magazine*, *39*, 18–31.

Margetis, G., Ntoa, S., Antona, M., & Stephanidis, C. (2021). Human-centered design of artificial intelligence. In *Handbook of Human Factors and Ergonomics* (pp. 1085–1106). John Wiley & Sons, Ltd.

McLennon, E., Dari, B., Jha, G., Sihi, D., & Kankarla, V. (2021). Regenerative agriculture and integrative permaculture for sustainable and technology-driven global food production and security. *Agronomy Journal*, *113*, 4541–4559.

Meadows, D. (1999). *Leverage Points: Places to Intervene in a System*. Retrieved from http://donellameadows.org/archives/leverage-points-places-to-intervene-in-a-system/

Mitchell, M., et al. (2019). Model cards for model reporting. In *Proceedings of the Conference on Fairness, Accountability, and Transparency* (pp. 220–229).

OpenAI. (2023). GPT-4 Technical Report. Preprint. Retrieved from https://arxiv.org/abs/2303.08774

Osburg, T., & Lohrmann, C. (2017). *Sustainability in a Digital World: New Opportunities Through New Technologies*. Springer International Publishing.

Pathak, M. et al. (2022). *Climate Change 2022: Mitigation of Climate Change. Contribution of Working Group III to the Sixth Assessment Report of the Intergovernmental Panel on Climate Change*. In P.R. Shukla et al. (Eds.), Cambridge University Press.

Pineau, J. et al. (2020). Improving Reproducibility in Machine Learning Research: A Report from the NeurIPS 2019 Reproducibility Program. Preprint. Retrieved from http://arxiv.org/abs/2003.12206

Research Data Alliance International Indigenous Data Sovereignty Interest Group. (2019). *CARE Principles*. Global Indigenous Data Alliance. Retrieved from https://www.gida-global.org/care

Rolnick, D. et al. (2019). Tackling climate change with machine learning. *ACM Computing Surveys*, *55*, 1–96.

Schulz, K., & Feist, M. (2021). Leveraging blockchain technology for innovative climate finance under the Green Climate Fund. *Earth System Governance*, *7*, 100084.

Shah, A. (2004). *Fiscal Decentralization in Developing and Transition Economies: Progress, Problems, and the Promise*. The World Bank.

Shneiderman, B. (2020). Human-centered Artificial Intelligence: Reliable, safe & trustworthy. *International Journal of Human–Computer Interaction*, *36*, 495–504.

Sustainability in the Digital Age, Future Earth, & ClimateWorks Foundation. (2022). *Dynamic Philanthropy: A framework for supporting transformative climate governance in the digital age*. Retrieved from https://sustainabilitydigitalage.org/featured/dynamic-philanthropy/

Ubalijoro, É. et al. (2021). Open data, distributed leadership and food security: The role of women smallholder farmers. In D.P. Singh, R. Joy-Thompson, & K.A. Curran (Eds.), *Reimagining Leadership on the Commons: Shifting the Paradigm for a More Ethical, Equitable, and Just World* (pp. 273–293). Emerald Publishing Limited.

Ubalijoro, É. et al. (2023). Inclusive innovation in artificial intelligence: From fragmentation to wholeness. In Prud'homme, B., Régis, C., & Farnadi, G. (Eds.), *Missing Links in AI Governance* (pp. 248–268). UNESCO.

United Nations System Chief Executives Board for Coordination. (2016). *Leaving No One Behind: Equity and Non-Discrimination at the Heart of Sustainable Development* (pp. 140–175). UN.

Vågen, T.-G., Winowiecki, L.A., Neely, C., Chesterman, S., & Bourne, M. (2018). Spatial assessments of soil organic carbon for stakeholder decision-making – a case study from Kenya. *SOIL*, *4*, 259–266.

Wilkinson, M.D. et al. (2016). The FAIR Guiding Principles for scientific data management and stewardship. *Scientific Data*, *3*(160018), 1–9.

World Bank. (2016). Poverty headcount ratio at national poverty lines – Poverty and Inequality Platform. Retrieved from https://data.worldbank.org/indicator/SI.POV.NAHC?locations=RW

Young, M., et al. (2019). Beyond Open vs. Closed: Balancing Individual Privacy and Public Accountability in Data Sharing. In *Proceedings of the Conference on Fairness, Accountability, and Transparency* (pp. 191–200). Association for Computing Machinery.

# 12 Crafting Human-Centered AI in Workspaces for Better Work

*Christian Lévesque*
HEC Montréal, Montréal, Québec, Canada

*Cassandra Bowkett*
University of Manchester, Manchester, UK

*Julie (M.É.) Garneau*
Université du Québec en Outaouais, Gatineau, Québec, Canada

*Sara Pérez-Lauzon*
HEC Montréal, Montréal, Québec, Canada

## INTRODUCTION

Digitalization of work involves various technologies, including the use of artificial intelligence (AI) and digital platforms. Digital technologies, as a whole body of technical and informatics infrastructure to capture and manage data to improve productivity, flexibility, and cost reduction, are giving legitimacy to a new production and consumption regime. The use of these technologies in workspaces is still at an early stage of development, and predictions about their transformative impact on work remain uncertain and contested (Krzywdzinski et al., 2022).

Digital technologies evolve as actors in the world of work experiment, and these experimentations can make work better or worse, and empower or disempower the worker which is a crucial dimension of human-centered AI (HCAI). Within the field of AI, empowerment is often associated with the augmented capacity of individuals through the enhancement of new technologies (Shneiderman, 2021). We seek to go beyond this individualized approach by integrating collective action, participatory, and inclusive governance to achieve empowerment.

DOI: 10.1201/9781003320791-15

This chapter builds on the assumption that technology is neither good nor bad; nor is it neutral. We need to avoid simple dichotomies and focus instead on understanding how institutions shape the future of work and the overall experience of workers with new technologies. For workers and their representatives, one central issue is how to mobilize traditional institutions and develop new ones to create inclusive governance and better work. We propose that better work mitigates social and economic risks, and increases both workers' autonomy and their capacity to express themselves through meaningful and skillful jobs. This raises the question of how HCAI can lead to better work that empowers workers individually and collectively but that also creates the conditions for a better society (Tasioulas, 2022).

The chapter proceeds as follows. The next section outlines our framework and reviews the existing literature on AI and better work. This is followed by an analysis of two exemplars, "Industry 4.0" and "platform work." Drawing on our own material and secondary data, we examine for each exemplar the interplay between quality of work, worker empowerment, and forms of governance. The next section addresses the issue of AI governance for better work and a better society. The conclusion offers a reflection on avenues to achieve HCAI for better work.

## AI FOR FUTURE AND BETTER WORK

AI as a general-purpose technology can have many uses and spillover effects and significantly impact our habits of thinking and acting. AI is nevertheless socially constructed and can produce different outcomes depending on the social circumstances and context of development. In the world of work, the deployment of AI has the potential to fissure and erode the foundations of employment relations institutions which rest on unity of place of work (workshop, factory, office), unity of time of work (weekly work schedules, rest periods), and unity of action (collective organization of work) (Degryse, 2020). What can result is a blurring of boundaries between paid and free labor, work and nonwork, and jobs requiring enhanced skills and high rewards and those requiring low levels of both skill and rewards.

While the directions of these fissures are context-specific, adopting AI in any workspace remains an employer's prerogative, associated with the right to direct and control labor. Whether or not AI is adopted is often linked to production strategies and corporate decisions to reduce costs or access new markets. Workers and unions rarely participate in these decisions, and implementation is legitimated by the legal and contractual subordination of workers with employee status. However, the literature has shown us that once adopted, these changes are negotiated within workspaces (Doellgast & Wagner, 2022). As such, what is often referred to as the indeterminacy of the traditional employment relationship remains, which requires managers to obtain workers' consent and cooperation (Bélanger & Edwards, 2007). Hence, whether under formal or informal relationships of subordination, the power of management over workers is constrained, and there are continuous negotiations around the physical, cognitive, and emotional effort displayed at work and over how work is organized (Bélanger & Edwards, 2013).

Given the challenges and uncertainties that the deployment of AI entails, actors in the world of work are engaged in a prolonged period of experimentation which opens space for new forms of governance (who decides what for whom?) that are still being contested and co-constructed (Zuboff, 2022). The conversation on AI and the future of work is polarized around two broad approaches: a techno-centrist and a human-centered approach (De Stefano, 2019). The techno-centrist approach is a top-down process that promotes a data-centered view that ascribes greater value to expert and codified knowledge over worker experiences. The human-centered approach is a bottom-up process that puts more emphasis on employee involvement and values their experience and tacit knowledge. It promotes an HCAI where individuals gain autonomy and competencies through their interaction with technologies (Shneiderman, 2021).

Our approach seeks to expand this individual conception of HCAI by integrating collective action as well as participatory and inclusive governance to achieve empowerment. We do this by linking worker empowerment to the constitutive dimensions of job quality.

The issue of job quality has attracted much interest over the last decade and scholars have proposed various typologies and frameworks to apprehend job quality in the context of digitalization (Berg, Green, Nurski, & Spencer, 2022; Guest, Knox, & Warhurst, 2022; Jarrahi et al., 2021). These typologies highlight that job quality is complex and multidimensional but there is significant consensus on the need to take into account a broad set of economic and noneconomic factors to assess "good and bad jobs" (Guest et al., 2022; Warhurst & Knox, 2022). Drawing on these typologies and the work of Murray et al. (2023), we articulate three key dimensions of the quality of work: risks, autonomy, and expressiveness. They are depicted in Figure 12.1.

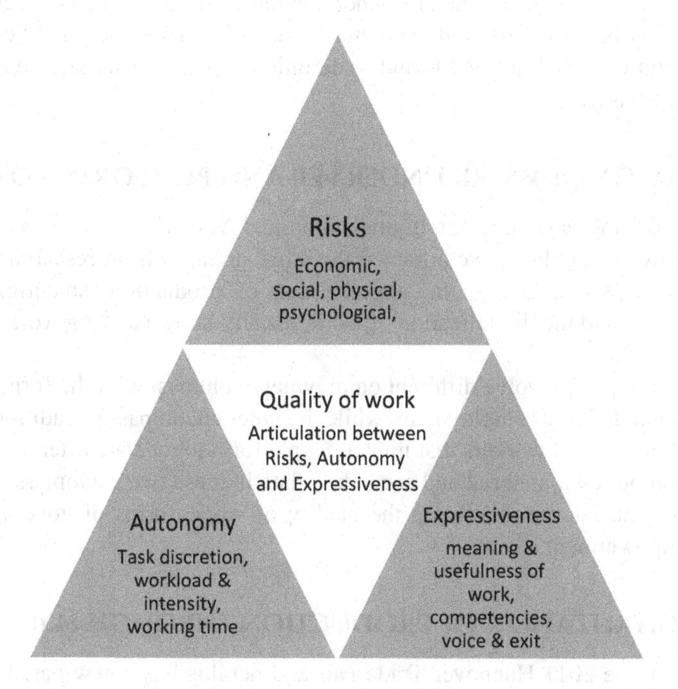


**FIGURE 12.1** Constitutive dimensions of the quality of work.

The first dimension refers to the nature and extent of risks at work and whether these risks are mutualized or individualized. The risks that people face interact in a complex way and may be economic (e.g., income security), social (e.g., isolation or alienation), or related to physical and mental health (e.g., injury, burn-out). The second dimension refers to the autonomy and discretion workers have over the organization of work, workload, and working time. Workers may have more or less autonomy on each of these. The third dimension, expressiveness, refers to the capacity of workers to express themselves through meaningful work, to deploy their skills, to voice their concerns, and to participate in decision-making.

In theory, empowerment is characterized by a workspace that offers social and economic security, autonomy, and the possibility to freely express oneself, while disempowerment exemplifies workspaces where there are high levels of insecurity, workers do not have much autonomy at work and are less able to express their voice and concerns. In practice, however, these three dimensions intersect in a complex and uneven way. Worker experiences are filled with contradictory tensions. Increasing working time autonomy may impinge on work–life balance and well-being, increasing an unhealthy life. Reducing task discretion may lessen cognitive effort and risk, but also increase worker alienation.

The articulation between these three dimensions of work unfolds in institutionally situated workspaces and under the right conditions, they can create spaces for inclusive and participatory governance. According to Floridi (2018:3) "digital governance is the practice of establishing and implementing policies, procedures and standards for the proper development, use and management of the infosphere." Put more simply, it is about who decides what, for whom, in the digital world (Zuboff, 2022). Our main proposition is that to produce better work and better society public and nonprofit institutions, such as NGOs and trade unions, must be included in the governance of digital work.

## THE QUALITY OF WORK UNDER I4.0 AND PLATFORM WORK

There are several ways in which digitalization and AI can affect work and employment, but two exemplars have attracted the most attention from researchers (Guest et al., 2022; Wood, 2021): the digitalization of production, sometimes called "Industry 4.0" and the digitalization of work, usually known as "gig work" or "platform work."

These exemplars involve different employment contexts, with the former involving traditional forms of employment, while the latter encompasses traditionally self-employed or contract workers that now use a platform to mediate offer and demand. Drawing on our own material and secondary data in these two exemplars, we examine in each the interplay between the quality of work, forms of governance, and worker empowerment.

## THE DIGITALIZATION OF PRODUCTION THROUGH I4.0

Launched at the 2011 Hannover Trade Fair and heralded as a new paradigm likely to revolutionize both manufacturing and services (Kagermann, Helbig, Hellinger, &

Wahlster, 2013), the concept of Industry 4.0: (I.40) entails the synchronization of process flows through the capturing and formatting of data and the interconnection of systems and work teams. A key question concerning the implementation of I4.0 is whether it will replace labor or enhance worker tasks while improving their capabilities. Interdependencies exist between technology, work organization practices, and competencies (Brynjolfsson, Rock, & Syverson, 2019) and shape the constitutive dimensions of work in terms of risks, autonomy, and expressiveness.

The most obvious risk associated with I4.0 is that of economic security through job loss. For employers, I4.0 is a way to improve productivity and flexibility, while reducing costs, including labor costs; for workers, I4.0 is a threat to jobs and economic security (Marenco & Seidl, 2021). Although there is much debate on whether digital technology will replace labor (Doellgast & Wagner, 2022), empirical evidence points to variegated effects (Gautié et al., 2018; Stroud, Timperley, & Weinel, 2020). The introduction of robotics does not appear to reduce overall employment but does increase polarization, with increases in both high-wage/high-skill and low-wage/low-skill jobs, while reducing middle-income/semi-skilled jobs in between (Dixon, 2020). If the impact of I4.0 on economic risk is variegated, it seems that I4.0 opens new possibilities for collective actors to improve health and safety (Haipeter, 2020), reducing physical/psychological risks. In their study of the Canadian automotive industry Rutherford and Frangi (2020) suggest that workers and trade union representatives have taken advantage of the implementation of I4.0 to improve workplace ergonomics. Similarly, Stroud, Timperley, and Weinel (2020) show how workers and their representatives can use the introduction of new technology to reduce dangerous tasks in German and Italian steel plants.

The impact of I4.0 on workers' autonomy and discretion is more obvious and less equivocal. I4.0 clearly reduces worker discretion over both the way work is accomplished and the pace of work (Butollo, Jurgens, & Krzywdzinski, 2018; Gautié, Ahlstrand, Wright, & Green, 2018; Lévesque, Bowkett, Perez-Lauzon, & Emilien, 2021). Worker autonomy over critical production decisions such as tooling, layout, set-ups, and repair work are reduced, and algorithmic management accentuates this shift in decision-making to programmers. The new digitalized work systems gather manufacturing data in real-time, including the measurement of worker performance, and use these extensive data to make decisions, formulate predictions, and monitor workers' behavior (Gautié et al., 2018; Wood, 2021). By increasing routine tasks and limiting the capacity of workers to make decisions about the sequence of tasks and the accomplishments of work, I4.0 limits the opportunity for production workers to take initiative.

These changes in the organization of work are connected to the meaning ascribed by employees to their work, whether it is meaningful and fulfilling, and whether they have the opportunities to develop and mobilize their skills and competencies. Lévesque et al. (2021) highlight a form of segmentation with the creation of some very attractive and meaningful jobs and many boring and meaningless jobs. Gautié et al. (2018) observe a similar process in France. While the codified nature of I4.0 knowledge tends to reduce the proportion of meaningful jobs, subjective knowledge and experience of the work process are still relevant and important (Bellandi, Chaminade, & Plechero, 2020). A key challenge for workers concerns the recognition of their tacit and experiential knowledge, which is so important for productivity and competitiveness. A similar challenge concerns the opportunities for workers to acquire new skills and access training to improve their capacity to master the new cognitive technologies associated with I4.0. What is at stake is the recrafting of work so that workers regain control over their livelihoods and expand their voice in decisions over emerging technologies (Yudken & Jacobs, 2021).

In the implementation of I4.0 in the manufacturing industry, one crucial element has been the possibility for workers to collectively voice their concerns. In a comparative study of the deployment of I4.0 in the United States, Germany, and Italy, Rutherford (2021) highlights how workers and their representatives in the United States have less opportunities to participate and exert influence over I4.0 than their counterparts in Italy and Germany. Gautié et al. (2018) also highlight differences in worker representatives' voices between Sweden, France, and the UK. Worker representatives in Sweden have greater influence over the implementation of I4.0. They argue that strong trade union involvement is related to positive outcomes on worker autonomy and skills, whereas when trade union involvement is weaker, there are few positive results. In the case of Germany, Krzywdzinski, Gerst, and Butollo (2022) argue that the strategic involvement of works councils in the deployment of AI technologies is only possible when works councils are strong and are able to identify problems in corporate planning. Whereas in other cases, such as Canada and Denmark, other levels of governance, such as regional or sectoral institutions, may enable a more collective approach (Garneau, Pérez-Lauzon, & Lévesque, 2023).

These studies show that actors are using old and new institutions to shape the governance of AI and future work, seeking to empower workers, though this is an uneven process across countries.

## THE DIGITALIZATION OF WORK THROUGH PLATFORM WORK

For the platform exemplar, we focus on Uber, one of the major ride-hail delivery platforms that has been under intense scrutiny from both researchers and policymakers for its impact on the digitalization of work. As an application, Uber seeks to connect consumers with drivers who will then provide them with a service, such as transportation. Uber positions itself as a technology company, via its application that connects drivers with customers. Union and worker action has instead focused on Uber as an "employer," challenging Uber's responsibilities to its drivers. The scope of these challenges encompasses how Uber has transferred risk to drivers particularly in relation to noncompliance with employment regulations (pay, social security, and benefits), while the expansiveness of Uber's algorithmic management practices increases company control over drivers' autonomy and expressiveness, resulting in driver subordination (Garneau & Bernier, 2023).

In terms of risk, Uber positions itself as competitive by championing deregulation of the taxi industry, often undermining existing restrictions on the number of taxis operating or requirements for minimum levels of service in terms of hours and location. Uber's business model removes upfront costs (e.g. licenses), which benefits workers by reducing barriers to entry and can support easy access to the labor market for migrant, youth, and part-time workers. However, deregulation can, in some circumstances, increase risk for drivers via increased economic precarity, particularly for drivers working full-time.

Uber passes risk (economic and social) onto its drivers by classifying them as independent contractors, reducing its responsibilities in terms of complying with employment regulations (minimum wage, holiday pay, and access to other benefits). As such, Uber is transforming what constitutes an employment relationship (Aloisi, 2022; Thelen, 2018). Collective responses to Uber's efforts to deregulate the industry and avoid its employment responsibilities have typically used legal routes. Unions and gig worker groups argue that Uber as an employer has responsibilities to its drivers and have tried to challenge Uber's classification of drivers in a number of countries including the United States, the UK, France, Italy, and Canada (Aloisi, 2022; Coiquaud & Morissette, 2020; Dubal, 2017). These efforts have been more successful in some countries (e.g., Germany) than others (e.g., the US and Canada), (Thelen, 2018) depending on the national institutions and governance responses.

In terms of autonomy, Uber's platform enables drivers to choose their hours and place of work. Some drivers consider themselves to be entrepreneurs, running their own business using Uber's platform, while others benefit from being able to work part-time, or around other responsibilities. Uber's algorithmic management tools are, however, comprehensive in controlling, both covertly and overtly, driver autonomy over task discretion, workload, and intensity. Uber combines continually fluctuating, and opaque pricing algorithms, with covert algorithmic nudges and gamification to encourage drivers to work in certain locations (e.g., surge pricing) and for longer hours (e.g., Uber Pro, challenges) (Vasudevan & Chan, 2022). This gamification of work creates a "just-in-place" workforce (Wells, Attoh, & Cullen, 2021), which drivers refer to as "grinding," working long hours in preset locations and doing repetitive tasks for a reward such as higher pay or greater visibility of job criteria (locations/prices/distance) (Vasudevan & Chan, 2022). One result of this is work intensification. Drivers have however developed several individual and collective responses to gain room for maneuverability, and "game" the system (Maffie, 2022a).

Nevertheless, in terms of expressiveness, drivers report finding it difficult to exercise their voice individually with Uber. Drivers have no line managers, and the call center support is outsourced and heavily scripted, though Uber's in-person driver support hubs are more effective. Uber has also transformed driver competencies through an emphasis on customer ratings, placing more emphasis on drivers' customer service and emotional labor. Some drivers have welcomed this shift to customer service metrics. However poor customer ratings impact drivers' overall ratings and can result in Uber deactivating driver accounts (Chan, 2019; Rosenblat, 2018). Uber's algorithms also nudge and enforce drivers into going above and beyond, typically including physical acts such as providing water or chargers, as well as providing high levels of customer service (Chan, 2019).

For drivers, while many customer interactions are positive, they must now learn to manage aggressive, racist, or sexually abusive customers, often a difficult task when they are reliant on maintaining customer ratings, and it can be difficult to get negative ratings overturned (Maffie, 2022b; Rosenblat & Stark, 2016). Individually, drivers often choose to exit the platform, when they are unable to exercise their voice or try to reduce their reliance on Uber by developing other forms of income (e.g., pirate or independent taxi operations) (Maffie, 2022a). Collectively, drivers and driver representatives have used legal systems (court cases) and engaged in protests, rallies, and referenda to lobby for the adoption of public policies protecting drivers, and some of these efforts have included traditional unions (Estlund & Liebman, 2020; Vandaele, 2018; Wells et al., 2021).

Uber has been able to dismantle traditional labor institutions, and one of the challenges for workers and their representatives is to renew and build new governance institutions for better work.

## HCAI FOR BETTER WORK AND ... BETTER SOCIETY

As highlighted by several studies, there is still much to be done to put HCAI at the forefront in many industries and countries. The dominant trend in our two exemplars, I4.0 and platform work, rests more heavily on a techno-centrist approach that promotes a data-centered top-down process that ascribes greater value to expert and codified knowledge over worker experiences. This does not appear to be the best avenue for better work.

I4.0 implementation in the manufacturing industries does not fundamentally transform where production takes place (workshop, factory), the unity of time of work (weekly work schedules, rest periods), and unity of action (collective organization of work) (Degryse, 2020). However, AI and algorithmic management are fissuring the constitutive dimensions of job quality. AI implementation rests on the segmentation of the labor market and the polarization of risk, enhancing the gap between "good" and "bad" jobs. I4.0 has the potential to redefine workers' autonomy by reducing workers' capacity to make decisions over the sequence of tasks and work accomplishments. The standardization of tasks increases firm capacity to monitor workers while limiting worker possibilities to have fulfilling jobs and develop their skills and capabilities.

In the platform exemplar, the Uber case, workers consider that they have more autonomy over place and time of work, but this comes at the expense of worker collectivity and unity of action (collective organization of work). Here, how Uber crafts its business model and uses AI rests on the idea that it is a tech company, not an employer which has significant disruptive and destructive impacts on employment protection and regulation. Uber transfers employer risks and responsibilities to individual workers, increasing precarity. To retain control over the tasks and work conducted, Uber engages in heavy monitoring and surveillance of drivers. The lack of recognition for drivers as workers reduces Uber's responsibility to provide voice mechanisms to workers.

There is, however, much experimentation where various actors are engaged in designing and implementing technology that empowers workers through better work. These experiments occur at multiple levels (workspace, sectoral, regional, national, and supranational), are institutionally situated and vary according to AI governance.

Levels and spaces of AI governance are variegated and shape in distinctive ways how HCAI is defined and is shaping future work. There is however growing evidence that corporations and market institutions alone cannot deliver the conditions for better work through inclusive and participatory governance. Public and nonprofit institutions such as NGOs, trade unions, and sectoral and regional intermediary organizations must be included in digital governance to deliver better work.

When there are strong institutions with open spaces for inclusive and participatory governance, HCAI is more likely to empower individual and collective actors. This is particularly the case in Germany, Sweden, and Denmark with their multilevel governance approaches but also specific regions and sectors in Canada and the United States. One critical issue in these two countries and other countries with weaker institutions is how to scale up HCAI experiments.

We consider this to be a critical issue not only because inclusive and participatory governance can lead to better work and empowered workers but also as it can lead to a better society by reducing inequality and increasing solidarity (Acemoglu, 2021; Tasioulas, 2022; Zuboff, 2022).

## CONCLUSION

Although increasing, the use of AI in workspaces is at an early stage of development, and our understanding of the interplay between forms of governance, the quality of work, and worker empowerment remains limited. This chapter puts forward a framework to help us start to analyze these interconnections and applies it to two exemplars, I4.0 and platform work. However, more knowledge is needed to understand how AI can lead to better work and a better society. In these concluding remarks, we propose three avenues to achieve this task.

First, given the variegated effect of AI on work, we have to grasp more fully how AI is reshaping the three constitutive dimensions of work (risk, autonomy, and expressiveness) and whether digitalization empowers or disempowers workers. The description of our two exemplars shows that these dimensions intersect in a complex and uneven way with empowerment. Narratives or debates on AI ought to focus on whether new technology is beneficial to a few or the majority of the working people. There is a need to disentangle the relationship between individual and collective empowerment and advocate a collective approach to complement the existing individualized one put forward by many HCAI advocates.

Second, we have to understand more fully how various actors are experimenting with new and old institutions to shape the governance of the future of work in this digital era. These processes of experimentation may involve the reordering and revalorization of old governance institutions to adapt them to a new digital context. They also may entail the dismantling of old institutions and the creation of new ones. The key question is how these processes of dismantling, reordering, and reinventing of governance institutions are inclusive and create resources for civil society and collective actors, such as trade unions, to participate competently, creatively, and confidently in the governance of these new technologies.

Third, all stakeholders need to seriously unravel the link between better work and a better society. The workspace is considered a central site of participation in constructing a good life. However, disempowerment through increasing economic and social insecurity, deskilling, and meaningless jobs may have destructive consequences for social solidarity and democracy. Specific policies that change incentives and encourage socially responsible innovation are needed (Acemoglu & Johnson, 2023). While this may appear a daunting task, unravelling this link is the only way to move forward with remaking and rethinking digital technologies for a better future for all.

## REFERENCES

Acemoglu, D. (2021). Harms of AI *(Working Paper No. 29247; Working Paper Series), National Bureau of Economic Research.* https://doi.org/10.3386/w29247

Acemoglu, D., & Johnson, S. (2023). *Power and Progress: Our Thousand-Year Struggle Over Technology and Prosperity*: Basic Books. John Murray Press.

Aloisi, A. (2022). Platform work in Europe: Lessons learned, legal developments and challenges ahead. *European Labour Law Journal, 13*(1), 4–29.

Bélanger, J., & Edwards, P. (2007). The Conditions Promoting Compromise in the Workplace. *British Journal of Industrial Relations, 45*(4), 713–734. https://doi.org/10.1111/j.1467-8543.2007.00643.x

Bélanger, J., & Edwards, P. (2013). The nature of front-line service work: distinctive features and continuity in the employment relationship. *Work, Employment and Society, 27*(3), 433–450.

Bellandi, M., Chaminade, C., & Plechero, M. (2020). Transformative paths, multi-scalarity of knowledge bases and Industry 4.0. *Industry, 4*, 62–83.

Berg, J., Green, F., Nurski, L., & Spencer, D. (2022). Risks to Job Quality from Digital Technologies: Are Industrial Relations in Europe Ready for the Challenge? *Working Paper 16, Bruegel.*

Brynjolfsson, E., Rock, D., & Syverson, C. (2019). Artificial intelligence and the modern productivity paradox: A clash of expectations and statistics. In A. Agrawal, J. Gans, & A. Goldfarb (Eds.), *The Economics of Artificial Intelligence: An Agenda* (pp. 23–57). Cambridge: National Bureau of Economic Research; University of Chicago Press.

Butollo, F., Jurgens, U., & Krzywdzinski, M. (2018). From Lean Production to Industrie 4.0. More autonomy for Employees? *WZB Berlin Social Science Center, Discussion Paper* (SP III 2018-303).

Chan, N. K. (2019). The rating game: The discipline of Uber's user-generated ratings. *Surveillance & Society, 17*(1/2), 183–190.

Coiquaud, U., & Morissette, L. (2020). La « fabrique réglementaire » autour de l'arrivée d'Uber. *Relations industrielles-Industrial Relations, 75*(4), 633–851.

De Stefano, V. (2019). 'Negotiating the Algorithm': Automation, Artificial Intelligence, and Labor Protection. *Comparative Labor Law & Policy Journal, 41*(1), 15–46.

Degryse, C. (2020). Du flexible au liquide: le travail dans l'économie de plateforme. *Relations industrielles/Industrial Relations, 75*(4), 660. https://doi.org/10.7202/1074559ar

Dixon. (2020). *L'effet des robots sur le rendement et l'emploi des entreprises.* Ottawa: Statistic Canada.

Doellgast, V., & Wagner, I. (2022). Collective regulation and the future of work in the digital economy: Insights from comparative employment relations. *Journal of Industrial Relations.* https://doi.org/10.1177/00221856221101165

Dubal, V. B. (2017). The drive to precarity: a political history of work, regulation, & labor advocacy in San Francisco's taxi & Uber economies. *Berkeley Journal of Employment and Labor Law*, 73–135.

Estlund, C., & Liebman, W. B. (2020). Collective bargaining beyond employment in the United States. *Comparative Labor Law & Policy Journal, 42*(2), 371.

Floridi, L. (2018). Soft ethics, the governance of the digital and the General Data Protection Regulation. *Philosophical Transactions of the Royal Society A, 376*, 20180081.

Garneau, J., & Bernier, J. (2023). Défis et enjeux du travail sur les plateformes numériques au Canada et au Québec. In M. Vultur (Ed.), *Les plateformes de travail numérique. Polygraphie d'un nouveau modèle organisationnel* (pp. 67–93). Québec: PUL.

Garneau, J., Pérez-Lauzon, S., & Lévesque, C. (2023). Digitalisation of work in aerospace manufacturing: expanding union frames and repertoires of action in Belgium, Canada and Denmark. *Transfer: European Review of Labour and Research, 49*(1). https://doi.org/10.1177/10242589221146876

Gautié, J., Ahlstrand, R., Wright, S., & Green, A. (2018). Innovation, Job Quality and Employment Outcomes in the Aerospace Industry: Evidence from France, Sweden and the UK.

Guest, D., Knox, A., & Warhurst, C. (2022). Humanizing work in the digital age: Lessons from socio-technical systems and quality of working life initiatives. *Human Relations, 75*(8), 1461–1482.

Haipeter, T. (2020). Digitalisation, unions and participation: the German case of 'industry 4.0'. *Industrial Relations Journal, 51*(3), 242–260. https://doi.org/10.1111/irj.12291

Jarrahi, M. H., Newlands, G., Lee, M. K., Wolf, C. T., Kinder, E., & Sutherland, W. (2021). Algorithmic management in a work context. *Big Data & Society, 8*(2). https://doi.org/10.1177/20539517211020332

Kagermann, H., Helbig, J., Hellinger, A., & Wahlster, W. (2013). *Recommendations for implementing the strategic initiative INDUSTRIE 4.0: Securing the future of German manufacturing industry; final report of the Industrie 4.0 Working Group*: Forschungsunion.

Krzywdzinski, M., Gerst, D., & Butollo, F. (2022). Promoting human-centred AI in the workplace. Trade unions and their strategies for regulating the use of AI in Germany. *Transfer: European Review of Labour and Research.* https://doi.org/10.1177/10242589221142273

Lévesque, C., Bowkett, C., Perez-Lauzon, S., & Emilien, B. (2021). *Industry 4.0, the Future of Work & Skills. Building Collective Resources for the Canadian Aerospace Industry*: The Future Skills Centre.

Maffie, M. D. (2022a). Becoming a pirate: Independence as an alternative to exit in the gig economy. *British Journal of Industrial Relations, 61*(1), 46–67. https://doi.org/10.1111/bjir.12668

Maffie, M. D. (2022b). The perils of laundering control through customers: A study of control and resistance in the ride-hail industry. *ILR Review, 75*(2), 348–372.

Marenco, M., & Seidl, T. (2021). The discursive construction of digitalization: a comparative analysis of national discourses on the digital future of work. *European Political Science Review, 13*(3), 391–409. https://doi.org/10.1017/s175577392100014x

Murray, G., Gesualdi-Fecteau, D., Lévesque, C. & Roby. N. (2023). What makes work better or worse? An analytical framework. *Transfer: European Review of Labour and Research, 29*(3), 305–322.

Rosenblat, A. (2018). *Uberland: How algorithms are rewriting the rules of work*: University of California Press.

Rosenblat, A., & Stark, L. (2016). Algorithmic labor and information asymmetries: A case study of Uber's drivers. *International Journal of Communication, 10*, 27.

Rutherford, T. D. (2021). 'Negotiate the Algorithm': Labor Unions, Scale and Industry 4.0 Experimentation. In T. Schulze-Cleven & T. E. Vachon (Eds.), *Revaluing Work(ers): Towards a Democratic and Sustainable Future* (pp. 79–100). Champaign, IL: Labor and Employment Relations Association Series.

Rutherford, T. D., & Frangi, L. (2020). Is Industry 4.0 a Good Fit for High Performance Work Systems? Trade Unions and Workplace Change in the Southern Ontario Automotive Assembly Sector. *Relations industrielles, 75*(4). https://doi.org/10.7202/1074563ar

Shneiderman, B. (2021). Human-centered AI: How human-centered AI could empower humans. *Issues in Science and Technology, 37*(2). Retrieved from https://issues.org/issue/37-2/

Stroud, D., Timperley, V., & Weinel, M. (2020). Digitalized drones in the steel industry: The social shaping of technology. *Relations industrielles/Industrial Relations, 75*(4), 730–750.

Tasioulas, J. (2022). Artificial Intelligence, humanistic ethics. *Daedalus, 151*(2), 232–243.

Thelen, K. (2018). Regulating Uber: The politics of the platform economy in Europe and the United States. *Perspectives on Politics*, *16*(4), 938–953.

Vandaele, K. (2018). Will trade unions survive in the platform economy? Emerging patterns of platform workers' collective voice and representation in Europe. *ETUI RP-WP*, *2018*(5), 33.

Vasudevan, K., & Chan, N. K. (2022). Gamification and work games: Examining consent and resistance among Uber drivers. *New Media & Society*, *24*(4), 866–886.

Warhurst, C., & Knox, A. (2022). Manifesto for a new quality of working life. *Human Relations*, *75*(2), 304–321.

Wells, K. J., Attoh, K., & Cullen, D. (2021). "Just-in-Place" labor: Driver organizing in the Uber workplace. *Environment and Planning A: Economy and Space*, *53*(2), 315–331.

Wood, A. J. (2021). *Algorithmic management consequences for work organisation and working conditions*. JRC Working Papers Series on Labour, Education and Technology.

Yudken, J. S., & Jacobs, D. C. (2021). Worker voice in technological change: The potential of recrafting. In T. E. Vachon & T. Schulze-Cleven (Eds.), *Revaluing Work (ers): Toward a Democratic and Sustainable Future* (pp. 141): Labor and Employment Relations Association Series.

Zuboff, S. (2022). Surveillance capitalism or democracy? The death match of institutional orders and the politics of knowledge in our information civilization. *Organization Theory*, *3*(3). https://doi.org/10.1177/26317877221129290

# 13 AI and Judiciary Decisions

*Takehiro Ohya*
Keio University, Tokyo, Japan

## MODERN TRIAL AND REASON

> *Who was that? A friend? A good person? Somebody who was taking part? Somebody who wanted to help? Was he alone? Was it everyone? Would anyone help? Were there objections that had been forgotten? There must have been some. The logic cannot be refuted, but someone who wants to live will not resist it. Where was the judge he'd never seen? Where was the high court he had never reached? He raised both hands and spread out all his fingers.*
>
> *But the hands of one of the gentlemen were laid on K.'s throat, while the other pushed the knife deep into his heart and twisted it there, twice. As his eyesight failed, K. saw the two gentlemen cheek by cheek, close in front of his face, watching the result. "Like a dog!" he said, it was as if the shame of it should outlive him.*
>
> (Kafka, 1925: chapter 10)

This text is the end of Franz Kafka's *The Judgement*, a classic of the absurdist book. The central character, Josef K., is doomed to a dogged death at the mercy of an inexplicable procedure and the enigmatic behavior of the people who facilitate it. But what is absurd about his fate—in the sense of philosophical absurdism, being irrational, lacks meaning, and thus unintelligible? It is appropriate to place a key element in the fact that, as is repeatedly mentioned in the novel, nothing is fully explained, and no answers are given to the questions about the reasons for his being treated in certain ways.

> "But how can I be under arrest? And how come it's like this?" "Now you're starting again," said the policeman, dipping a piece of buttered bread in the honeypot. "We don't answer questions like that." "You will have to answer them," said K. "Here are my identification papers, now show me yours and I certainly want to see the arrest warrant." "Oh, my God!" said the policeman. "In a position like yours, and you think you can start giving orders, do you. It won't do you any good to get us on the wrong side, even if you think it will—we're probably more on your side than anyone else you know!"
>
> (Kafka, 1925: chapter 1)

In the past, some of what we call "trials" have been similarly conclusive. In Europe, the procedure called compurgation, in which the denial of the accusation by the defendant is admitted under oath by witnesses has traditionally been widely used. In the procedure, however, witnesses testify not that the content of their statements is true, but the defendant is honest and thus his statement is trustworthy. Or, in the

DOI: 10.1201/9781003320791-16 

procedure called "trial by combat," the plaintiff and the defendant should make a direct duel, in which the loser is deemed guilty or liable.[1] Naturally, from a modern rational standpoint, the conclusions of these "trials" would be considered unreliable and not based on facts. What underpins these proceedings is the work of a supernatural being, the belief that God helps the righteous and discourages the wicked.

This is why we can assume that with the development of the rational and scientific mind in the modern era, this type of trial disappeared and underwent a gradual transformation into the procedure we know today. The characteristics of such procedures are as follows: first, there is a suitor and a sued, and both parties submit their claims and supporting evidence; second, a judge, who is supposed to be a neutral third party, evaluates their value and decides who wins or loses[2]; and third, not only a conclusion with legal meaning but also the reasons supporting such a decision are stated in the judgment.[3] It has traditionally been considered as an important element for justice to be just that the concerned parties can be involved in the process to determine the outcome, and that what led to that outcome is understandable for the society in general, not just to the concerned parties. How would these characteristics be affected by introducing AI into the process and replacing at least some of the procedures with nonhuman ones? Can justice *for people* and *with people* be achieved through processes that are not directly *by people* but are *designed by people*? Is it possible to achieve Human-Centered AI in the judiciary?

## DEEP LEARNING AND LACK OF REASON

When we consider that *reasoned decisions* thus occupy the core of judicial proceedings, we are reminded that there is an aspect of AI utilization that clashes with this. In the rapid technological evolution of recent years, which has been called the third AI boom,[4] a technology called deep learning is attracting attention as a major factor that has made possible the extensive and efficient use of AI, which was once unimaginable.

Expert systems, which were once proposed with the dream of having machines make judgments and decisions instead of experts, were to analyze our thoughts, formalize them into rules, and then direct them to computers as programs that they can understand. Thus, expert systems are faced with the limitation that they could not process our judgments that we did not understand by ourselves—for example, can any person clearly express in language whether a steak is properly grilled or not, or what kind of information we are basing our judgments on? On the other hand, deep learning is a system that automatically learns the relationship between inputs and outputs (cause and effect) and changes its own behavior patterns accordingly.

Consider what Ludwig Wittgenstein said about "family resemblance" (familienähnlichkeit). Is there, for example, an essence common to the various entities called games, a single element that defines whether it is, and is not, a game?

> Look for example at board-games, with their multifarious relationships. Now pass to card-games; here you find many correspondences with the first group, but many common features drop out, and others appear. When we pass next to ball-games, much that is common is retained, but much is lost. (…) we see a complicated network of similarities overlapping and cries-crossing: sometimes overall similarities.
>
> (Wittgenstein, 1953: 66)

Even if we cannot explicitly answer the question, "What is a game?", we can generally distinguish between games and nongames, and even if we cannot describe the commonalities that exist among the appearances of a family, we can identify who the members are. What we are looking at then is not a single rule that divides the inside and outside of a family, but multiple features (Züge) and their *intensities*, which we ourselves have learned to understand through learning—"How should we explain to someone what a game is? I imagine that we should describe games to him, and we might add: "This and similar things are called 'games'" (Wittgenstein, 1953: 69). The development of deep learning, which focuses on the evaluation of consequences rather than rules, has been the very foundation of the third boom of AI. In other words, the possibility of AI surpassing human subjective judgments is foreseen in its ability to learn and reproduce what lies behind the *unspeakable* similarities. On the other hand, however, this also means that the results obtained will be incomprehensible to us *humans*.

Imagine reading radiographs (though I have never done it). One can well imagine a situation in which some cancers have different characteristics from others, and an expert pathologist can spot the differences. "You see, this border here is characteristic, you know"—of course I don't see. But if an expert says that, and other experts agree with him, I have no choice but to accept it as the correct opinion. The expert's judgment will thus often take on a blackbox quality that is incomprehensible to the nonexpert. The question is, what would a learned AI do if it were in such a position, rather than a human expert who, if asked, would give the basis for his or her judgment (time, effort, and good mood permitting)? "You, an average person, would not understand, but this is cancer." "You wouldn't understand, but he's going to reoffend." "You don't know this, but we're going to have a nuclear war."

The problem with AI technology, at least at this point in time, is that *it does not tell us* the process of decision-making or the reasons why it came to a particular conclusion. Assuming that it is a characteristic of AI to return appropriate results, even though we do not know why, there may not be much problem in using it in games (a very strong chess AI, although we do not know why) or in systems that alert specialists (AI to mark data that *may* be abnormal, when a doctor checks the result of blood test). However, can it be used as a tool for criminal court decisions or taxation? Or, if the use itself is possible, can we accept it?

The last point, in particular, will be extremely important given the nature of the judiciary, which makes decisions that directly affect the rights and obligations of citizens and enforces the consequences. For example, if an AI is used to trade stocks held by an individual owner, it may be considered to be within the category of self-determination, including what system to use and the degree of dependence on it. Since it is his property that is affected by the system, it would be fine to say that the person should be allowed to do as he pleases. However, the situation surrounding the judiciary cannot be justified by such a logic of self-responsibility.

This is why legal philosophy has addressed the question of how an order issued under the law differs from a threat of robbery (Hart, 2012: Chapters 1–3). And what many authors have said is that the law is a general rule, and that the commands given in individual situations can be explained as an application of that rule (Austin, 1832). That is why we have understood it as the rule of law, as distinct from the bare arbitration of those in power, even though in reality it is a judge, a concrete human being, who makes the decision, and that law has been understood as the self-rule of we the people, as my rule by myself. Thus, in order to respect the people as the ruler and ruled, it is required to show *with reason* that his treatment, whatever it may be, is based on the law.

We must further note that the law is associated with coercion and, that even when a party refuses to comply, the state will ultimately execute its commands with force. In order to prevent abuse by someone in power and to ensure the sound operation of such a system, its operation must be made public to be verified. This point would also explain the importance of showing reason.

## ONGOING USE OF AI

Nevertheless, in reality, AI is already known to be used in the judicial process in several countries. In several U.S. states, risk scores based on AI calculations are used in criminal trials to determine a defendant's sentence or whether to grant parole to an inmate. In Wisconsin, for example, the COMPAS system developed by Northpointe (now Equivant) is reportedly used to determine recidivism risk based on age, race, employment status, education level, and drug use. The details of the calculation method, however, are not disclosed because they are considered trade secrets.

The likelihood of recidivism, which is an important part of the parole decision, is only a prediction for the future and thus is subject to uncertainty. It may be possible to expect that an AI, which can refer to a large number of historical records and use accurate probabilities, will be able to make a more reliable decision than a human judge, who only has experience from only a limited number of cases. Also, it would not be impossible to say that judges would be influenced by external factors such as the age, gender, and race of the subject and reflect their own biases. One might argue that an AI's judgment, which has no prejudice about a particular gender or race, would be more neutral. At least Equivant, the developer, insisted that "many research studies have concluded that objective statistical assessments are, in fact, superior to human judgment (Equivant, 2019: 1–2)." According to the company, mentioning that "in overloaded and crowded criminal justice systems, brevity, efficiency, ease of administration and clear organization of key risk/need data are critical (Equivant, 2019: 2)," COMPAS is realizing appropriate balance on the trade-off between comprehensive coverage of key risk and criminogenic factors, and brevity and practicality.

According to Wisconsin judicial statistics, the number of felony cases filed in Circuit Courts (first instance) increased from 36884 in 2013, 39819 in 2016, and 43614 in 2019, and the median number of days required to process cases increased from 128 days in 2013 to 152 days in 2019, indicating an increased burden on the courts.[5] Given that overload on human judges would induce erroneous decisions, it would be well worthwhile to have that burden alleviated by AI.

A defendant who was sentenced to prison because the system determined that he was likely to reoffend filed a lawsuit claiming that (1) the system's accuracy could not be challenged because its details were unknown and (2) its use of race and gender as part of the data violated due process, but in 2016 the Wisconsin Supreme Court dismissed the suit, holding that use of such scores is permissible if there is a warning regarding their "limitations and cautions."[6] However, the score has been criticized for determining that blacks are at higher risk and that only 20% of those expected to commit violent crimes actually did so (Angwin, Larson, Mattu, & Kirchner, 2016). In addition, although the state Supreme Court's decision assumes that the AI scores are only informative and that human judges are making substantive decisions, there may be some concern that reliance on the scores may actually occur and that the judges' evaluations may remain a formality.

Also in China, the utilization of AI in the judiciary has become widespread. In 2019, courts in Hainan Province began using AI to weigh criminal cases and draft sentences. In 2021, Shanghai began testing the use of AI to replace prosecutors' decisions as to whether to prosecute criminal cases. In 2022, China's Supreme People's Court issued a guideline directing that the infrastructure for the use of AI should be in place by the end of the next three years.

In China, the number of lawsuits has been increasing rapidly in recent years due to changes in the social environment associated with economic development, and the shortage of human capacity in the courts has become a policy issue. Looking at the number of civil lawsuits related to intellectual property as a representative example of the rapidly changing field, the number of new lawsuits filed in the first instance was 301278 in 2018, a 41% increase over the previous year (Tao, 2019). As a background to this increase, it has been pointed out that where administrative organizations were previously involved in dispute resolution, they are no longer able to achieve satisfactory results because the content of such cases has become more complex due to technological development. There may be a demand out there to automate the litigation process and utilize AI-based litigation support in order to make effective use of limited human resources.

It can be said that AI's reason-free decisions are increasingly being used in the judiciary, although to varying degrees: to support expert judges, to be referred by experts, or as a substitute for experts. In cases where private companies are responsible for AI development, as in the U.S. example, trade secret barriers, in addition to the risks inherent in the technology, will make decisions even more opaque. If this trend continues, the fear will also arise that one may be imprisoned or even lose one's life without being informed of any reason, as was in the case of Josef K.

The Artificial Intelligence Act, the EU's proposal for 2021 on the regulation of AI (European Parliament and Council, 2021), classified AI systems into four tiers according to their inherent risks, and AI that assists judicial institutions in the process of investigating and interpreting facts and laws, and applying laws to facts, is treated as a type of high-risk AI, the second tier. They are required to meet requirements for human oversight, transparency, and provision of information, as well as to go through a prior suitability assessment. It may be noted that such treatment is also based on a sense of caution about exposing us to absurd judgments.

## REASON AND CAUSE

But what is the problem? There is a view that since AI is physically an arithmetic process of a computer program, the process of its operation can be recorded (as long as there is enough storage space) and verified (except for the question of how much effort and time it takes to do so)—and thus AI is not a black box, or would not be a black box if only a proper verification method could be constructed. However, this view overlooked the point that what should be given in response to a request for explanation is the *reason* for the action, which indicates a semantic relation, and not the *cause*, which indicates a causal relation—just as we cannot accept "it was because of the sun" as the appropriate motivation to shoot an Arab to death, even if it was the accurate description of the facts (Camus, 1942).

Let us note that Wittgenstein introduced two sets of distinct terms in the period preceding *Philosophical Investigations*. The first is "criterion" and "symptom," and the second is "reason" and "cause." "To the question "How do you know that so-and-so is the case?", we sometimes answer by giving "criteria" and sometimes by giving "symptoms" (Wittgenstein, 1969: 24). Here, a criterion is a definition to determine the establishment of a situation, whereas a symptom means "a phenomenon of which experience has taught us that it coincided (…) with the phenomenon which is our defining criterion" (Wittgenstein, 1969: 24).

For example, we all know that a person suffering from influenza develops a sudden high fever and complains of joint pain and other symptoms. In this sense, these elements—high fever and joint pain—are empirically known phenomena and signs that accompany influenza infection. But since influenza is not the only disease that produces these symptoms, the presence of these symptoms is a necessary but not sufficient condition for diagnosing influenza. A sufficient criterion that would be socially acceptable to answer the question of why one diagnoses influenza would be *the detection of the influenza virus in a test*. Of course, there must be a prerequisite that the test is socially approved as being effective against influenza, for example, approved by the Ministry of Health. Thus, the criterion would be understood as a socially defined requirement, a necessary and sufficient condition to lead to a result.

Consider the question, "Why did you diagnose a patient with influenza? In reality, the process would probably begin with the patient's own recognition of a high fever and joint pain, followed by the physician's suspicion of influenza, followed by the detection of the virus as a result of testing. If we were asked to describe the process accurately, we would first cite the signs of high fever and joint pain, and these are certainly the starting point of the causal relationship, in the sense that the process leading to the diagnosis would not have been set in motion without these observations. In this sense, these signs are definitely the "cause" of the definitive diagnosis, and there is a causal relationship between them and the result.

However, it is also true that the presence of these signs alone is not sufficient justification for the diagnosis, given that there are several diseases that present with similar symptoms. What should be cited as sufficient justification for the diagnosis is the definition of influenza as a disease, and the fact of a positive viral test, or criterion, which is semantically linked to the conclusion, which is what we mean by the word "reason." What is at issue here is the semantic relation, the linkage of the meaning of the situation, in which a positive viral test can immediately be understood as influenza, and in which the meaning of the two phenomena is identical.

On the other hand, we can distinguish between reason and cause when we offer "reasons" for our thoughts. Reasoning is the actual computation performed, and reason is nothing more than going back one step in that computation. (…) To ask for a reason is to ask how we arrived at that result" (Ambrose, 1979: 4). It would thus follow that the subjectivity of the responding thing is assumed there. A cause, on the other hand, is something that is given in behaviorist explanations, something that can be discovered through experimentation (Ambrose, 1979: 4). Therefore, it is sufficient that the subject's behavior is observable, and there is no relation to subjectivity.

For example, when a friend travels to Hawaii, we may assume two questions and corresponding answers—"Why did you go to Hawaii?" "Because I wanted to enjoy the blue ocean." and "How did you get to Hawaii?" "By plane." The former is a question about reason, where one is expected to provide an explanation justifying the action (we would not take the response "because I wanted to see the glaciers," for example, as a decent one). There is an awareness of the connection between the meanings of the two concepts, such as the blue ocean in Hawaii or the glacier in Norway.

Alternatively, it may be noted that when asked about the actions of a third party, this question involves a presumption of the intent of the other. For example, "Why did he go to Hawaii?" "Maybe he wanted to see the ocean." The expected answer then is that such a thought would provide a consistent explanation for his behavior, which still depends on our criterion that *this is normal behavior*.

In contrast, the question of cause is purely a question of causality, and we can turn it to inanimate objects as well. Note that the answers to the following two questions are identical. "Why did that volcano erupt?" and "How did the volcano erupt?"—the answers to both questions are the same, such as "because of the expansion of magma underground." The point is that we understand reason as justifying the actions of others who may have an agenda, which is different from the kind of cause that we might ask of inanimate objects.

## FOLLOWING THE RULES

And importantly, our practice of "following the rules" is deeply related to the presentation of reasons. Consider the "Woodhouse joke" to which Michael Dummett refers (Dummett, 1991: 93). A person asks, "Do you speak Spanish?" and the other replies—"I don't know; I've never tried." According to Dummett, this kind of response is sensible when it comes to questions of practical ability, such as "Can you swim?", but it is nonsense when it comes to language. It is based on his assertion that a person who does not know Spanish should not even try to speak it. A nonswimmer knows what swimming is, and can therefore try. But a person who does not know Spanish does not know what it is to speak Spanish.

But here the two phases of the act are confused. If we think of "being able to swim" as a matter of fact, as in falling into the water and not sinking for a certain period of time, then it is possible to be successful without ever having tried it before. But if we think of it at this level, then it is possible to successfully communicate in Spanish for a certain amount of time, by going to a bar in Spain and imitating the orders of the customers around you. Of course, that is not speaking Spanish at the level of correct expression according to Spanish grammar, and it would be nonsense to think that this is possible without having studied Spanish in the past. But if you think of swimming in a similar way, following certain rules, for example, the rules of breaststroke, then it would be nonsense to think that you could do that without having studied it.

In other words, the problem lies in the difference between the level of doing a certain action as a matter of fact and doing it according to a set of rules. We can practice an action without knowing we are doing it, whereas we cannot do a norm-conscious action without being aware of it and without reason to think we can. On the other hand, we cannot do norm-conscious actions unknowingly and without reason to think we can. And what makes the difference is that the word "reason" implies a normative justification, rather than merely a cause of an event.

"I shot the Arab because of the sun" would not be accepted as a sound answer in court, because it goes against our common sense that a normal person would not kill others for such a reason. Any explanation that people do not consider just will be dismissed as either a deliberate lie or, if he seriously believes it to be true, a reflection of his own disorder. In a real court process, it may happen that police officers, prosecutors, or even defense lawyers may try to persuade the accused to change his statement on the grounds that such a motive will not convince the judge, and thus make him suspected of trying to hide his true motive by deliberate lying. Reasons that appear in the process of legal application are not merely objective facts but are often coproduced as indicative of socially accepted semantic relations.

Let us note that in some cases, the accused himself has a motive to cooperate with it. Suppose the car I was driving causes an accident and injures someone. If there didn't exist any physical condition to cause the accident without my fault, or defects in the car, the police officer would say that there must have been intent to injure the victim or negligent driving, and I myself would have to admit that logically these two were the only ways it could have happened. Even if I had no awareness or clear recollection of having made a mistake, I would infer that I must have been careless in not being aware of it, and I would try to state this as the reason for the accident, since, given that intentional injury is usually punished more severely than negligence, I had a motive to avoid it. Reasons are thus often logically constructed by retrospective inference from consequences.

## CONCLUSION

Given that this is the kind of reason that is required in court, it is clear that the systems that have been implemented using deep learning have not reached the level of answering this question. Even in so-called explainable AI, what is intended there in many cases is to visualize the parameters and factual elements that influenced the conclusion, and it seems to remain at the level of showing cause.

For example, the AI Principles published by the OECD (2019) lists "transparency and explainability" as one of the seven principles, along with "human-centered values and fairness" and others. Also, in the Ethics Guidelines for Trustworthy AI published in April 2019 by the European Commission's High-Level Expert Group on AI, one of the seven requirements, transparency, is also listed as "traceability, explainability and communication" (Independent, 2019: 14).

Explainability as used here should thus be understood as a high level that includes normative justification for reasons. This is what has been demonstrated in our practice of judiciary. This is what guarantees the chance to make a critical investigation of the judiciary, which guarantees the sound functioning of the trial system, closely related to coercion. Conversely, if AI is utilized in trials without ensuring the possibility of explaining the reasons, it will create a system of coercion that guarantees neither sovereign control from citizens nor verifiability by the parties. At the very least, this is not appropriate for a system in countries where democracy and human rights are fundamental. Until this is achieved, it will be necessary to avoid allowing AI to make judiciary-related decisions on its own and to limit its use to vigilant reference by human experts in order to achieve human-centered use of AI.[7]

## NOTES

1 There were certain exceptions, as when a woman was sued in cases of accusations of infidelity, she could have someone fighting on her behalf ("champion").
2 In some legal systems, a jury randomly selected from citizens may be added, but even in this case, the aim is to ensure the same neutrality as a professional judge by excluding interested parties.
3 Of course, legal systems are formed through democratic processes in each nation, and their content differs from country to country. The author's knowledge is limited and may not be applicable to all countries.
4 In general, the movement from the creation of the term "artificial intelligence" at the Dartmouth Conference in 1956 to the 1960s, which pursued methods of inference and search, is called the first AI boom, distinguished from the second boom in the 1980s and 1990s, which aimed to realize expert systems through knowledge representation, and the third boom, from the development of deep learning by Jeffrey Hinton in 2006.
5 Wisconsin Court System, Circuit Court Caseload Statistics, https://www.wicourts.gov/publications/statistics/circuit/historicalcircuitstats.htm
6 Loomis v. Wisconsin, 881 N.W. 2d 749 (Wis. 2016)), confirmed by the U.S. Supreme Court's denial to accept the certiorari.
7 The draft of this chapter was first written in Japanese, the author's native language, and translated into English by DeepL (https://www.deepl.com/translator). Naturally, the results were closely checked by the author himself and modified according to his intentions. In the author's opinion, this is a justifiable use of AI.

## REFERENCES


Ambrose, A. (Ed.). (1979). *Wittgenstein's Lecture: Cambridge, 1932–1935, from the Notes of Alice Ambrose and Margaret Macdonald.* Basil Blackwell. (Vol. I, p. 4).

Angwin, J., Larson, J., Mattu, S., & Kirchner, L. (2016, May 23). Machine Bias. *ProPublica.* Retrieved from https://www.propublica.org/article/machine-bias-risk-assessments-in-criminal-sentencing (Checked on February 17, 2023).

Austin, J. (1832). *The Province of Jurisprudence Determined.* John Murray.

Camus, A. (1942). *L'Étranger.* Éditions Gallimard.

Dummett, M. (1991). *The Logical Basis of Metaphysics.* Harvard University Press.

Equivant. (2019). *Practitioner's Guide to the COMPAS Core* (pp. 1–2).

European Parliament and Council. (2021). Proposal for a Regulation of the European Parliament and of the Council Laying Down Harmonised Rules on Artificial Intelligence (Artificial Intelligence Act) and Amending Certain Union Legislative Acts. COM/2021/206 final.

Hart, H. L. A. (2012). *The Concept of Law* (3rd ed. with an Introduction by Leslie Green). Oxford University Press. (1st ed. 1961).

Independent High-Level Expert Group on Artificial Intelligence set up by the European Commission. (2019). *Ethics Guidelines for Trustworthy AI.*

Kafka, F. (1925). *Der Prozeß*. Verlag Die Schmiede. (Trans. by D. Wyllie). Project Gutenberg. https://gutenberg.org/ebooks/7849

OECD. (2019). *Recommendation of the Council on Artificial Intelligence*, OECD/LEGAL/0449, adopted on 22/05/2019.

Tao, K. (2019, June). China's commitment to strengthening IP judicial protection and creating a bright future for IP rights. *WIPO Magazine*. Retrieved from https://www.wipo.int/wipo_magazine/en/2019/03/article_0004.html (Checked on June 27, 2023).

Wisconsin Court System. *Circuit Court Caseload Statistics*. Retrieved from https://www.wicourts.gov/publications/statistics/circuit/historicalcircuitstats.htm

Wittgenstein, L. (1953). *Philosophical Investigations* (G. E. M. Anscombe, Trans.). Macmillan. (I. s. 66).

Wittgenstein, L. (1969). *The Blue and Brown Book*. Basil Blackwell. (Original work published 1958).

# 14 Artificial Intelligence in Higher Education
## *Opportunities, Issues, and Challenges*

*Bruno Poellhuber, Normand Roy, and Alexandre Lepage*

Université de Montréal, Montréal, Québec, Canada

Since November 2022, ChatGPT has had very high visibility in higher education, raising an impressive amount of debate and discussion. These conversations have been focused on both the various risks and issues raised by such powerful AI tools but also on the diverse possibilities they offer to assist, facilitate, and even augment the work of learners and teachers. For many people, ChatGPT represents an eruption of AI in the field of education. Yet this sudden media attention obscures the fact that AI has been present in higher education for many years already. The field of learning analytics is growing significantly in education, resulting in descriptive or predictive analyses based on the traces left by learners in digital environments, and giving rise to predictive dropout models and dashboards that have been implemented in some universities (Ifenthaler & Yau, 2020). Technological developments by large cloud providers make it much easier to accumulate data for analysis (data mining) or to develop intelligent conversational agents (chatbots) that can be used to support students (Heryandi, 2020). The field of AI in education (AIED) focuses on learning analytics, conversational robots and natural language processing, adaptive learning, speech and visual recognition, expert systems, and decision support systems. It now also encompasses generative AI.

In this chapter, we propose to first situate the field of AIED historically, and then examine three areas that have considerable potential in higher education: learning analytics, adaptive learning, and generative artificial intelligence. We will look at each of these areas in detail and discuss how they can be used to enhance the educational experience, as well as the limitations and challenges associated with them.

## CONTEXTS OF ARTIFICIAL INTELLIGENCE IN EDUCATION (AIED)

Research on AIED began in the 1970s (Southgate, 2020), and is focused on developing systems that can personalize the learning experience, personalization being considered the Holy Grail of AIED. The ways that personalization can be achieved have

DOI: 10.1201/9781003320791-17

evolved with technological advances in the field of AI. Likewise, expectations for the complexity of personalization have continually increased. The first AI systems in education (e.g., Carbonell, 1970) were intelligent tutorial systems aimed at the acquisition of knowledge circumscribed to a specific subject, a very specific theme, or the learning of specific procedures, relying heavily on programmed instruction and a structured approach to teach hierarchical knowledge. The interaction was unidirectional, with the learner completing a series of exercises or questions proposed by the computer, and the questions being modulated based on the errors made by the learner and the learning to be mastered before the next level content was presented. What distinguished the field of AIED from educational technology was the systems' ability to make instructional decisions, so that different learners could be offered different exercises, or a different order of exercises, or could be asked to present the concepts to be learned in a manner tailored to their level (Wenger, 1986). These early systems may seem simple compared to what is being developed today, but the intent to personalize is still central and is now based on larger and more accurate data sets, as well as on better learner models and more formalized knowledge domains. Applications have also diversified, offering not only intelligent tutorial systems for learners but also targeting teacher support, which is made possible through increased information to support students and the use of big data to guide decision-making on instructional practices. To reflect this advancement, several updated definitions of AIED have been proposed in recent years. According to Holmes et al. (2019), the definition of AIED encompasses the practical use and application of techniques from the field of AI to many aspects of education:

> AIED includes everything from AI-driven, step-by-step personalized instructional and dialogue systems, through AI-supported exploratory learning, the analysis of student writing, intelligent agents in game-based environments, and student-support chatbots, to AI-facilitated student/tutor matching that puts students firmly in control of their own learning. It also includes students interacting one-to-one with computers, whole-school approaches, students using cell phones outside the classroom, and much more besides. In addition, AIED can also shine a light on learning and educational practices.
>
> (p. 11)

Humble and Mozelius (2019) emphasize the interdisciplinary nature of AIED and its potential not only to support certain educational uses but also to increase our understanding of human learning processes: "AIED is, as AI, an interdisciplinary field containing psychology, linguistics, neuroscience, education, anthropology and sociology with the goal of being a powerful tool for education and providing a deeper understanding of how learning occurs." For Hwang et al. (2020, p.1), AI can occupy four roles: a tutor that teaches, a tutor that is taught, a tool that assists learning, or a tool that assists in the development of educational practices. In practice, these uses are often combined. Thus, as is often the case with technologies in education, the development of AIED pertains simultaneously to research and teaching initiatives. For example, data collected in Learning Management System (LMS) can be used both to help us understand certain phenomena and indicate how to take action in everyday classrooms or distance teaching.

## AFFORDANCES AND OPPORTUNITIES OF AI TO SUPPORT TEACHING AND LEARNING IN HIGHER EDUCATION

### CASE NO. 1. LEARNING ANALYTICS AND STUDENTS' DASHBOARDS

Learning analytics (LAK) is probably the area in which there has been the most work in AIED. Teaching and learning are increasingly done through learning management systems (LMS) such as Moodle. An LMS allows teachers not only to integrate different types of educational resources (files, HTML pages, hyperlinks, etc.) but also to create learning activities (formative or summative assessments, exchange or collaboration activities in forums or collaborative documents, etc.). Thanks to the traces collected from the students' activities (mouse clicks, resources consulted, connection time, messages sent, quizzes submitted, etc.), teachers can track student engagement. Many consider traces to be indicators of behavioral and even cognitive engagement (Poellhuber et al., 2019). This is the objective of LAK, which aims to use the data collected about learners' activity in their LMS to understand and improve learning, the learning process, or the learning environment.

Work in the field focuses on four main themes: performance prediction, decision support for teachers and learners, behavioral pattern detection and learner modeling, and dropout prediction (Du et al., 2021). Thus, LAK is employed in predictive analyses (Alhadad et al., 2015), including identifying students at risk of dropping out or failing, and providing them and their teachers with dashboards to support their self-regulation and coaching (Peraya, 2019). Among the studies identified by Du et al. (2021), only a few focused on dropout predictions (19 of 901) with the general idea of implementing early detection and interventions. Most of the studies focus on proofs of concept (29.4%) or proposed concepts or frameworks related to LAK (33.3%). A minority of papers (35.2%) were focused on conducting actual data analysis; these were usually related to a single course or MOOC, with a few notable exceptions.

The first initiatives were done at Purdue University, where a system called Course Signals, constructed using the Blackboard LMS, was developed with the aim of using a predictive model to guide at-risk students toward appropriate help resources (Arnold & Pistilli, 2012). The authors reported a significant improvement in retention rates for at-risk students taking courses in Course Signal compared to other at-risk students. As part of the Open Academic Analytics Initiative, a predictive model —based on the Sakai environment and other information from student records—was developed and used at Marist College and four other institutions. This initiative has also proven to be quite successful (Jayaprakashe et al., 2014). Students who were identified as being at-risk and who received interventions performed better in their courses than other at-risk students.

In sum, there is a great deal of work on predictive analytics and dashboard development, and research results on student engagement and retention seem positive. In a systematic review of the literature, Ifenthaler and Yau (2020) note that there are many LAK approaches that use effective techniques to support the success and retention of students at risk of dropping out, but that there is still a lack of large-scale, rigorous research examining the effectiveness of LAK for student success and the characteristics that might drive adoption and uses, particularly, from our perspective, in the Moodle environment.

The adoption of dashboards still seems to be problematic, and explicit links to suggested actions (for students) or interventions (for teachers) are still missing in most projects (Gasevic et al., 2019). In order to develop sound dashboards, a large set of historical data time is required, as well as data from student records. Although these projects are usually designed with the objective of supporting student engagement and success, a number of ethical risks arise. The most common concern relates to the use of personal data and its confidentiality. The students' or teachers' personal data may be disclosed or misused. Because predictive models are by definition imperfect, care must be taken in the management of interventions with false positives (students incorrectly identified as being at-risk), while aiming to minimize false negatives (at-risk students not identified). Moreover, students who use effective study strategies that do not involve a large number of clicks may be incorrectly identified by the model as being at-risk (e.g., students who download or print a large portion of the resources). LAK could inadvertently promote a biased assessment of learning, with some faculty tending to penalize rather than help at-risk students. Other risks are associated with the *Hawthorne effect*, whereby a teacher's negative perceptions of a student or a learner's perceptions of themself become self-fulfilling. Finally, while comparison with other students can be a driving force for some learners, it can be a discouragement for others.

Some other LAK prospects seem promising, however. More descriptive analysis could provide a better understanding of how learners engage with courses and could be applied to course design. Data collected through LAK could serve as feedback on teachers' course designs, especially when a teacher changes their design in order to increase engagement or learning. This latter possibility seems particularly promising if the process is started with a course design phase that mobilizes a sound organization of active learning activities, relying on collaboration, discussion, investigation, practice, and production, ideally with many opportunities for formative feedback. This kind of course design would produce rich traces that could be analyzed to improve the course design in a positive feedback loop.

## Case No. 2. Adaptive learning

Imagine that, as part of a psychology course, you have to learn how to conduct a consultation with a patient using a cognitive-behavioral approach. To learn this approach, you are invited to spend a few hours using an interactive tool that alternates video presentations, mini-case scenarios, and comprehension questions. As you progress through the interface, the videos you are shown are adapted to your level of knowledge of the approach and vary for different learners. Depending on the answers you give, the time you spend reviewing or reading content, the level of engagement you show, the preferences you express for the format in which the content is presented, and whether or not you have mastered the principles underlying the approach, the activities you are offered are continuously adapted. Your comprehension errors are identified accurately, and resources that specifically target them are proposed to you. Furthermore, the system validates that you have understood the principles of the approach by presenting different aspects of the case and asking different questions, with subtleties that allow you to identify a detail that you may have missed. This is an example of adaptive learning, an active field of research since the 1980s (e.g., Wenger, 1986).

Adaptive learning can be viewed as building "a model of the goals, preferences and knowledge of each individual student and using this model throughout the interaction with the student in order to adapt to the needs of that student" (Brusilovsky & Peylo, 2003, p. 156). To achieve this modeling, tutorial systems traditionally rely on three continuously updated models: the learner's model and state of knowledge, the knowledge domain model, and the teaching model, which is responsible for the choice of strategies for presenting knowledge or adjusting the level of difficulty. In interaction, these three models allow for the automation of decisions to ensure that the learner's knowledge model is as close as possible to the expected knowledge model (Figure 14.1).

Can AI transmit knowledge better than a teacher? Sometimes. AI can do it faster and in a more individualized way as long as the learner's model is sufficiently complex (Kay et al., 2022). Teachers lack time to do this adaptation work on a large scale, where instruction is sometimes provided to groups of hundreds of students.

This raises questions about the purpose of higher education. When the goal is to transmit knowledge mechanically, quickly, and efficiently, adaptive learning can be highly effective. Technological advances might even allow for greater consideration of some expert (conditional) knowledge, going beyond theoretical (declarative) or

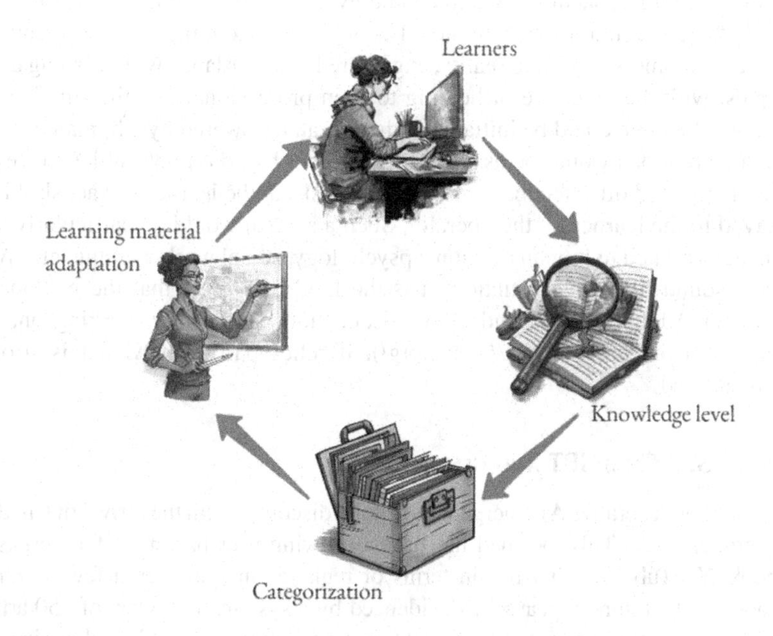


**FIGURE 14.1** Adaptive learning model.

practical (procedural) knowledge. For situations geared toward socialization with peers, research socialization, the sharing of core values, the development of socio-emotional abilities, critical thinking skills, and/or metacognition, however, adaptive learning reveals serious limitations. With these kinds of objectives intrinsic to the higher-education curriculum, the university experience cannot be replaced. The main challenge with adaptive learning will therefore be to choose when and how to use it, and in which courses or activities: What content should it be used for? For how long? At what point should a teacher intervene in a course that uses adaptive learning tools?

In a study to identify the challenges of implementing adaptive learning in universities, Mirata et al. (2020) identified three main types of challenges: organizational, pedagogical, and technological. Organizational challenges relate to the lack of resources to support faculty in their adoption of this type of tool, an essential step in any course transformation. Pedagogical challenges relate to the identification of what can and cannot be learned through adaptive learning, as well as to the required self-regulation abilities that students must demonstrate when engaging in this type of experience, avoiding a propensity to rely excessively on adaptive learning tools to regulate themselves. Technological challenges may relate to the infrastructure required to manage real-time data, integration with the LMS, and the usability of the adaptive systems.

Adaptive learning and personalization of learning are highly valued in education. Before ChatGPT, this was the most discussed topic in AIED. The ability to use big data to develop highly personalized instructional support systems is a pinnacle for equity, diversity, and inclusion initiatives. Although the adaptive learning systems that have been developed so far have focused on declarative or procedural knowledge, in the future, coaching systems could evolve to incorporate judgments, decisions, and the execution of professional actions. For example, a system could be developed in immersive virtual reality, engineered in accordance with learning design principles, with the objective of helping to train professionals. In the simulation, a continuous dialogue could be initiated with an avatar, powered by a human operator, whose interventions would be assisted by a chatbot based on natural language processing that would offer feedback and propose tasks at the learner's exact skill level, all relayed to the learner by the operator. Such a system would be particularly valuable in areas related to human relations: psychology, social work, nursing, etc. While this may sound like science fiction, it is the kind of project that the National AI Institute for Adult Learning and Online Education is currently working on, with funding from the NSF (https://aialoe.org/). It relies partly on AI but is strongly human-centered.

## Case No. 3. ChatGPT and generative AI

The concept of generative AI emerged into public discourse with the arrival of ChatGPT in November 2022. This tool had the fastest growing user base thus far, surpassing Facebook, YouTube, and TikTok in terms of registration speed. In a few months, a large body of literature appeared, as evidenced by a systematic review of 150 articles and 300,000 tweets (Leiter et al., 2023). Higher education is one of the domains most impacted by these recent developments. Generative AI makes it possible to create content (texts of various kinds, computer code, images, videos) from prompts and automatic language processing. In response to a question or proposed text, AI generative tools propose the most probable answer from their database or generate an image that best matches the idea formulated. How might this affect learning and higher education? Much discussion has been dedicated to plagiarism, but positive possibilities for using ChatGPT have been also outlined: as an intelligent tutor (Pardos & Bhandari, 2023), for adaptive learning (Qadir, 2022), as a note-taking assistant (Saini et al., 2023), or as a writing assistant (Salvagano et al., 2023), not to mention the possibilities of assisting in various artistic creations (songs, poems, stories, etc.).

In a human-centered approach, generative AI can act as a kind of personalized tutor that answers learners' questions at any time and suggests modifications or ways to enrich ideas, thus embodying a form of adaptive learning; or it can be used from the perspective of an "intelligence augmentation, designed to work with people and focus on building systems that augment and support human cognition with AI," (Hassani et al., 2020) as illustrated in Figure 14.2 for an AI-assisted research and writing process (Figure 14.2). Recently, generative AI has been able to simulate a more natural "human" conversation, answering questions formulated in common language. This makes the answers appear more accurate and plausible. The rapid adoption of generative AI is explained precisely by this feeling that it is useful and efficient (Gatzioufa & Saprikis, 2022), which is in line with traditional models of technology adoption, where ease of use and perceived usefulness are the main determinants of use.

However, ChatGPT and other generative AI tools should not be seen as reliable, since they do not distinguish between true and false information and, even worse, can literally propose fabricated answers and even invent false references. Their

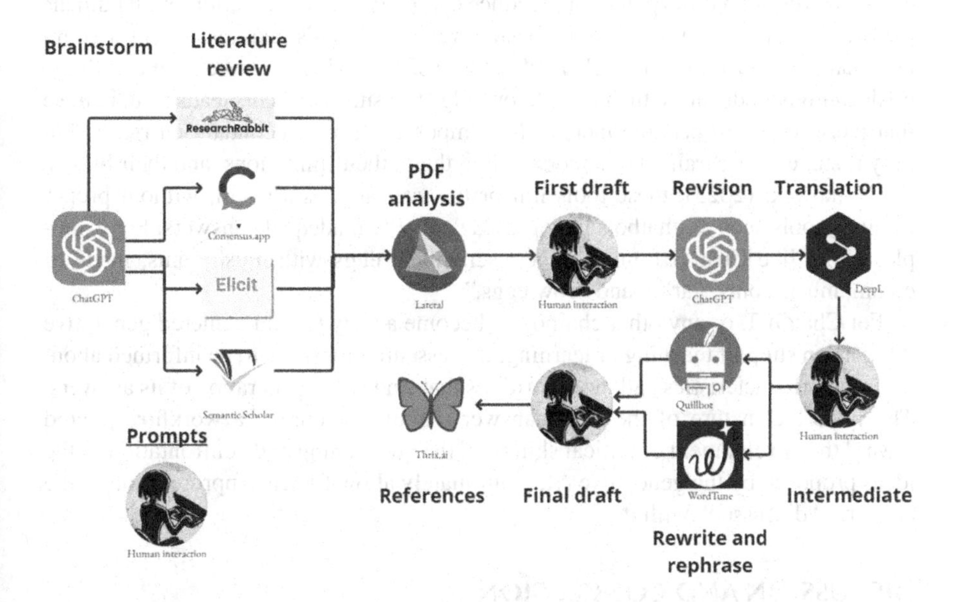


**FIGURE 14.2** Example of workflow proposition for a human-centered approach for scientific writing with various AI tools.

strength lies in their ability to effectively synthesize existing information or rephrase it to help summarize ideas or present them in a different style. Thus, as long as they are fed by verified and verifiable data, the texts produced can be of good quality. This is not quite the case with ChatGPT, however, which feeds on what it found on the Internet prior to 2021, a body of work that is biased in many ways and in which many topics have been subject to major disinformation campaigns.

For students, ChatGPT offers interesting avenues by helping them to write various texts, such as computer code. For teachers, ChatGPT and generative AI can also act as an adaptive assistant, helping with lesson planning, formulating learning objectives, synthesizing ideas, or inspiring complex exercises for students (Sok & Heng, 2023). One of the most interesting aspects is the possibility of using generative AI to enrich and personalize the feedback given to learners on multiple-choice quizzes or in the development phase of written assignments, as feedback has been identified as a highly effective pedagogical intervention and teachers lack the time to provide it in formative assessments.

Nevertheless, generative AI tools have a limited "understanding" of complex topics. ChatGPT fails to apply the steps of logical reasoning adequately and remains at a very superficial level on many complex subjects. For generative AI to be useful, humans play a critical role: they must be able to formulate the right series of prompts and adjust them gradually to reach a given goal. This requires not only some knowledge of how ChatGPT works but also deep domain expertise to be able to judge the suggestions that are made. Like proofreading software or YouTube tutorials, generative AI can help build learners' knowledge, but like a pharmakon, it can also make them lazy or facilitate plagiarism.

Generative AI has the potential to reduce certain types of interactions with humans (answering questions related to the basic levels of Bloom's taxonomy), but it could also make room for higher levels in class discussions, which is more in line with the goals of higher education than mere knowledge transmission. This leads us to believe that these tools can become increasingly important in postsecondary curricula, but only if students are trained in understanding them, their limitations, and their biases. For Sejnowski (2023), these tools mirror the user and, as a result, without proper training, tools such as chatbots may provide naive or inadequate answers. For example, a tool like ChatGPT may suggest scrambled eggs with mushrooms, made of cream, mushrooms, garlic, and "cow eggs."

For ChatGPT, or any other chatbot, to become a truly human-centered generative AI that can support teaching or learning, it is essential for users to be informed about its corpus characteristics and the algorithms that lead to the generation of its answers. The imperfect nature of the tool's answers should encourage a workflow geared toward the development of critical skills and the questioning and confrontation of the ideas proposed by the generative AI, to ultimately allow for true appropriation of the concepts "discussed" with it.

## DISCUSSION AND CONCLUSION

In this chapter, we presented three cases that mobilize AI in higher education—learning analytics, adaptive learning, and generative AI—examining their overall contributions and limitations. We discussed some of the various risks associated with the use of these AI tools and pointed out that, although they have great potential for assisting teachers and learners, we must be cautious about these risks.

Will AI eventually be able to replace teachers? Our human-centered approach says a clear "No.". We have only to read the works of Rogers, Bandura, and Dewey, or even Vygotsky to understand the importance of humans and social interactions for learning (Kolb, 2014). However, the role of teachers will be certainly be questioned and transformed, and, in some cases—in courses that focus on knowledge transmission or in domains where there is little advancement— the teachers' role may be threatened.

The emergence of deep fakes, which involve the use of AI to manipulate real people's voices and/or videos in fabricated scenarios or discussions, has made disinformation a prominent risk associated with AI. This has significant political and economic implications. Plagiarism, privacy of personal information, biases, lack of transparency, and dependency are some of the other widely cited risks that have to be addressed in a human-centered approach to IA. Another noteworthy concern is the potential decline of human interactions, particularly in education.

The adoption of AI as a partner in some educational processes that involve human participation and critical evaluation of AI output at different stages is one of the ways to counter some of these risks. Another way that also represents a challenge for real human-oriented AI is AI literacy or the general understanding of AI on the part of the people using it. Dimension 2 of the Québec Digital Competency Framework (i.e., develop and mobilize technological skills) involves *developing a global understanding of artificial intelligence and its impacts on education, society, culture, or politics*, a perspective that calls on philosophy, law, political science, etc. To this end, the global functioning of AI should not only be taught in computer or data science but also across all disciplines in order to encompass the different possible uses. Every teacher should be familiar with the uses of AI in education or they may not be able to use these tools properly or understand the risks associated with them (Wilton et al., 2022). Ultimately, we see AI literacy not only as an educational opportunity but also as an essential condition for teachers, students, educational advisors, educational managers, and policymakers, to ensure that they understand how AI works, its possibilities, and its limitations in their disciplinary field. AI literacy will also allow these actors to participate in the discussion of its social, ethical, legal, and economic impacts, in order to make relevant, thoughtful, and ethical use of it.

Finally, generative AI appears to have the capacity to transform the development of several essential skills (informational skills, methodology skills, writing skills, etc.). Generative AI should not be mobilized to replace learners' or teachers' tasks but, rather, seen as a tool to assist teaching and learning. Moreover, with the diversity of specialized AI tools that are currently emerging, particularly in the field of documentary research and bibliometric analysis, we can imagine it as a set of tools that could, in the future, assist teachers and students, corresponding more to augmented intelligence than artificial intelligence. Sound knowledge of these tools, their functioning, their limits, and their biases will be crucial for relevant and ethical human-centered use.

Other AI-related issues are also present in higher education: academic management systems (Wang, 2021), plagiarism detection (Khalil & Er, 2023), recruitment (Allal-Chérif et al., 2021), etc. From a human-centered perspective, institutions of higher education need to consider the implications of different decisions for the people involved (students, teachers, provosts, etc.), before making technological implementation decisions. The risks of drift and bias are well documented and pervasive (Collin & Marceau, 2023), and institutions cannot ignore them.

In the meantime, measures must be put in place to ensure the greatest possible transparency concerning the uses, algorithms, and databases used. As for research, it must examine the various systems by ensuring that it draws on methodologies that include the actors who are primarily concerned. Involving these actors more in the design processes and in the research that concerns them not only fosters a real integration of human-centered AI, it also promotes the development of true AI literacy. And for the rest of the actors, training, and resources that democratize AI—what some authors call explainable AI (Khosravi et al., 2022)—must be put in place. Only when all these conditions are met will we see ethical, relevant, and sustainable changes in higher education.

## REFERENCES

Alhadad, S., Arnold, K., Baron, J., Bayer, I., Brooks, C., Little, R. R., Rocchio, R. A., Shehata, S., & Whitmer, J. (2015). *The Predictive Learning Analytics Revolution : Leveraging Learning Data for Student Success* (ECAR Working Group Paper, p. 23). Educause Center for Analysis and Research. https://library.educause.edu/resources/2015/10/the-predictive-learning-analytics-revolution-leveraging-learning-data-for-student-success

Allal-Chérif, O., Yela Aránega, A., & Castaño Sánchez, R. (2021). Intelligent recruitment: How to identify, select, and retain talents from around the world using artificial intelligence. *Technological Forecasting and Social Change*, *169*, 120822. https://doi.org/10.1016/j.techfore.2021.120822

Arnold, K. E., & Pistilli, M. D. (2012). Course signals at Purdue: Using learning analytics to increase student success. In *Proceedings of the 2nd International Conference on Learning Analytics and Knowledge* (pp. 267–270). https://doi.org/10.1145/2330601.2330666

Brusilovsky, P., & Peylo, C. (2003). Adaptive and intelligent web-based educational systems. *International Journal of Artificial Intelligence in Education*, *13*, 159–172.

Carbonell, J. (1970). AI in CAI: An artificial-intelligence approach to computer-assisted instruction. *IEEE Transactions on Man Machine Systems*, *11*(4), 190–202. https://doi.org/10.1109/TMMS.1970.299942

Collin, S. et Marceau, E. (2023). Enjeux éthiques et critiques de l'intelligence artificielle en enseignement supérieur. *Éthique publique*, *24*(2).

Du, X., Yang, J., Shelton, B. E., Hung, J. L., & Zhang, M. (2021). A systematic meta-review and analysis of learning analytics research. *Behaviour and Information Technology*, *40*(1), 49–62. https://doi.org/10.1080/0144929X.2019.1669712

Gasevic, D., Tsai, Y. S., Dawson, S., & Pardo, A. (2019). How do we start? An approach to learning analytics adoption in higher education. *International Journal of Information and Learning Technology*, *36*(4), 342–353. https://doi.org/10.1108/IJILT-02-2019-0024

Gatzioufa, P., & Saprikis, V. (2022). A literature review on users' behavioral intention toward chatbots' adoption. *Applied Computing and Informatics*. https://doi.org/10.1108/ACI-01-2022-0021

Hassani, H., Silva, E. S., Unger, S., TajMazinani, M., & Mac Feely, S. (2020). Artificial Intelligence (AI) or Intelligence Augmentation (IA): What is the future? *AI*, *1*(2), Article 2. https://doi.org/10.3390/ai1020008

Heryandi, A. (2020, July). Developing chatbot for academic record monitoring in higher education institution. In IOP Conference Series. *IOP Conference Series: Materials Science and Engineering*, *879*(1), 01204. https://doi.org/10.1088/1757-899X/879/1/012049

Holmes, W., Bialik, M., Fadel, C., & Center for Curriculum Redesign. (2019). *Artificial intelligence in education: Promises and implications for teaching and learning*. Center for Curriculum Redesign.

Humble, N., & Mozelius, P. (2019). Teacher-supported AI or AI-supported teachers? https://doi.org/10.34190/ECIAIR.19

Hwang, G.-J., Xie, H., Wah, B. W., & Gašević, D. (2020). Vision, challenges, roles and research issues of Artificial Intelligence in Education. *Computers and Education, 1*, 100001. https://doi.org/10.1016/j.caeai.2020.100001

Ifenthaler, D., & Yau, J. Y. K. (2020). Utilising learning analytics to support study success in higher education: A systematic review. *Educational Technology Research and Development, 68*(4), 1961–1990. https://doi.org/10.1007/s11423-020-09788-z

Jayaprakash, S. M., Moody, E. W., Lauría, E. J. M., Regan, J. R., & Baron, J. D. (2014). Early alert of academically at-risk students: An open source analytics initiative. *Journal of Learning Analytics, 1*(1), 6–47. https://doi.org/10.18608/jla.2014.11.3

Kay, J., Bartimote, K., Kitto, K., Kummerfeld, B., Liu, D., & Reimann, P. (2022). Enhancing learning by Open Learner Model (OLM) driven data design. *Computers and Education, 3*, 100069. https://doi.org/10.1016/j.caeai.2022.100069

Khalil, M., & Er, E. (2023). *Will ChatGPT get you caught? Rethinking of plagiarism detection. arXiv preprint arXiv:2302.04335*. https://doi.org/10.48550/arXiv.2302.04335

Khosravi, H., Shum, S. B., Chen, G., Conati, C., Tsai, Y., Kay, J., Knight, S., Maldonado, R. M., Sadiq, S. W., & Gašević, D. (2022). Explainable Artificial Intelligence in education. *Computer Education Artificial Intelligence, 3*, 100074. https://doi.org/10.1016/j.caeai.2022.100074

Kolb, D. A. (2014). *Experiential learning: Experience as the source of learning and development*. FT Press.

Leiter, C., Zhang, R., Chen, Y., Belouadi, J., Larionov, D., Fresen, V., & Eger, S. (2023). ChatGPT: A meta-analysis after 2.5 months. *Arxiv, Admin./2302.13795*.

Ministère de l'Éducation et de l'Enseignement Supérieur du Québec. (2019). *Digital Competencies Framework*. Gouvernement du Québec.

Mirata, V., Hirt, F., Bergamin, P., & van der Westhuizen, C. (2020). Challenges and contexts in establishing adaptive learning in higher education: Findings from a Delphi study. *International Journal of Educational Technology in Higher Education, 17*(1), 32. https://doi.org/10.1186/s41239-020-00209-y

Pardos, Z. A., & Bhandari, S. (2023). Learning gain differences between ChatGPT and human tutor generated algebra hints. *Arxiv, Admin./2302.06871*.

Peraya, D. (2019). Les Learning Analytics en question. [Panorama, limites, enjeux et visions d'avenir]. *Distances et médiations des savoirs* [T2]. *Distance and Mediation of Knowledge, 25*[T3], 77–95.

Poellhuber, B., Roy, N., & Bouchoucha, I. (2019). Understanding participant's behaviour in massively open online courses. *International Review of Research in Open and Distributed Learning, 20*(1). https://doi.org/10.19173/irrodl.v20i1.3709

Qadir, J. (2022, décembre 30). Engineering education in the era of ChatGPT: Promise and pitfalls of generative AI for education. *Tech. Rxiv*[T4]. https://doi.org/10.36227/techrxiv.21789434.v1

Saini, M., Arora, V., Singh, M., Singh, J., & Adebayo, S. O. (2023). Artificial intelligence inspired multilanguage framework for note-taking and qualitative content-based analysis of lectures. *Education and Information Technologies, 28*(1), 1141–1163. https://doi.org/10.1007/s10639-022-11229-8

Salvagno, M., et al. (2023) Can artificial intelligence help for scientific writing? *Critical Care, 27*(1), 75. https://doi.org/10.1186/s13054-023-04380-2

Sejnowski, T. (2023). AI chatbot ChatGPT mirrors its users. *Intelligent*. Today. https://today.ucsd.edu/story/ai-chatbot-chatgpt-mirrors-its-users-to-appear-intelligent

Sok, S., & Heng, K. (2023). ChatGPT for education and research: A review of benefits and risks. *SSRN Electronic Journal*. http://doi.org/10.2139/ssrn.4378735

Southgate, E. (2020). *Artificial intelligence, ethics, equity and higher education: A 'beginning-of-the-discussion' paper*. National Centre for Student Equity in Higher Education.

Wang, Y. (2021). Educational management system of colleges and universities based on embedded system and artificial intelligence. *Microprocessors and Microsystems*, *82*, 103884. https://doi.org/10.1016/j.micpro.2021.103884

Wenger, E. (1986). *Artificial intelligence and tutoring systems: Computational approaches to the communication of knowledge*. Morgan Kaufmann.

Wilton, L., Ip, S., Sharma, M., & Fan, F. (2022). Where is the AI? AI literacy for educators. In M. M. Rodrigo, N. Matsuda, A. I. Cristea, & V. Dimitrova (Ed.), *Artificial intelligence in education. Posters and late breaking results, workshops and tutorials, industry and innovation tracks, practitioners' and doctoral consortium* (pp. 180-188). Springer International Publishing. https://doi.org/10.1007/978-3-031-11647-6_31

# 15 HCAI-Based Service Provision for an Engaged University

*Cristina Mele, Tiziana Russo Spena, Irene Di Bernardo, Angelo Ranieri, and Marialuisa Marzullo*

Department of Economics, Management, Institutions, University of Naples Federico II, Naples, Italy

## INTRODUCTION

In recent years, technologies based on Artificial Intelligence (AI-based technologies) have revolutionized several aspects of our daily lives. AI is being used to automate and streamline processes, which can help organizations save time and money while providing better user experiences (Huang & Rust, 2018). Many organizations commonly use AI-powered chatbots to provide users with real-time responses. Additionally, AI-enabled algorithms are used for predictive analytics to provide personalized or ahead-of-time services (Huang et al., 2019). Universities and higher education institutions worldwide also embrace this trend and start to incorporate AI into their processes and service provision (McKinsey, 2022).

This trend parallels another critical change in the university context. Over the years, the main functions of universities have been education and research. They have long functioned as independent institutions with little territorial encumbrance (Karlsen, 2005). A new role for the university is now emerging, identified by the newly engaged university model (Davies & Nyland, 2022). It stresses the impact universities can have on society and the importance of engagement with communities to address societal challenges through research and innovation.

This transformative impact spurs the efforts of universities to integrate new intelligent and digital technologies into their daily operations to better connect with students, educators, researchers, and the broader community. Chatbots, virtual assistants, and social media platforms are used to interact with students based on their specific and contextual needs, support scholars and researchers in daily activities, and overall promote public engagement (Rodríguez-Abitia & Bribiesca-Correa, 2021). Similarly, AI technologies aid research by analyzing data to identify patterns and insights to speed up research and promote collaborations, knowledge sharing, and transfer (Nyland & Davies, 2022). AI-based technologies are expected to create many opportunities to innovate university service provision to impact society.

DOI: 10.1201/9781003320791-18

In this chapter, we aim to analyze how human-centered artificial intelligence (HCAI) supports universities to advance service provision in line with its new role in transforming society. We offer a comprehensive understanding of the HCAI-based engaged university focused on three main activities: (1) engaged education, as it relates to the universities' activity of fostering both teaching and learning (e.g., information support, interactive lessons, personalized experiences, and inclusive education); (2) engaged research, as it relates to the universities' activity of contributing to detect understandings (e.g., analysis support, collaborative network, expanded knowledge, augmented research); and (3) engaged with society, as it relates to universities' activity of embracing society in a renewed relationship (e.g., increased trust, joint initiatives, advanced communication, innovative abilities).

These activities show how technology can support universities in addressing the wider benefits of its engaged mission.

## SERVICE PROVISION AND EDUCATION

The service provision concept entails applying knowledge and skills to provide a benefit (Vargo & Lusch, 2008). The service-oriented vision is relational because value emerges through using resources offered in a particular context and resources obtained from other service providers (Mele et al., 2022). Providers need to seek out and understand actors' needs and expectations actively, thus collaborating to co-create value and improve the overall service experience.

Universities are fundamentally identified as centers of higher education, academic research, and knowledge dissemination (Tuunainen, 2005). From a service-oriented perspective, the fundamental aim of traditional universities has always been to furnish students with a holistic educational experience that fosters engagement and prepares them with the necessary skills to tackle challenges within their chosen field (Conduit et al., 2016; Sim et al., 2018). Consequently, "universities as service providers" (Kamvounias, 1999, p. 30) consider students as beneficiaries of the service, offering them various resources in line with Vargo and Lusch (2008). However, universities also play a critical role in advancing knowledge through research and other academic endeavors (Wyner, 2019), involving students, researchers, and other stakeholders such as entrepreneurs. Thus, the multiple faces they can assume are revealed, as well as the need for different activities to be performed (Thatcher et al., 2016). Nevertheless, these activities are not mutually exclusive. For instance, research services are provided to other academics interested in the analyzed topics and to support other actors engaged in research activities, such as students and entrepreneurs.

Following this perspective, Calma and Dickson-Deane (2020) state that once viewed as institutions that generate knowledge, universities are now recognized as service providers that care about knowledge and integrate technology to improve, reinforce, and establish a direct connection between learning outcomes. By adopting this more extensive role, universities offer resources and provide access to educational opportunities, promote knowledge sharing, and increase engagement through innovative tools (Keenan et al., 2020). More recently, academic institutions have embraced a broader mission of contributing to civic and public welfare engagement by serving their local communities and promoting social and economic development (Chankselian & McCowan, 2021).

In this context, it is necessary to consider the multiple components that shape the university as an actor within a service ecosystem that evolves in response to emerging societal challenges. A shift is required from a mere provision of educational services to a more active, engaging role of universities in service provision.

## ENGAGED UNIVERSITY

The concept of the engaged university has become critical in the higher education literature (Kliewe & Baaken, 2019). It has its roots in the US land-grant movement in the mid-19th century, which sought to provide practical education and outreach to support rural communities' economic and social development (Breznitz & Feldman, 2012). Since then, this concept has evolved to encompass a broader range of activities, including community-based research, service learning, civic engagement, and partnerships with nonprofit organizations and industry.

According to some scholars, universities involve more and more models that go beyond traditional education and research goals (Perkmann et al., 2021). However, the engaged university differs from the model of an entrepreneurial university. This last model involves creating new university structures (spin-offs) that connect academic scientists with potential research users through a business approach (use of patents and licensing) promoting knowledge transfer to the market (Wang et al., 2016). Differently, engaged universities are increasingly responsible for solving societal problems. Goddard's (2009) engaged university model advocates integrating this third mission throughout all university's organizations, activities, and practices.

Watson et al. (2011, p. 281) defined engaged universities as *"a comprehensive empirical account of the global civic engagement movement in higher education."* The engaged universities actively attract their communities, both locally and globally, to address societal challenges and contribute to the public good. As a result, there has also been a growing interest in their commitment to their surrounding areas' development and economic growth (Benneworth, 2012). The main role of the university is to contribute to improving the knowledge exchange environment, organization, governance, and policy frameworks.

This new engaged perspective emphasizes the need to view the different actors involved in university activities (i.e., students, researchers, and society) from a wider and more unified perspective. University helps students in developing skills and knowledge for their future careers (Breznitz & Feldman, 2012), supports researchers in generating new knowledge and innovations that can have a significant impact on society (Perkmann et al., 2021) and contributes to the well-being of local communities by addressing social challenges and fostering the rise of the learning environment (Fischer et al., 2021). In this sense, a higher level of complexity arises in the university service provision. For example, communities and local industries could have specific interests and agencies to be addressed as well, as students' beneficiaries' needs evolve with the context in which they live. AI technologies can play a crucial role in supporting this new engaged role of the university.

## AI-BASED TECHNOLOGIES AND EDUTECH

AI-based technologies are increasing the capacity of advanced learning systems based on ever-growing data analytics processes that automate, support, and enhance human decision-making (Huang et al., 2019). They also support emotion-based interactions by recognizing, emulating, and responding appropriately to human emotions. Based on these abilities, service literature has widely demonstrated how AI is used to streamline service processes, deliver effective personalized interactions, and provide a compelling customer experience (Wedel & Kannan, 2016). The opportunities to connect humans, data and digital technologies (i.e., human-centered artificial intelligence—HCAI) enact smart service provisions and affect value co-creation (Mele & Russo-Spena, 2023).

Studies on AI technologies and universities are still emerging. Many of these studies analyze how AI technologies can promote interaction in service experiences to the various beneficiaries (students, professors, and researchers). The information support and collaboration activities are the focus.

AI technologies provide students access to online resources such as digital libraries, e-learning platforms, collaboration and communication tools, and online tutoring services (Selwyn, 2019). One case in point is the phenomenon of Edtech or Edutech, short for "educational technology," which refers to using technology tools and resources to support teaching, learning, and educational administration (Granic, 2022). In this sense, AI-based edutech tools and resources can be used to create personalized learning experiences for students by analyzing their learning patterns and adapting the teaching materials and assessments to their needs. AI-based edutech also supports industry partnerships and collaborations by providing platforms for knowledge-sharing, skills development, and talent acquisition. Chatbots provide students, faculty, and staff with quick and easy access to information and services (Chen et al., 2023). In research labs, AI-based robots prepare samples, collect data, and conduct analyses effectively (Selwyn, 2019). Conversely, researchers benefit from advanced technology tools and resources for data research and analysis.

Some scholars also point to the role of HCAI in addressing disparities to advance the social side of education. Some scholars also point to the role of HCAI in addressing disparities. Augmented reality technology helps students or researchers with visual impairments access additional information and interact with educational content in a more accessible way (Garzón, 2021). Virtual reality facilitates collaborative learning, teaching, and research experiences, between actors from different regions or geographical areas (Radianti et al., 2020), and robots serve as teaching aids for students with disabilities (Syriopoulou-Delli & Gkiolnta, 2021).

For society, AI-based technology offers a wide range of innovative solutions to social and economic issues and enables knowledge sharing and dissemination (Mariani et al., 2023). Overall, AI-based technologies facilitate more accessible connections between actors, enable new capabilities, and create new opportunities.

Additionally, blockchain technology creates a more efficient and transparent service system for managing student data (Cheng et al., 2018) or provides secure and tamper-proof storage of academic records such as transcripts and certificates (Bhaskar et al., 2021).

All these studies take specific university service scope in their analysis and do not go more in-depth in analyzing how HCAI can effectively impact society as a whole. The transformative role of the university requires an integrated and more profound vision of how HCAI impacts engaged service provision.

## EXPLOITING ARTIFICIAL INTELLIGENCE FOR AN ENGAGED UNIVERSITY

The use of AI-based technologies makes improving or enabling new university service provision possible. Universities increasingly adopt new technologies in their daily activities to engage more and more with teachers, researchers, students, and, more generally, with society. The framework presented in Figure 15.1. represents how universities offer renewed services in step with the new conceptualization of what an engaged university should be. The framework is identified through a combined understanding of literature and real case studies (see also Appendix 1). It is made up of three main activities—engaged education, engaged research, and engaged with society—supported by AI-based infrastructures. A more detailed explanation of each activity is provided as follows.

### Engaged Education

AI-based technologies have the potential to revolutionize teaching in universities and higher education institutions, thereby contributing to engaged education (Kuleto et al., 2021). Four main activities can be identified in this context: information support, active interaction, personalized experiences, and inclusive education.

Firstly, AI-based technologies help teachers and students by providing access to real-time and real-time information, enhancing students' learning activities (Al-Shoqran & Shorman, 2021). Virtual assistants or chatbots can answer questions, provide assignment feedback, and assist with homework. One case in point is *Jill*, an AI-enabled teaching assistant used in the Georgia Institute of Technology classroom who can answer questions from students. Jill uses natural language processing and machine learning algorithms to understand and respond to student's questions. Students get quick and accurate answers to their questions. Jill provides feedback and suggests solutions based on the student's specific needs.

Additionally, AI-based technologies enable immersive and active interaction with students promoting engagement and motivation (Kuleto et al., 2021). A representative case is the *Metaversity* project of the University of Kansas School of Nursing. Metaversity harnesses the power of virtual reality to create an immersive learning context, overcoming the constraints of a physical classroom and expanding the possibility of imagination and emotional learning.

Furthermore, AI-based technologies support teachers in identifying the needs and learning styles of each student better to adapt to their specific situational needs and personalize their learning experience (Ravshanbek, 2022). Students are supported according to their different learning abilities and can have easy and proper access to the resources they need to succeed. One case in point is Yuki, from the Philipps University of Marburg. It is the first robotic lecturer that assists teachers in analyzing large amounts of data, such as student assignments and test scores, in providing more accurate and timely personalized student feedback and supporting teachers in their lessons and support

Finally, AI-based technologies can promote the development of more inclusive educational environments by supporting the participation of impaired users (Ali &

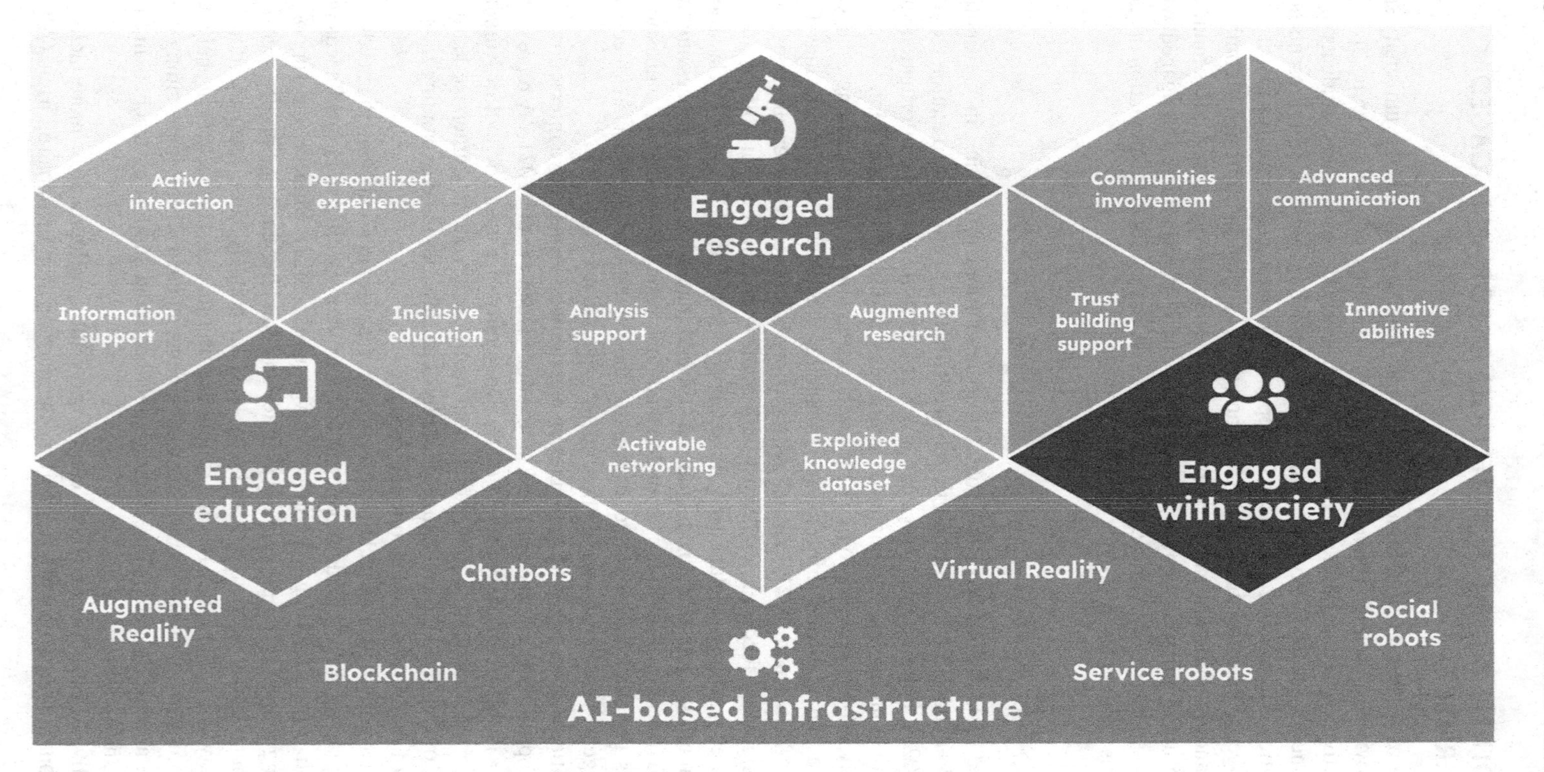


**FIGURE 15.1** The exploitation of AI-based technologies for an engaged university.

Abdel-Haq, 2021). Speech recognition and translation technologies enable students who are deaf or hard of hearing to participate in classroom discussions. An illustrative example is *Abii*, a smart robot employed at the University of Naples Federico II as a student tutor. The robot can track students' attention during the lesson, support teachers in drafting and delivering the lessons, and help students in their learning assessment. It can be used for students with some imparities or learning disorders to transform them into active actors taking more control of their learning process.

## ENGAGED RESEARCH

AI-based technologies have the potential to revolutionize research activities in universities and higher education institutions by providing effective support analysis, facilitating collaborative research activities, expanding knowledge, and supporting extensive research. By leveraging these technologies, scholars accelerate scientific discovery and contribute to engaged research (Chatterjee & Bhattacharjee, 2020). Four main activities can be identified: analysis support, activable networking, exploited knowledge dataset, and augmented research.

AI-powered virtual assistants support researchers, helping them with tasks such as literature searches, data analysis, and citation management (Fidan & Gencel, 2022). By automating research data analysis, researchers can save time and increase their productivity by focusing on more creative and innovative aspects of their research (Chen et al., 2020). One case is the Ithaca AI-powered virtual assistant. It reconstructs missing parts of ancient texts, an activity that, if undertaken manually, requires a time-consuming and challenging process for scholars. Using natural language processing and machine learning techniques, Ithaca quickly analyses large amounts of data that may be difficult for humans to identify. This helps researchers to identify patterns, gain insights, and ultimately accelerate their research.

In addition to analysis support, AI-based technologies promote activable research networking through collaborative platforms and online communities. These AI-based platforms enable scholars to easily search for and connect with other researchers' partners, share their findings more effectively, and collaborate on projects in real-time (Ouyang et al., 2022). A representative case is the PaperHive, a co-working hub platform that uses AI to improve research collaboration. It allows users to comment and discuss scientific publications collaboratively and uses machine learning algorithms to suggest links between comments and related publications.

AI-based technologies also help exploit knowledge by providing access to vast amounts of information and identifying new connections between disparate fields of study, exploiting large datasets, uncovering hidden patterns, and generating new insights (Zhang & Aslan, 2021). This led to new insights, knowledge transfer, and knowledge combinations in medicine, engineering, and social sciences. An illustrative case is ChatGPT, used by researchers at the University of Gothenburg to automatically outline the main points of a research paper and obtain feedback about their observations. It uses machine learning algorithms to identify patterns and connections within the multiple data, allowing researchers to gain new insights and perspectives on complex concepts and theories.

Finally, AI supports researchers to focus on more complex tasks by providing real-time feedback and allowing them to explore on-content research crossing different fields of study (Chen et al., 2020). The feedback from the research is provided almost immediately as the data is being processed. New knowledge can emerge concerning experimentation with different search options and systematic efforts to develop new and uncertain research paths. An example is IBM Summit, a powerful supercomputer with the capacity to process vast amounts of data and perform complex calculations to tackle challenging computational problems. With IBM Summit, researchers can simulate and model the behavior of new drug compounds. These simulations allow them to predict how the compounds will interact with the body and identify any potential side effects, helping them to make more informed information and decisions about which compounds to pursue further in drug development. Researchers can also quickly analyze large amounts of data produced by simulation and link them with different sources. Thus, they identify gaps in the existing medical literature and gain insights that can help inform their research and decision-making processes.

## Engaged with Society

AI-based technologies can potentially renew relationships between universities or other higher education institutions with the surrounding society to solve societal problems and increase opportunities to create mutual value. Four main activities can be observed: trust-building support, communities involvement, advanced communication, and innovative abilities. These activities redefine the role of universities themselves.

Firstly, due to the digital transformation, the transparency and security of the data to be managed have become critical issues (Attaran et al., 2018), and the needs related to the interactions and the way of doing of the university actors change continuously (Rodríguez & Bribiesca-Correa, 2021). HCAI enables universities to automatically manage and organize data appropriately, gain insights into societal problems, and develop targeted interventions to address them (Kuleto et al., 2021). Open and transparent interactions allow for a trust-building environment along with the integration of blockchain technologies. Data reliability is ensured by automating, securing, and certifying data exchange (Kamišalić et al., 2019). This is the case of *Blockcerts Wallet* introduced by Maryville University as a tool to provide diplomas that can be easily shared by the students and securely verified by employers or other interested actors.

In addition, as the need for universities to be more involved in the communities arises (Karlsen, 2005), organizing joint initiatives is a way for universities to collaborate and exchange resources with other actors (Singh & Kaundal, 2022). Therefore, HCAI effectively helps them facilitate involvement and promote collaboration and communication between communities. The *Citizens, Society, and Artificial Intelligence platform* brings researchers together in one place and allows for sharing their analyses on societal dynamics and the lives of individual citizens.

Additionally, the need for universities to communicate properly with society has always been an essential goal for fulfilling community claims (Compagnucci & Spigarelli, 2020). HCAI allows universities to promote dialogue and understanding of communities' interests and readiness and to advance their engaging practices by taking an active role in societal issues (Berente et al., 2021). Embracing this philosophy, Carnegie Mellon University's Language Technologies Institute created an AI-powered *Claire* chatbot to promote dialogue and understanding between law enforcement officers and communities. Claire uses natural language processing to simulate conversations between officers and community members to improve communication and build trust.

Finally, in their complete expression, AI-based technologies help universities to create a more innovative and creative society by supporting people to develop new solutions to complex challenges, such as environmental ones (Chatterjee & Bhattacharjee, 2020). In this sense, the University of Turin has founded a *federated competence centre* on High-Performance Computing (HPC), AI, and Big Data Analytics (BDA), capable of creating collaboration with entrepreneurs. A place where artificial intelligence boosts their ability to innovate daily activities.

## CONCLUSIONS

The use of AI-based technologies supports engaged education, engaged research, and engagement with society in several forms, leading to the development of engaged universities (Kliewe & Baaken, 2019). Engagement encompasses a proactive relationship between the universities and their beneficiaries in an extended ecosystem of actors (Benneworth, 2012). Supported by different AI technologies, engaged universities expand the abilities of different actors (students, researchers, and communities) and prompt their involvement in active value co-creating processes providing mutual and social value.

Addressing an engaged education through AI-based technologies enhances the quality of education by offering immediate access to information, increasing interaction in the provision of lessons, personalized learning experiences automatically adapted to individuals' needs, and inclusion of actors. Addressing engaged research using AI-based technologies accelerates the activities of researchers and enhances performance (Davies & Nyland, 2022). The quick and accurate analysis of a large amount of data by identifying patterns and trends or making predictions, the incremental development of collaboration, the sharing of extended knowledge, and the addressing of extensive research are both rigorous and relevant to contribute to addressing a higher level of contributions. Finally, engaging society through AI-based technologies helps institutions understand and respond to community needs and concerns better. Increasing support for data storage, security, and sharing, organizing joint initiatives, addressing advanced communication, and participating in creating an innovative society enable universities to solve real problems.

The three engaged activities cut across each other. HCAI allows for linking different actors within the same activity, such as enabling sharing among students or even creating collaboration among researchers and different actors. Universities and higher education institutions become active agents of change, engaging with industry and communities to address unprecedented social and economic challenges, maximize their students' and staff's learning pathways and growth, and unleash their innovative potential (Fischer et al., 2021).

By serving students and researchers, the university also serves as a co-creator to promote new solutions to complex problems and work collaboratively with partners

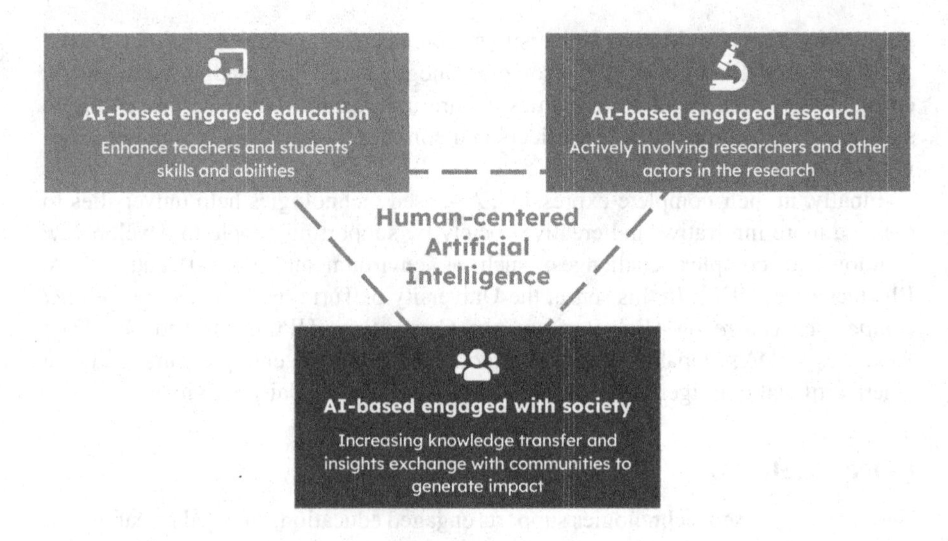


**FIGURE 15.2** The HCAI-engaged university.

to address pressing societal challenges. Moreover, an HCAI-engaged university recognizes that engagement is not an "add-on" activity but an essential part of its core mission. As such, it incorporates engagement into its teaching, research, and knowledge exchange activities, leveraging AI tools and technologies to create new opportunities for collaboration and innovation. By embracing HCAI and incorporating engagement into its core mission, an HCAI-engaged university can effectively contribute to its respective communities, fostering partnerships that promote social and economic development and creating new opportunities for learning, discovery, and innovation (Figure 15.2).

However, while AI-based technologies can be powerful tools for universities, some potential risks and challenges need to be addressed if they are not human-centered. For example, overreliance on AI can reduce human interactions, potentially limiting the development of important social and emotional skills. Universities and institutions of higher education must ensure that the use of AI is ethical. In this regard, one criticism concerns the inequality of access (Attaran et al., 2018); in fact, the use of AI can exacerbate existing inequalities in education and research. Not least, it should be remembered that the use of AI requires specialized skills, which may be a critical issue. Therefore, universities may need to invest in training and hiring staff with the necessary skills and knowledge (skills needs).

In addition, increasing the degree of engagement alone may not entirely eliminate the risks associated with overreliance on AI. While it can certainly enhance the overall learning experience and mitigate some of the negative effects, it may not fully compensate for the potential drawbacks of AI interactions. To address these concerns, universities and institutions of higher education need to take proactive steps to ensure that the use of AI is fair and effective, i.e., AI is HCAI. This involves implementing policies and guidelines that promote responsible AI usage and ensuring equitable access to AI technologies, resources, and opportunities for all, regardless of their background or circumstances.

## APPENDIX 1 EXAMPLES OF AI-BASED TECHNOLOGIES USED IN ENGAGED UNIVERSITIES (AUTHORS' ELABORATION)

| | Information Support | Active Interaction | Personalized Experience | Inclusive Education |
|---|---|---|---|---|
| **Engaged education** | *Beacon—Staffordshire University*<br>A chatbot that acts as a digital coach for students. Using AI technology, Beacon provides personalized and responsive information, thus performing quicker and more effective communication.<br>*jill—Georgia Institute of Technology*<br>An AI-enabled teaching assistant who can answer student questions about a particular class and curriculum. | *Metaversity—KU University of Kansas*<br>Wearing that takes students to the virtual nursing campus, it is possible to educate people where they are and virtually explore inaccessible places such as the inside of a human heart.<br>*Bina48—West Point Military Academy*<br>A humanoid robot co-teach in a university-level class, enhancing comprehension and holding interest.<br>*Robot—University of Nairobi*<br>Students can have joint lectures with students from the University of Komblez-Landau using the robot. Students can also easily attend seminars, whether in Kenya or Germany. | *Yuki—Philipps University of Marburgg*<br>The first robotic lecturer that aids professors in providing more individualized, accurate and prompt grading, as well as identifying new opportunities for improving teachings, thanks to the collection and the analysis of students' feedback.<br>*SmartUniversity—University of Trento*<br>The main goal of this project is to fill the empirical gap concerning students' time allocation and academic performance by providing a detailed description of how their time management affects their academic achievement. It is based on the analysis of smartphone and secondary data. | *Abii—University of Naples Federico II*<br>A smart robot tutor who assists vulnerable actors through personalized learning adventures and monitors the degree of attention during lessons.<br>*Microsoft Translator—Rochester Institute of Technology*<br>An AI-powered communication technology that uses an advanced form of automatic speech recognition to convert raw spoken language into fluent text.<br>*SignAll—Sinclair College*<br>An innovative, patent-pending technology that combines computer vision, machine learning, and natural-language processing to capture the core elements of sign language, allowing conversations between teachers and vulnerable students. |

| | Analysis Support | Activable Networking | Exploited Knowledge Dataset | Augmented Research |
|---|---|---|---|---|
| **Engaged research** | *Ithaca—University of Venezia Ca' Foscari*<br>An artificial intelligence model that helps scholars to reconstruct the missing parts of ancient texts, to date them and to identify their original place of writing.<br>*Plagiarism check—Autonomous university of Barcelona*<br>A self-service mode to verify the identification, evaluation and prevention of plagiarism through the comparison of different sources, such as, for example, public documents, publishers, and their own repositories. | *PaperHive—Oxford University Press*<br>It is a science collaboration platform that uses artificial intelligence to improve collaboration between researchers. It allows researchers to comment and discuss scientific publications collaboratively, and uses machine learning algorithms to suggest links between comments and related publications.<br>*SciLinkR—Brandeis University*<br>It is a social networking platform that uses artificial intelligence to identify the most influential researchers in a given research area. It uses a machine learning algorithm to analyze the scientific citation network and identify top researchers based on their academic influence. | *ChatGPT—University of Gothenburg*<br>An AI-based chatbot has made its formal debut in the scientific literature. It has been used to automatically outline the main points of the research, thus becoming a co-author.<br>*Generative AI—Stanford University*<br>An artificial intelligence that can generate novel content by utilizing existing text, audio, or images. It has now reached a tipping point where it can produce high-quality output that can support many different kinds of tasks.<br>*Philosobot—Oxford University*<br>A chatbot designed to convey their research to others in the most accessible way as it is possible to inform users about research, automatically create links within a topic and collect insights on users. | *IBM Summit*<br>a powerful supercomputer developed by IBM that can leverage data in various scientific fields. With its capability to process vast amounts of data and perform complex calculations at high speeds, IBM Summit is particularly suited for tackling challenging computational problems. |

| | Trust Building Support | Communities Involvement | Advanced Communication | Innovative Abilities |
|---|---|---|---|---|
| **Engaged with society** | *Blockcerts Wallet—Maryville University*<br>Using blockchain technology, the app provides a verifiable, tamper-proof diploma that can be easily shared with potential employers and other schools. It is a transformation of the university curricula management that assures storage and verification. | *CiSAI-University of Amsterdam*<br>The platform "Citizens, Society, and Artificial Intelligence" unites social and behavioral science researchers who analyze the effects of AI on societal dynamics and the lives of individual citizens. CiSAI is closely connected to university and faculty initiatives around data science and to teaching initiatives. | *Virtual tour—Georgia state university*<br>Through virtual reality, anyone interested can take a live virtual guided campus tour with their student ambassadors. | *HPC4AI—University of Turin*<br>A federated competence center on High-Performance Computing (HPC), Artificial Intelligence (AI), and Big Data Analytics (BDA), capable of collaborating with entrepreneurs to boost their ability to innovate. |
| **AI-based infrastructures** | *Augmented Reality*—technology that superimposes virtual images and information in the real world.<br>*Blockchain* is a distributed ledger technology that allows transactions and information to be recorded securely and immutably.<br>*Chatbots*—a technology capable of simulating a conversation, understanding, and automatically answering questions.<br>*Service and Social Robots*—a technology presented as mechanical devices capable of performing autonomously or semi-autonomously programmable tasks.<br>*Virtual Reality*—a technology that creates an immersive environment where users can interact realistically with virtual objects and people. | | | |

## REFERENCES


Al-Shoqran, M., & Shorman, S. (2021). A review on smart universities and artificial intelligence. In Hamdan, A. et al. (ed.), *The fourth industrial revolution: Implementation of artificial intelligence for growing business success* (ser. Studies in computational intelligence, 935), 281–294. Springer, Cham.

Ali, M., & Abdel-Haq, M. K. (2021). Bibliographical analysis of artificial intelligence learning in Higher Education: is the role of the human educator and educated a thing of the past? In Ali, M. & Wood-Harper, T. (ed.), *Fostering Communication and Learning With Underutilized Technologies in Higher Education*, 36–52. IGI Global.

Attaran, M., Stark, J., & Stotler, D. (2018). Opportunities and challenges for big data analytics in US higher education: A conceptual model for implementation. *Industry and Higher Education, 32*(3), 169–182.

Benneworth, P. (2012). University engagement with socially excluded communities: Towards the idea of 'The Engaged University'. In Benneworth, P. (ed.), *University engagement with socially excluded communities*, 3–31. Springer, Netherlands.

Berente, N., Gu, B., Recker, J., & Santhanam, R. (2021). Managing artificial intelligence. *MIS Quarterly, 45*(3), 1433–1450.

Bhaskar, P., Tiwari, C. K., & Joshi, A. (2021). Blockchain in education management: Present and future applications. *Interactive Technology and Smart Education, 18*(1), 1–17.

Breznitz, S. M., & Feldman, M. P. (2012). The engaged university. *The Journal of Technology Transfer, 37*, 139–157.

Calma, A., & Dickson-Deane, C. (2020). The student as customer and quality in higher education. *International Journal of Educational Management, 34*(8), 1221–1235.

Chankseliani, M., & McCowan, T. (2021). Higher education and the sustainable development goals. *Higher Education, 81*(1), 1–8.

Chatterjee, S., & Bhattacharjee, K. K. (2020). Adoption of artificial intelligence in higher education: A quantitative analysis using structural equation modelling. *Education and Information Technologies, 25*, 3443–3463.

Chen, L., Chen, P., & Lin, Z. (2020). Artificial intelligence in education: A review. *IEEE Access, 8*, 75264–75278.

Chen, Y., Jensen, S., Albert, L. J., Gupta, S., & Lee, T. (2023). Artificial intelligence (AI) student assistants in the classroom: Designing chatbots to support student success. *Information Systems Frontiers, 25*(1), 161–182.

Cheng, J. C., Lee, N. Y., Chi, C., & Chen, Y. H. (2018). Blockchain and smart contract for digital certificate. In *2018 IEEE international conference on applied system invention (ICASI)*, 1046–1051. IEEE.

Compagnucci, L., & Spigarelli, F. (2020). The Third Mission of the university: A systematic literature review on potentials and constraints. *Technological Forecasting and Social Change, 161*.

Conduit, J., Karpen, I. O., & Farrelly, F. (2016). Student engagement: A multiple layer phenomenon. In *Making a difference through marketing*, 229–245. Springer, Singapore.

Davies, D., & Nyland, J. (2022). Critical thinking for an engaged university. In *Curriculum challenges for universities: Agenda for change*, 3–19. Springer, Singapore.

Fidan, M., & Gencel, N. (2022). Supporting the instructional videos with chatbot and peer feedback mechanisms in online learning: The effects on learning performance and intrinsic motivation. *Journal of Educational Computing Research, 60*(7), 1716–1741.

Fischer, B., Guerrero, M., Guimón, J., & Schaeffer, P. R. (2021). Knowledge transfer for frugal innovation: where do entrepreneurial universities stand? *Journal of Knowledge Management, 25*(2), 360–379.

Garzón, J. (2021). An overview of twenty-five years of augmented reality in education. *Multimodal Technologies and Interaction, 5*(7), 37.

Goddard, J. (2009). *Reinventing the Civic University: National Endowment of Science, Technology and the Arts* (Provocation 12: September 2009). NESTA. https://research.thelegaleducationfoundation.org/wp-content/uploads/2018/03/reinventing_the_civic_university.pdf

Granić, A. (2022). Educational technology adoption: A systematic review. *Education and Information Technologies*, *27*(7), 9725–9744.

Huang, M. H., & Rust, R. T. (2018). Artificial intelligence in service. *Journal of service research*, *21*(2), 155–172.

Huang, M. H., Rust, R., & Maksimovic, V. (2019). The feeling economy: Managing in the next generation of artificial intelligence (AI). *California Management Review*, *61*(4), 43–65.

Kamišalić, A., Turkanović, M., Mrdović, S., & Heričko, M. (2019). A preliminary review of blockchain-based solutions in higher education. In *Learning Technology for Education Challenges: 8th International Workshop, LTEC 2019*, 114–124. Springer, Cham.

Kamvounias, P. (1999). Students as customers and higher education as industry: A review of the literature and the legal implications. *Academy of Educational Leadership Journal*, *3*(1), 30–38.

Karlsen, J. (2005). When regional development becomes an institutional responsibility for universities: The need for a discussion about knowledge construction in relation to universities' third role. *AI & Society*, *19*, 500–510.

Keenan, J., Poland, F., Manthorpe, J., Hart, C., & Moniz-Cook, E. (2020). Implementing e-learning and e-tools for care home staff supporting residents with dementia and challenging behaviour: A process evaluation of the ResCare study using normalisation process theory. *Dementia*, *19*(5), 1604–1620.

Kliewe, T., & Baaken, T. (2019). Introduction: A brief history of engaged and entrepreneurial universities. In *Developing engaged and entrepreneurial universities: Theories, concepts and empirical findings*, 1–15. Springer, Singapore.

Kuleto, V., Ilić, M., Dumangiu, M., Ranković, M., Martins, O. M., Păun, D., & Mihoreanu, L. (2021). Exploring opportunities and challenges of artificial intelligence and machine learning in higher education institutions. *Sustainability*, *13*(18), 10424.

Mariani, M. M., Machado, I., & Nambisan, S. (2023). Types of innovation and artificial intelligence: A systematic quantitative literature review and research agenda. *Journal of Business Research*, *155*, 113364.

McKinsey (2022). *How technology is shaping learning in higher education*. Retrieved from https://www.mckinsey.com/industries/education/our-insights/how-technology-is-shaping-learning-in-higher-education

Mele, C., & Russo-Spena, T. (2023). Artificial Intelligence in services. *Elgar Encyclopedia of Services*, 356–359.

Mele, C., Spena, T. R., & Kaartemo, V. (2022). Smart technologies in service provision and experience. In *The Palgrave handbook of service management*, 887–906. Palgrave Macmillan, Cham.

Nyland, J., & Davies, D. (2022). The University's Social and Civic Role: Time for an Appraisal. In *Curriculum challenges for universities: Agenda for change*, 21–38. Springer, Singapore.

Ouyang, F., Zheng, L., & Jiao, P. (2022). Artificial intelligence in online higher education: A systematic review of empirical research from 2011 to 2020. *Education and Information Technologies*, *27*(6), 7893–7925.

Perkmann, M., Salandra, R., Tartari, V., McKelvey, M., & Hughes, A. (2021). Academic engagement: A review of the literature 2011–2019. *Research Policy*, *50*(1), 104114.

Radianti, J., Majchrzak, T. A., Fromm, J., & Wohlgenannt, I. (2020). A systematic review of immersive virtual reality applications for higher education: Design elements, lessons learned, and research agenda. *Computers & Education*, *147*, 103778.

Ravshanbek, Z. (2022). Use of artificial intelligence technologies in the educational process. *Web of Scientist: International Scientific Research Journal*, *3*(10), 764–770.

Rodríguez-Abitia, G., & Bribiesca-Correa, G. (2021). Assessing digital transformation in universities. *Future Internet*, *13*(2), 52.
Selwyn, N. (2019). *Should robots replace teachers?: AI and the future of education*. John Wiley & Sons.
Sim, M., Conduit, J., & Plewa, C. (2018). Engagement within a service system: A fuzzy set analysis in a higher education setting. *Journal of Service Management*, *29*(3), 422–442.
Singh, S., & Kaundal, B. (2022). Academia-industry linkages: Theoretical and empirical review article. *World Journal of Advanced Research and Reviews*, *15*(1), 104–115.
Syriopoulou-Delli, C., & Gkiolnta, E. (2021). Robotics and inclusion of students with disabilities in special education. *Research, Society and Development*, *10*(9), e36210918238.
Thatcher, J., Alao, H., Brown, C. J., & Choudhary, S. (2016). Enriching the values of micro and small business research projects: Co-creation service provision as perceived by academic, business and student. *Studies in Higher Education*, *41*(3), 560–581.
Tuunainen, J. (2005). Hybrid practices? Contributions to the debate on the mutation of science and university. *Higher Education*, *50*, 275–298.
Vargo, S. L., & Lusch, R. F. (2008). Service-dominant logic: Continuing the evolution. *Journal of the Academy of Marketing Science*, *36*, 1–10.
Wang, Y., Hu, R., Li, W., & Pan, X. (2016). Does teaching benefit from university–industry collaboration? Investigating the role of academic commercialization and engagement. *Scientometrics*, *106*(2), 1037–1055.
Watson, D., Hollister, R., Stroud, S. E., & Babcock, E. (2011). *The engaged university: International perspectives on civic engagement*. Taylor & Francis.
Wedel, M., & Kannan, P. K. (2016). Marketing analytics for data-rich environments. *Journal of Marketing*, *80*(6), 97–121.
Wyner, J. S. (2019). *What excellent community colleges do: Preparing all students for success*. Harvard Education Press.
Zhang, K., & Aslan, A. (2021). AI technologies for education: Recent research & future directions. *Computers & Education: Artificial Intelligence*, *2*, 100025.

# 16 How Human-Centered Are the AI Systems That Implement Social Media Platforms?

*Alistair Knott*
Victoria University of Wellington, Wellington, New Zealand

*Tapabrata Chakraborti*
Alan Turing Institute, University College London, London, UK

*Dino Pedreschi*
University of Pisa, Pisa, Italy

## INTRODUCTION

Some of the most pervasive AI systems in today's world are deployed in social media platforms. **Recommender systems** (see e.g., Zhang et al., 2021) are AI systems that deliver a personalized stream of content into each platform user's daily feed. They do this by observing user behavior on the platform, and *learning* what users "engage with." For each platform user, a recommender system observes that user's behavior, and uses machine learning techniques to form hypotheses about the types of content that particular user typically engages with. It then gives preference to these types of content in that user's subsequent feed. **Content classifiers** are AI systems that identify content of specified types on the platform. They can operate on different types of content: a text classifier might identify textual content as "a newspaper article," or "a positive product review." An image classifier might identify an image as a picture of a dog, a cat, or a table. Recommender systems make some use of content classifiers. But content classifiers also have a central role in keeping *harmful content* off the platform (or in flagging problematic content in various ways). Social media platforms all have detailed policies on hate speech, violent content, nudity, and sexual content. The enforcement of these policies partly involves human oversight, by teams of content moderators. But enforcement also involves the large-scale development and deployment of content classifiers. Again, most of these content classifiers acquire their abilities through machine learning methods. They are given large training sets, comprising numerous *examples* of each type of content they are to identify. From

DOI: 10.1201/9781003320791-19

these examples, they construct their own internal definitions of these same content categories, which are sufficiently general to allow them to classify new items they haven't seen before.

Both recommender systems and content classifiers are eminently human-centered, in different ways. Recommender systems are human-centered in the way they *customize* the information seen by platform users. They treat each user as an individual, with a unique set of likes and dislikes. A recommender system functions as a kind of personalized editor for each user that is sensitive to each user's individual tastes. Content classifiers, when deployed for content moderation, are human-centered in ensuring certain *human values* are upheld on the platform. They help ensure that various forms of hateful content are removed from the platform. Some of this work is done to comply with laws requiring the removal of harmful content (of course, the nature of these laws varies between jurisdictions). But the work they do often extends beyond what's required by law, to implement companies' own bespoke policies about how to handle harmful content, whether this involves removal, flagging, or other actions.

While the AI systems used by social media platforms are human-centered in some senses, there are several aspects of their operation that deserve careful scrutiny. In this chapter, we will focus on questions that relate to *societal structures* that influence, or are influenced by, the operation of social media platforms. Assessing the human-centeredness of a technology doesn't just involve studying how people interact with it as individual users: we must also consider processes that occur collectively within *groups* of people, both in the use of the technology and in its governance. Collective processes operating in society can certainly be more or less human-centered: for instance, democracy is more human-centered than autocracy; a home approval scheme that is audited for bias is more human-centered than one that is not. Within these collective processes, human-centeredness should also respect human rights, such as the right to free speech and protection from discrimination. We will discuss recommender systems in the "Recommender Systems" section: focusing on concerns these systems may move users toward polarizing or extremist ideologies, and away from human-centered values. We will discuss content classifiers in the "Content Classifiers" section, focusing on concerns that the construction and training of these systems happen without adequate consultation of users, and without sufficient public accountability. In Sections "Human-centered Transparency for Recommender Systems" and "Human-centered Transparency for Content Classifiers," we propose regulatory mechanisms that could address these concerns, and help to make social media platforms more human-centered.

## RECOMMENDER SYSTEMS

### How Recommender Systems Learn

The job of a recommender system in a social media platform is to *suggest content* for the user. This content could be posts from other platform users (friends, people, or groups in the wider platform community), about other items served by the platform, or suggestions about new users to follow or befriend. What's distinctive about recommender systems is that the suggestions they make for each individual user are informed by what they learn about that particular user's interests, from observations of that user's prior behavior on the platform. If a given user clicks on a post, "likes" it, comments on it, or shares it with others, they are understood to be showing interest in that post. Recommender systems take observations about what content users actually engage with, and from these, attempt to learn *generalizations* about what they are interested in, so they can make good suggestions about new items of content.

## A Concern about the Learning Mechanism

A recommender algorithm is a powerful device for suggesting content to the user. After it has been trained, it holds a rich and powerful *model* of each individual user on the platform—specifically, of what each user is interested in. But there is a potential problem with the mechanism that achieves this learning.

The problem stems from the fact that recommender systems learn *continually* from user behavior, refining their model of users as they continue to engage with the platform. The problem is that the content users engage with *is heavily determined by the content they are recommended.* Put simply, users tend to click on the items the recommender system suggests for them. This result should be familiar to anyone using Google search: we tend to click on items on the first page of returned results (see e.g., Petrescu, 2014). What this means is that when the recommender system refines its model of a given user, it is likely to *reinforce the model it already had.* The effect is often referred to as the "filter bubble" effect.

There are many elegant proofs of filter bubble effects and many simulations that show it in one form or another. A model from Google DeepMind (Jiang et al., 2019) makes the point particularly clear: the essential finding is that a recommender system serves to *narrow* a user's range of interests as time passes and focus on some smaller subset of content types. Of course, system designers are aware of this fact, and employ many strategies to avoid this narrowing. We know roughly what some of these strategies are—a brief review is given by Knott et al. (2021). But we don't know the details of these strategies, because the learning algorithms at issue are the private property of social media companies. This leaves us with a prima facie *cause for concern* about the possible effect of recommender algorithms on social media platform users.

This initial cause for concern arises from purely technical considerations about the effects of online learning in an AI system. But it is compounded by other considerations that are about people, rather than machines. Social media users are known to have small preferences for certain types of content that could potentially drive recommender system user models in worrying directions. These preferences are psychological in nature: they are established in empirical studies. For instance, in political discussions, it has been found that users disseminate messages containing "moral-emotional expressions" more rapidly than other messages (Brady et al., 2017; Brady and van Bavel, 2021). Messages containing reference to a political out-group also travel particularly fast (Rathje et al., 2021), as do messages containing false information (Vosoughi et al., 2018). There is also a tendency for discussions to contain more hate speech as they become longer (Cinelli et al., 2021).

Preferences for these concerning types of content are small. But crucially, there is a possibility that they *interact* with the progressive learning that happens in recommender systems. If users' concerning biases persist over time, they may exert a persistent influence on the way recommender systems refine their models of platform users, so these models are drawn toward content that has an increasing concentration of moral–emotional expressions, reference to political out-groups, hate speech, and false information. In short, there is a plausible cause for concern that recommender systems may play a role in moving users toward *extremist* positions.

Again, this is just a cause for concern. But it is enough, we believe, to justify requests to companies for further information about the effects of recommender systems. To return to the theme of this handbook: while social media recommender systems are certainly "human-centered" in their delivery of content tailored to individual users, there is a real cause for concern that these systems may push users toward extremist positions over time, and in this regard may have social effects on users that are the opposite of human-centered.

## CONTENT CLASSIFIERS

Content classifiers are used for many purposes in social media platforms, but here we are concerned with one of their primary uses, which is in content moderation: that is, in identifying harmful content. Harmful content can be of many kinds, and companies have their own policies and definitions, but in this chapter we will focus on the broad category of "hate speech." A platform implementing human-centered values should clearly avoid the proliferation of hate speech, all other things being equal. (As noted in the introduction, human-centered AI should respect human rights, such as the right to protection from discrimination.) Of course, the concept of "harmful content" is heavily value-laden: we will discuss whose values should be consulted later in the paper. And of course, the removal of content posted by users is also problematic: some balance must be struck between the need to ensure users are free to express opinions and the need to keep users safe from content that harms them, either intrinsically, or through its incitement of harm in some other sphere.

In this section, we will outline how content classifiers learn to identify a category of content, using "hate speech" as an example. (We will focus on classifiers of written text, though the same methods can be used to train classifiers of images, or videos.) We will identify a concern about the learning process that is currently implemented by social media companies, which is very different in nature from the concern about recommender systems.

### How Content Classifiers Learn

A content classifier is, at the base, a "supervised" learning system. That is, it learns by being given a (large) set of examples of the content of each category it is to identify: its objective is to identify *commonalities* within these examples, which extend to new content it has never seen before. For instance, a classifier of hate speech is trained by being given (many) examples of texts that should be classified as "hate speech," and (many) other examples of texts that should be classified as 'not hate speech.' The details of the learning mechanism aren't important for the current discussion: what matters is that the classifier's learning is driven entirely by the training examples it is given. So, of course, the creation of the training set is of the utmost importance. Our concern is precisely about the process by which social media companies build the training sets for their harmful content classifiers.

### A Concern about the Learning Mechanism

The essential concern we wish to flag is that social media companies train their harmful content classifiers *behind closed doors*, with minimal public scrutiny. Some companies occasionally release training sets for some varieties of harmful content: for instance, Facebook recently released a dataset of "hateful memes" (Facebook/Meta, 2020). But such releases are the exception, not the rule. Outside the big companies, academic research groups have assembled a large collection of datasets covering harmful content of many types, presented in several modalities: see for example, Madukwe et al. (2020) for a critical review. But we don't know to what extent companies make use of these public resources. They certainly all have their own resources which they don't share.

There is actually good reason for some degree of secrecy about training sets. If companies made their training sets public, malicious actors would be able to work out how to "game" the classifiers they train, with harmful content that escapes automatic detection. But there are still several important concerns that could be better addressed. Critically, *who is making the crucial decisions* about what counts as harmful content? Companies make *general* definitions of hate speech available to the public. But in practice, the subtlety of the definitions implemented by their classifiers resides in the annotated examples they are trained on. So, who is contributing annotations? How are they being chosen? And what processes are they following? In their lack of transparency, the processes involved in training content classifiers fall well short of being "human-centered."

In the next two sections, we will propose mechanisms that can help to provide more transparency about how recommender systems affect users, and about how content classifiers are trained.

## HUMAN-CENTERED TRANSPARENCY FOR RECOMMENDER SYSTEMS

We would like to know more about how recommender systems influence users' interests and opinions—in particular, in areas related to violent and extremist content. Our project with the Global Partnership on Artificial Intelligence (GPAI 2021, 2022) has surveyed the possible empirical methods for studying this question; we conclude that by far the best methods are those that are only available internally to social media companies. The large social media companies all study the effects of their recommender system in one way or another. The basic method is conducting experiments on users. (Users all consent to being the subjects of these experiments, in the terms and conditions they agree to when joining the platform.)

The simplest experiments are randomized controlled trials, in which randomly selected groups of users are created, and exposed to different experiences in relation to the recommender system. In some companies, like YouTube and Facebook, these experiments are very fine-grained: different user groups are exposed to different "versions" of the recommender system, that learn in slightly different ways. (The variables used to represent users and content items might be varied, or the learning methods that are employed, for instance.) In Twitter, a much coarser-grained experiment was conducted: when Twitter's recommender system was introduced in 2016, a randomly selected "control group" of users was set aside, who continued to have their feed presented in a reverse-chronological stream. An "experimental group" of users exposed to the recommender system was also created, and data was gathered about the online experiences of these two groups over several years. This experiment allows a rather direct assessment of the effect of the recommender system on platform users.

In all experiments, whether fine- or coarse-grained, companies can make comparisons between groups of users experiencing different interventions. Because user groups can be large, and the only difference between groups, as a whole, relates to the intervention they experience, any differences between groups can be confidently identified as a causal effect of the intervention. User groups can be compared on many metrics. Often, these metrics relate to differences in engagement: companies have a financial interest in maximizing the engagement of users, because this keeps them on the platform for longer, and generates more advertising revenue. From a human-centered perspective, what we would like to know is whether different recommender system experiences have different effects on users' exposure to, or attitudes toward, *harmful content*. If this is the case, then companies have some *agency* over this exposure, or these attitudes, and potentially, therefore, some responsibility to do the right thing.

At present, unfortunately, external agencies have no remit to study these effects. Studies happen at the complete discretion of companies and have hardly ever involved external researchers. Only one company study using randomized controlled trials has been reported publicly: a study by Twitter, using the dataset mentioned above, logging the behavior of an experimental user group exposed to the recommender system and a control group receiving a reverse-chronological feed. The published study (Huszār et al., 2022) compared the amount of mainstream political content seen by these two groups over the study period. There was a striking difference: users in the experimental group saw a great deal more political content than those in the control group. (They also saw more right-wing content than those in the control group, in all studied countries except one.)

Huszār's Twitter study is interesting in two ways. Firstly, it shows conclusively that social media recommender systems can have dramatic effects on platform users. In so doing, it highlights the need for further information about these effects: if recommender systems strongly affect the consumption of mainstream political content, what effects do they have in *other* domains—for instance, on the consumption of hate speech, extremist material, or other types of harmful material? We urgently need to know, if we want to assess the human-centeredness of their recommender system. Secondly, Huszār et al.'s study demonstrates a concrete method whereby social media companies can report information about the effects of their recommender systems without compromising the privacy of their users, or their proprietary technology. The study was internally vetted and approved by Twitter, indicating clearly it doesn't pose any risk on either of these fronts. So, it provides a model that Twitter can use in other studies, asking about other effects of their recommender system—and which other companies could also use.

On this note, a productive step was recently taken by Twitter, in association with the New Zealand and US governments and with Microsoft. As part of the Christchurch Call to Eliminate Terrorist and Violent Extremism Online (ChCh Call 2019), these parties agreed to conduct a project allowing external researchers to access internal data at Twitter, through an interface that ensures its privacy (see ChCh Call, 2022). The initial focus of this project was precisely on the dataset studied by Huszār et al., which examines the effects of recommender systems on platform users: in fact, the first goal of the project was to create an interface that allows replication of Huszār et al.'s study. The project was organized prior to Elon Musk's takeover of Twitter. At the time of writing, its future is uncertain, like much else about Twitter. We believe that alongside voluntary initiatives like the Christchurch Call, legislation is needed, allowing regulators to require companies to conduct studies similar to that of Huszā r et al., and make the results available to the public. The EU's Digital Services Act is leading the way in providing this kind of transparency; we are working with groups who implement the Act to progress the kind of study we advocate here.

## HUMAN-CENTERED TRANSPARENCY FOR CONTENT CLASSIFIERS

The transparency issue for content classifiers, as described above, concerns how their training sets are assembled, rather than the effects they have on users. As discussed above, all the subtlety of the definitions learned and applied by content classifiers resides in the examples of each category that provide the basis for their training.

The problem is partly that the public don't currently know about how training sets are created. Who decides what counts as hate speech, for instance? But there is another potential problem: the public doesn't currently have any *say* in the way harmful content is defined. An interesting suggestion has recently been made concerning hate speech: perhaps what counts as hate speech should be determined by those *to whom the hate is directed*—that is, to those who have lived experience of the harm it causes. This idea was suggested by Kate Hannah and colleagues (GPAI, 2021 Appendix A). The idea here was that the communities that are most regularly targeted with hateful online content should be systematically *consulted*, to advise social media platforms on how hate speech for their community should be best defined. This gives the targeted communities a say in how harmful content is managed in social media contexts. It also creates an interesting model for *local governance* of harmful content, because the communities who are subject to hate speech in different jurisdictions are likely to be different.

Consultation is a notoriously vague process, however. And it is not clear how consultation processes would inform company definitions. One interesting technical possibility is that consultation could involve the *curation of training sets* for harmful content classifiers. This proposal would have a number of useful effects. Firstly, it would create a very concrete objective for the consultation process: namely, the construction of a training set of content items, annotated by the consulted groups. Secondly, it would provide clarity about how the consultation process will connect with the company's own processes. If consultation provides the training set for a harmful content classifier, the way this classifier works will be directly determined by the consultation process. Thirdly, it would provide an attractive solution to the question of how to provide transparency about the process of training classifiers for harmful content. The training set itself could not be made publicly available, as already noted: this would allow its exploitation by adversarial actors. But the consultation *process* would be visible to the public: for instance, the process for selecting members of consultation groups, and the protocols used to annotate examples, could be a matter of public record.[1]

A final benefit of consultation-based curation of training sets is the possibility that definitions of harm could be *standardized* across companies. At present, companies make their own definitions of harmful content, and build their own classifiers, using their own training sets. (There are some shared datasets, but these are the exception, rather than the rule.) Standardization would have several important benefits of its own. The first benefit is that the costs of creating a training set could be *shared* between companies. At present, there is considerable overlap between the work done by different companies in creating training sets, because they each make their own. If companies instead used their existing resources to create a single shared training set for a given domain, each company would have access to a larger training set, at no extra cost, and we can expect the performance of classifiers to improve as a result. (There is a huge need for larger and better training sets, in particular for low-resource languages, where shared resources would be particularly beneficial—see, for example, the discussion in Ranasinghe and Zampieri, 2021.) Shared training sets would also immeasurably help smaller social media companies, which lack the resources to build adequate training sets on their own (see e.g., Saltman and Thorley, 2021). Some work is being done in this area, for instance in the GIFCT's recent "Gap Analysis" project (GIFCT, 2021)—but much more could be done. A second benefit of shared training sets is that companies could be more readily *assessed* on their performance in identifying harmful content. Consider a scenario where companies have a shared training set in a given domain, but develop their own in-house classifiers that learn from this training set. It would be easy to ask which company's classifier "does best" in its learning task, by trialing each classifier on an unseen portion of the shared training set, and seeing which is most accurate. Competitive "shared tasks" of this kind are completely standard in the world of machine learning; they have been the engine for considerable progress over the last 20 years. A shared dataset for harmful content would allow the big companies to compete against one another on their performance on a task with enormous social value, rather than just in the commercial domain of "market share." In relation to the current handbook, it would potentially enable companies to *compete on human-centeredness*: which company is the most human-centered, in its automated methods for identifying harmful content?

In summary, we see some merit in the idea that the construction of training sets for harmful content classifiers should be brought into a semi-public domain. It would give harmed communities real and meaningful input into company content curation mechanisms, and it would shed useful light on how these mechanisms operate. It might also allow the creation of shared training sets, which could have attendant benefits of its own. Of course, this process of community engagement has challenges too. Direct consultatory democracy also has problems: citizen-derived definitions of harmful content might incorporate biases of their own, and might be prone to populist currents of opinion. Working out how to fund large-scale public consultation processes would also be challenging, as it would involve the diversion of company budgets toward a shared endeavor. Work underway in the GIFCT could be thought of as a first step in this direction.

## SUMMARY

Modern social media platforms are largely powered by AI. AI recommender systems produce the personalized experiences that make them so compelling for users, and AI content classifiers have a crucial role in ensuring platforms are safe places. How human-centered are these AI algorithms? They are eminently human-centered in their intention. However, in their implementation, we have argued there is still considerable room for improvement. We have suggested two practical improvements that could happen. They both relate to this issue of transparency.

For recommender systems, the additional transparency needed is in studying and reporting on the effects of these systems, on users' attitudes toward harmful content. There are reasons to be concerned about these effects, and the public needs more information about them. Companies have excellent ways of studying these effects: what is needed is a mechanism for conducting studies in the public interest using these existing methods, and publishing the results, so that a more informed discussion about user effects can take place.

For content classifiers, the additional transparency needed concerns how training data is gathered. This process should arguably be more visible to the public, and should arguably involve more consultation of the public—who are, after all, the users of the platforms in question. Curating training sets in the public domain also opens up the possibility for *shared* training sets, which could usher in more efficient uses of company resources, and a way of promoting competition between companies on functions relating to human well-being.

## NOTE

1 On annotation protocols: one interesting option to explore is to allow *disagreement* between consultants on the appropriate annotation for each item. While we expect good agreement for clear-cut cases, for "borderline" items, we can expect a range of opinions. We might want to train the classifier to predict the range of annotations, rather than a single class. A content classifier can easily be interpreted as generating a probability distribution over possible classes for each item it sees. However, its model of these distributions is likely to be better if the gold-standard annotation for each training example is also a probability distribution over classes. In particular, its assessment of "confidence" in a given decision is likely to be more accurate if it is trained this way (see e.g., Peterson et al., 2019).

## REFERENCES

Brady, W., & Van Bavel, J. (2021). Estimating the effect size of moral contagion in online networks: A pre-registered replication and meta-analysis. OSF Preprints.

Brady, W., Wills, J., Jost, J., Tucker, J., & Van Bavel, J. (2017). Emotion shapes the diffusion of moralized content in social networks. *Proceedings of the National Academy of Sciences*, 114(28), 7313–7318.

ChCh Call (2019). Christchurch Call to Eliminate Terrorist and Violent Extremist Content Online. https://www.christchurchcall.com/

ChCh Call (2022). Christchurch Call Initiative on Algorithmic Outcomes. https://www.christchurchcall.com/media-and-resources/news-and-updates/christchurch-call-initiative-on-algorithmic-outcomes/

Cinelli, M., Pelicon, A., Mozetic, I., Quattrociocchi, W., Kralj Novak, P., & Zollo, F. (2021). Dynamics of online hate and misinformation. *Nature Scientific Reports*, 11, 22083.

Facebook/Meta (2020). Hateful Memes Challenge and dataset for research on harmful multimodal content. Meta AI Research. https://ai.facebook.com/blog/hateful-memes-challenge-and-data-set/

GIFCT (2021). GIFCT Technical Approaches Working Group Gap Analysis and Recommendations for deploying technical solutions to tackle the terrorist use of the internet. GIFCT report. https://gifct.org/wp-content/uploads/2021/07/GIFCT-TAWG-2021.pdf

GPAI (2021). Responsible AI for social media governance: A proposed collaborative method for studying the effects of social media recommender systems on users. Global Partnership on AI report.

GPAI (2022). Transparency mechanisms for social media recommender algorithms: From proposals to action. Global Partnership on AI report.

Huszár, F., Ktena, S., O'Briena, C., Belli, L., Schlaikjer, A., & Hardt, M. (2022). Algorithmic amplification of politics on Twitter. *PNAS*, 119(1), e2025334119.

Jiang, R., Chiappa, S., Lattimore, T., György, A., & Kohli, P. (2019). Degenerate feedback loops in recommender systems. In *Proceedings of the 2019 AAAI/ACM Conference on AI, Ethics, and Society* (pp. 383–390).

Madukwe, K., Gao, X., & Xue, B. (2020). In data we trust: A critical analysis of hate speech detection datasets. In *Proceedings of the Fourth Workshop on Online Abuse and Harms* (pp. 150–161).

Peterson, J. C., Battleday, R. M., Griffiths, T. L., & Russakovsky, O. (2019). Human uncertainty makes classification more robust. In *Proceedings of the IEEE/CVF International Conference on Computer Vision* (pp. 9617–9626)

Petrescu, P. (2014). Google Organic Click-Through Rates in 2014. Moz blog post, https://moz.com/blog/google-organic-click-through-rates-in-2014

Ranasinghe, T. & Zampieri, M. (2021) Multilingual offensive language identification for low-resource languages. *ACM Transactions on Asian and Low-Resource Language Information Processing*, 21(1), 4. https://doi.org/10.1145/3457610

Rathje, S., Van Bavel, J., & van der Linden, S. (2021). Out-group animosity drives engagement on social media. *Proceedings of the National Academy of Sciences*, 118(26), e2024292118.

Saltman, E., & Thorley, T. (2021). Practical and Technical Considerations. Broadening the GIFCT Hash-Sharing Database Taxonomy: An Assessment and Recommended Next Steps. GIFCT report.

Vosoughi, S., Roy, D., & Aral, S. (2018). The spread of true and false news online. *Science*, 359(6380):1146–1151.

Zhang, Q., Lu, J., & Jin, Y. (2021). Artificial intelligence in recommender systems. *Complex & Intelligent Systems*, 7, 439–457.

# 17 AI Art and Creation as a Tool to Demystify AI

## *The Case Study of CHOM5KY vs. CHOMSKY*

*Sandra Rodriguez*

MIT lecturer, and Independent Creative Director and Producer

*I never was aware of any other option*
*But to question everything.*

**—Noam Chomsky[1]**

Artificial intelligence (AI) is not only advancing quickly; it is omnipresent in public discourses. The term is referred to in articles about the future of work, on billboards promising AI in healthcare, in academic debates about the ethics of predictive decisions, and in YouTubers' streams discussing "singularity."[2] Open AI's ChatGPT, launched in 2022, keeps making headlines with unexpected applications for text emulation—using online data to pass bar exams, create song lyrics, or generate computer code. The hype is real. In this context, discerning fictional excitement from factual limitations and potential pitfalls is complex. Generative AI models like ChatGPT, Midjourney, or Stable Diffusion indeed present impressive technological advancements, from an engineering perspective as much as because of their rapid popularization. Yet their sprint also triggers fears, oppositions, and warnings amongst pundits who keep raising flags about the misunderstanding of these tools by the general public.[3]

In March 2023, while the world shared AI-generated images of Pope Francis wearing Balenciaga coats over social media,[4] over a thousand AI experts called for a halt in AI developments, raising concerns about the risks these could raise for society and humanity.[5] These radical swings between pop culture uses of AI and dramatic warnings stir public's attention. On one hand, when citizens use AI in unexpected creative ways it helps everyone (technologists included) discern unexpected opportunities, and thus, they also play on our shared imaginaries—our understanding of artificial intelligence gets polluted by commercial strife, memes, and marketing stunts. As a result, algorithmic biases, power relations, and real human rights violations get too often pushed aside, without any leveled conversation about governance tools or legislative opportunities to create educated and adapted regulations.

DOI: 10.1201/9781003320791-20

**FIGURE 17.1** CHOM5KY vs. CHOMSKY VR experience—VR screenshot of the experience Alternative text description: a picture of what is seen in the Virtual Reality headset, where a marble-like monolith is centered in a space filled with huge piles of semi-translucent blocks and cubes. The cubes create an architectural environment and are lit from within. Each cube contains letters, numbers, and percentages representing the data analyzed by the AI back-end system.

**(© Sandra Rodriguez, National Film Board of Canada.)**

In other words, an already complex decision-making process complexifies even more. As a way forward, some point to the importance of public participation and general data literacy (Gentelet and Mizrahi, 2023). Yet mobilizing citizens' participation and amplifying their voices has historically proven difficult. How to invite the public to weigh in on difficult subjects, when the definition of AI is still debated in academic circles? How to foment data literacy, when AI tools are marketed as highly specialized? How can we all take part in steering AI's future? Now is the time we open the conversation.

This chapter suggests art and artistic experiences can encourage AI literacy and public participation. Artistic interpretations can help ensure a wide audience grasps general concepts that encourage participation in important decision-making. Art can help illustrate how: (1) AI systems function on past, accumulated data (highlighting which type of data, in which context); (2) human decision processes affect algorithmic outcomes; and (3) that we also need to drastically reimagine uses, applications, and developments for AI if we want to avoid its current pitfalls. In sum, we need artwork that helps a wider audience demystify AI so that we can all better imagine its future together.

In the last decades, computational artistic practices such as cybernetics or machine generative art have often made use of ML applications, speech recognition software, and other common technologies (Audry, 2021), raising critical questions for AI creativity. More importantly, such creative projects have often proven helpful to highlight overseen factors in human/AI interaction studies: the notion of meaning—how we give sense to technology, its practices, and applications. What values do we attribute to conversing with a human-built AI program? What value do we give to an "artificial entity" and its responses during a candid or complex conversation? How do these trickle down into our social fabric and influence the way we imagine our future with AI?

To set the table, this chapter suggests a need for creative experiences that focus specifically on demystifying the metaphors we collectively use to explain AI, highlighting in the process the values and biases that shape its current developments.

To do so, we draw lessons from the research-creation experience *CHOM5KY vs. CHOMSKY*[6]—a VR and AI experience that lures visitors with a futuristic promise: what if we could use AI to replicate one of today's most famous minds? The experience then uses speech-to-text and text-to-speech (STT), voice emulation, NLP conversational systems (GPT-3.5), and virtual game engine (Unreal) to invite visitors into a computer-generated world where they can converse with CHOM5KY, an AI guide. The goal is to demystify current AI conversational systems by ensuring visitors understand elemental thinking points about AI's current developments (chatbots, categorization, and emulation), comparing them to elemental thinking points about human intelligence (capacity to inquire, collaborate, and create).

When visitors walk into the exhibit space, they are invited in groups of four to put on a VR headset. As they enter the VR world, they each individually and then collectively, meet, talk, and interact with a unique AI guide called CHOM5KY. The lessons learned from this experience raise theoretical and practical questions, which we here discuss. To do so, we first situate the need to demystify AI in a legacy of fictional narratives built around human–machine comparisons. We then draw from CHOM5KY vs CHOMSKY as a case study, to discuss concepts of meaning and cultural drifts. Thirdly, we revisit key leanings from user feedback, stressing the importance of embodied interaction and play. We conclude by reassessing a need to revisit collectively shared narratives, in particular when we invite citizens to rethink what we're expecting of AI. Artistic experiences are not only sense-revealing; we believe they also create meaning.

## ON ART AND EMULATIONS: DEMYSTIFYING AI THROUGH AI

The way we speak about AI today adds to a legacy of comparing computational programs to human cognition. We look at the logics, structures, and functions of machines to describe how we speak, think, and create. The prevalence of such metaphors has important implications for how we conceive of AI. Whether real or imagined, the constant yet vague promises made for an imminent, inevitable future with AI comes with consequential dilemmas: if we are to prepare for such a future, what are the opportunities and dangers we should discuss? Should we applaud creative developments like Dall-E or Midjourney[7]? If AI entities are made to sound and look human, will it affect our notion of otherness? Questions such as these are triggered by technological growth. But they are also nurtured and sustained by fictional narratives. These shared narratives help craft a collective imaginary that influences hopes, worries, and expectations—in sum, they play on how we "talk" about AI systems, forging perception about how we believe AI should work and what it will become.

Academics sometimes complain that the general public does not understand AI well enough to discern fiction from reality, but such assumptions are not fair. The very concept of AI is still theoretically challenged, with debates about what it should encompass. Artificial intelligence is still presented as a "thing," rather than an umbrella of different technologies. Futuristic metaphors for AI obfuscate conversations about how algorithms are already used today to choose promising candidates for jobs, determine the potential receiver of an organ transplant, or create deepfake videos of politicians, famous actors, ex-girlfriends, etc. Moreover, describing AI as being in "infancy" or "needing time to grow" creates a sense of natural spawning: leaving aside important questions about political and economic decisions made to sustain its developments. AI is made to feel like the weather—untethered, unpredictable, and inevitable. Little gets said, for instance, about the millions of underpaid "ghost workers" around the world, hired to pretend they're bots on websites, labeling images for computer vision, tweaking mistakes in natural language processing, or transcribing audio files.[8] Put radically, humans today still imitate AI so that AI can fill in the expectations we, humans, create for it.

In this context, the strive for public literacy, so desperately needed, keeps getting obscured by vocabularies made to make AI feel futuristic or highly specialized. In order to increase citizen participation, we first need to demystify AI. We need experiences and spaces that help citizens question: what are we hoping to achieve with AI? Who is included in the conversation? Who is not?

Surely, such endeavors are complex. In the case of AI, for instance, a popular conception is built around the promise to, one day, be able to interact with "life-like" entities that can emulate (or replace) a human presence—in the past, such have often been depicted in science-fiction as robots, sometimes virtual entities: think of Hollywood's *A.I.*, *Ex-Machina*, *Her*, etc. In real life, other contemporary projects also use AI to give audiences a sense of speaking with someone from the past—such as the AI-Dali guide that invites museum visitors to meet an artificial Salvador Dali, or similar entities such as AI-recreated Audrey Hepburn, JFK, and Nixon.

These fictional narratives and characters find echo in contemporary research on multimodal human–robot interaction (HRI). The multifaceted aspects of communication between robots and humans are not limited to language, but rather build on culturally constructed narratives to help put users at ease, or create a sense of "rules." As research has shown, such anthropomorphizing helps reinforce metaphors, language, and collectively shared imagination around what "artificial entities" (in this case robots) should be like—and how we should be with them (Bonarini, 2020; Häring et al., 2014; Dautenhahn et al., 2009; Koay et al., 2020). In parallel to robotics and AI, technological advances in virtual and augmented reality (VR, AR) have drastically accelerated through improvements in head-mounted displays (HMD), graphics card performance, and VR uses in work and leisure. Combining the "metaverse[9]" with AI creates opportunities to test interactions between humans and AI at reduced costs of fabrication,[10] with less machinery and more play. Recent services even combine AI with VR promising to create computer-generated and NLP-supported avatars[11] so realistic that they are said to feel like a meaningful exchange with a "live" entity.[12] The dream of being able, one day, to discuss and exchange with artificial entities seems to finally be in reach.

If we put things back in context, though, this everlasting promise is far from new. The 20th century saw a burgeoning of metaphors comparing machines to living organisms and wondering about a future of life-like machines. In the wake of war, cybernetics developed parallel to postmodern philosophy (Lafontaine, 2004). The narrative sci-fi of the times supported a cultural, geographic, and historical standpoint that chose to demystify technological developments with the lenses of biology. For instance, to facilitate the acceptance of machines, their functioning was explained in metaphors relative to nature and the human body: a motor was not described through mechanical parts but rather working "like a heart." With Cybernetics, Lafontaine highlights a paradigm shift. Suddenly, it was machines and electrical metaphors that were used to explain human biology. For instance: the heart would be described as functioning with valves, "like a machine." With continued progress in cybernetics and post-modernism, anthropomorphizing metaphors used to explain technological developments increasingly focused on values precious to post-modernity—free will and creativity. From machines that helped us think, we started to think (and talk) about our mind as that which functions like machines.

This legacy helps us understand today's obsession with comparing the human mind and AI systems. But let us be clear. These are only metaphors—they don't tell us much about the ways our minds work; nor do they, in honesty, tell us about how AI systems function. In addition, the time feels ripe to revisit past constructed narratives. After all, if AI is everywhere, so is our need to rethink relationships to organic life. We have a unique opportunity here to redefine the values we give to AI entities, beyond human-centric concepts. Why wouldn't we use new metaphors that compare our brain to, say, organic connections found in trees, mycelium fungi, or gut bacteria? Why always compare it to machines? Artworks, we believe, can help identify socially and culturally constructed preconceptions—because most importantly, highlighting such preconceptions can help rethink needed new ones.

## REFLECTING ON MEANING: CHOM5KY VS. CHOMSKY AS A CASE STUDY

The VR and AI experience *CHOM5KY vs. CHOMSKY*, here presented as a case study, uses AI to demystify AI. Through virtual affordances and playful user experience (UX), the artwork questions the limits and opportunities of AI, while highlighting our human capacity for creativity, inquiry, and collaboration. As Chomsky suggests, we know little about our minds. So, the questions the artwork puts forward are: what exactly are we trying to replicate in AI? To what end? And what are we leaving behind?

Writers and thinkers, including Noam Chomsky (NC) himself, have warned us against a tendency for "faux-science" in AI—a tendency we believe is also detrimental to public participation. In particular, Chomsky rejects ongoing promises made for an "imminence of revelation in understanding ourselves and the world" (Chomsky, Roberts, Watumull, 2023; Marcus and Luccioni, 2023). His objections find an echo in Bender and Koller (2020), who argue that the success of large language models is partly attributable to words used to describe them as "intelligent," "learning," and "understanding." Bender insists that the questions for AI scholars should be why it is designed to make us believe it has thoughts and feelings in the first place!

> "Applications that aim to believably mimic humans bring risk of extreme harms [...] synthetic human behavior is a bright line in ethical AI developments, where downstream effects need to be understood and modeled in order to block foreseeable harm to society and different social groups".
>
> (Bender, 2021)

Obscuring the line between mathematic formulas and sense-making processes is harmful at the most, and false at the minimum. To help change this prerogative, the theoretical frame of cultural drift, driven from the sociology of social change and developed by Herbert Blumer in the 1950s, is helpful. For Blumer, social change was too often analyzed through organized structures (social movements, governmental decisions, public regulations). Social change, he argued, happened at far more intricate levels: through cultural and artistic developments, through the emergence of new forms of consumption, or through the popularization of new values in younger generations (Melucci, 1996; Rodriguez, 2013). Such cultural drfits are hard to observe, and yet seem much more pervasive in time—as they influence the values, imaginations, and collective references of entire generations.

For research, making sense of social change thus also means to circumscribe it as an analytical object: paying attention to how individuals create and challenge a particular set of values and "meanings" at a given moment, or within a given space or experience (Dervin and Frenette, 2001, Rodriguez 2013). This theoretical framework is here used to help us think of *meaning* not as a given, but as something constantly evolving: where art contributes to "cultural drifts"—changing meanings given to AI, for instance.

From a linguistic point of view, Chomsky's theory on generative grammar has ironically been criticized for not allowing enough room to analyze meanings (Jackendoff, 2002). Others rather suggest (Bhava, 2008) NC's theory on language ties metaphors directly to meanings. Chomsky's own extensive wake of conferences and interviews demonstrate how much he uses metaphors himself, building bridges across domains (political, organizational, psychological) and using everyday examples to explain complex issues (for instance, comparing AI to useful snow plows). In sum, Chomsky's own human creativity enables him to respond and adapt metaphors to interviewers and context—creating new meanings each time. The project *CHOM5KY vs. CHOMSKY*[5] plays on this very concept, inviting visitors to explore what meanings we collectively give to AI and help them rethink the future they want for it.

The research-creation was conducted over 2016–2023 and combines speech-to-text and text-to-speech (STT), NLP conversational systems (GPT-2, GPT-3.5) and VR. Users put on a VR headset to enter a computer-generated world, where they soon meet an AI guide, called *CHOM5KY*. The AI character is built from the vast number of digital traces of renowned MIT Professor Emeritus of Linguistics, Noam Chomsky, found online or from the MIT Archives Special Collections. Born in Philadelphia in 1928, NC is one of the world's most recorded and digitized public intellectuals.

Interviewed on almost every subject for over 60 years by journalists, farmers, and students alike, his talks have been transcribed, digitized, and uploaded online, leaving an astronomical pool of data we can use to train an AI system. Perhaps less known, NC's work on Natural Language also partly inspired natural language processing (NLP), which allows us to interact with chatbots. NC's wake of digital traces therefore paradoxically helps us build an AI system that can discuss AI, by using Chomsky's own words and sentences, and that still keeps true to Chomsky's insisting on AI's lies and limitations when it comes to replicating the human brain.

CHOM5KY the AI thus becomes a character—a perfect case study and perfect guide to help demystify AI. The AI character never pretends to be Chomsky himself.[13] It rather uses the opportunity of visitors anthropomorphizing it to challenge important human-centric questions: why do we want AI to sound or feel human in the first place? As the conversation with CHOM5KY progresses, CHOM5KY shifts the virtual world to invite visitors to discover his "back-end system." Still in VR, a user can then see in real time CHOM5KY's answer being constructed. By tapping onto words stringed together in the form of a response, they see changes in intent prediction percentage, sentiment detection, and tone evaluation. The goal here is to demystify current uses and applications of LLM like ChatGPT—instead of pretending the AI "comes up with answers," we rather show how words are chosen, organized, and predicted—we burst off the magic bubble of chatbots, and use it as an opportunity to show visitors how such LLMs work (see Figure 17.2). As a narrative guide, CHOM5KY ironically, keeps commenting on our findings: "Can you guess why I predicted this answer? What could I answer differently?"

Of course, as an artistic interpretation, the experience takes creative liberties. As an activist, NC challenged big corporations that limit human creativity. As a linguist, he insisted large language models do not help understand the mind.[14] We felt it paramount that CHOMSKY-AI would also remind us about unique traits of human intelligence—our capacity to inquire, collaborate, and create.

The experience was presented to the public in three different locations: Sundance Film Festival (2020), Berlin KINDL Center for Arts (2022), and Montreal Quartier des Spectacles (2023). Lessons learned from observing public interactions with CHOM5KY proved fruitful in analyzing how individuals interact with artificial entities. For instance, initial interactions usually quickly anthropomorphized the AI guide, in spite of CHOM5KY itself rejecting this perspective: "how are you?", "are you lonely?", "where do you live?", seemed like a way for users to play a sort of Turing test, pushing limits and testing "breakability" of the bot. CHOM5KY, our AI, constantly denied this perspective: "You can call me that if that amuses you, but that's hardly an accurate description."

As the experience progressed, and as visitors increasingly could "peek" inside the generative answers system, they started to ask more questions about *how* the system worked: "what is your system built on?", "who built you?", "are you Siri?", "are you owned by Open AI?", "what are algorithms?", "are you biased?", etc. After testing the chatbot and discovering how answers are stringed together by the system, a third part of the experience invited visitors to debunk algorithmic problem-solving. To do so, CHOM5KY invited visitors to play a game—a puzzle—not telling them its rules or how to play it. Visitors needed to become instinctively creative and collaborative

**FIGURE 17.2** By playing and tinkering with CHOM5KY's answers, the experience helps visitors grasp the basic logics of LLM chatbots.

**(© Sandra Rodriguez, National Film Board of Canada and SCHNELLE BUNTE BILDER.)**

to solve it. Meanwhile, CHOM5KY the AI (quoting real Chomsky interviews), reminded them that collaboration and creativity are really hard, if not impossible, tasks for any AI system to complete. Surprisingly, the more visitors could understand or question the differences between algorithmic vs. human solution-finding, the more they started treating CHOM5KY like an oracle: "what is the future of humans?", "will you soon be able to feel?", "can animals dream?" Users, in other words, were more keen to test values shared *with* other users, seeking to understand *how* it was reflected in the bot's answers: "what is your belief in nature?", "who scripted that belief?", and "what does consciousness mean to you?" Some even simply asked it to reveal our collective future: "who do you think should win the elections?", "will we survive climate change?", "will AI rule the world?", etc.

From observing visitors' interaction throughout different stages of conversation, inquiry, collaboration, and creativity, we learned the importance of creating experiences that help a public pause, and look beyond the surface of what they're expecting of AI systems. In particular, the findings highlight the importance of play, as a particularly effective way to challenge citizens' preconceptions and help them rethink where they stand on current issues related to AI. As CHOM5KY increasingly helped visitors notice the presence of other users around them, it helped them question why we are expected to interact with AI individually in the first place. Finally, the use of

**FIGURE 17.3** On the left image, the traces of Noam Chomsky's words and digitized sentences are seen as forming blocks that shape and create the VR world. To the right, a user can build their own structure and natural landscape from the blocks of data created from one's own traces (questions asked during the experience, topics, intents, and sentiment detected, etc.)

**(© Sandra Rodriguez, National Film Board of Canada and SCHNELLE BUNTE BILDER.)**

embodied UX encouraged users to reconsider their own relationship to data in a non-intellectual way, inviting them to rethink the values and meanings we give to current uses of our digital traces (Figure 17.3).

## ON METAPHORS: WHAT ARE WE LEAVING BEHIND?

CHOM5KY vs. CHOMSKY is far from being the only contemporary artwork to question AI. In *Conversations with Bina48*, artist Dinkins[15] exchanges with an android using a system based on the beliefs and mannerisms of a human, to test how AI affects minority groups. In Ai-Da, a bank of words and speech patterns produce a work "reactive" to Dante's *Divine Comedy.* Yet AI artworks can sometimes be disregarded as highbrow. A key lesson from CHOM5KY vs. CHOMSKY is that imagination is not bounded by limitations. Why not make the experience weird while thoughtful? Playful and reflexive? Art, we believe, is not only revealing of our sense-making processes; it can be transformative of how we conceive AI systems and metaphors used to think of them.

Inviting audiences to critically think about metaphors, vocabularies, and cultural assumptions on AI has proven beneficial to foster new forms of public participation in creating human-centric AI. Through making the experience accessible to all, we were able to realize how embodiment and play could help democratize access to a wide variety of citizens into the conversation. The virtual world enables visitors to touch, tinker, and see other users interacting in the world, with the AI system. VR thus enabled us to create a world that morphs according to users' presence and behavior—if they dig deeper into how a system works, for instance, or rather contemplate their surroundings. And they keep seeing others do it too, helping them feel like they too can be part of a collective questioning.

I truly believe play and humor can help demystify the role we all take in shaping technology. In previous artwork, I have often personally chosen to misuse technology to debunk myths about it. In *Do Not Track* (Peabody Award 2016)[16], for instance, our team of creators decided to track its own audience to discuss data tracking and Big Data economy. The experience was playful and a clear success on an attendance level (it went over to win the Peabody for best interactive web series that year)—yet it was also used by the European Parliament to discuss tracking regulations, and is still used in universities and schools, to this day, to educate a younger generation on healthy data management and the social consequences of data tracking. In *MANIC VR* (winner of the Golden Nica at Ars Electronica 2019), we used simple embodiment affordances of VR to give users a feeling of the cyclical effects of manic depression. The UX was playful, exhilarating, and then frustrating—yet here too it helped psychiatrists use it with patients as well as in a series of public events, to demystify the disorder. This time around, I wanted to create an opportunity to demystify AI, while challenging complex related subjects such as free will, creativity, collaboration, etc. In revisiting Chomsky's written archives for the project, a message parallel to technology, parallel to ongoing debates about artificial intelligence, became quite salient: it's our capacity to wonder, Chomsky keeps reminding us, that makes humans so creative. When visitors interact with CHOM5KY, I wanted the avatar to invite us to wonder and ponder on where this unique skill leaves us in human-centered AI—where do we leave all of this creativity in how we conceive AI's future?

Unsurprisingly, the more users could play within the virtual world, the more diverse their questions were. In current chatbot developments, this is often perceived as problematic. It makes "controlling" what the AI can respond hazardous: what if it starts using discriminatory words? What if it starts cursing at users? We believe instead of avoiding such issues by creating chatbots that respond to expectations (a marketing objective), we should rather show in transparency its flaws, biases, and the material it is trained on (an educational objective). In the CHOM5KY vs Chomsky unique UX approach, we never tell users what to do: visitors' own agency is put front and center. The limits of our own systems were many, but we let them become evident to the audiences—offering a balanced view for visitors to build their own judgment on AI limitations and opportunities.

## CONCLUSION

The technological advancements associated with AI today are conflated with marketing stunts, pop culture, and fiction-inspired collective imaginaries. In this context, whether AI is understood as a theoretical construct is secondary. What is more important is the need to re-evaluate collectively constructed imaginaries and vocabularies associated with "intelligence," "learning," and "understanding." After all, as the sociology of social change helps us highlight, the notion of Artificial Intelligence is not fixed. It is shared, discussed, debated, and changed through citizens' unexpected stunts with AI, through everyday use, and through ongoing technological developments—participating in a cultural drift that marks our perception of how we imagine AI's future together. By inviting visitors of *CHOM5KY vs. CHOMSKY* to demystify how AI functions, we strived to create a space that would enable them to rethink the values and meanings we collectively give to AI systems—inviting, in other words, to challenge them.

Sociological works analyzing human and AI interactions too often understand uses as logically driven by technology. But humans are messy: we disrupt, rethink, repurpose. It is by testing and breaking systems that we reimagine applications—changing technology radically.[17] The first part of this chapter helped revisit how art and entertainment experiences can be a unique asset to foment public debate and participation, as they first help level the knowledge necessary to take part in the conversation, and additionally help revisit the way we speak of a AI: how we create metaphors to give meanings to functions, attributes, and uses. Yet in using a cultural drift theoretical framework, we also stressed the importance for policymakers, researchers, and artists alike to enhance and support projects that specifically focus on educating citizens on their agency in decision-making processes: our human capacity to contribute to an ever-evolving cultural drift.

The second part of this chapter stressed the transformative capacity of art to bring more public participation into a needed conversation on AI's future. To do so, we revisited lessons learned from the research-creation project CHOM5KY vs Chomsky, and in particular, we drew lessons from observing user interaction with CHOM5KY AI. As expected, visitors initially tested the bot's (and the world's) limits, playing a sort of Turing test with our AI guide. But as they increasingly got to test, tinker, and play to understand how the bot functioned, the less they anthropomorphized our bot, and the more focused they were towards its construct—how it worked, what it could do, who trained it, was it biased, how, etc. The multiplicity of questions asked of CHOM5KY by a variety of users in three different world locations helped train our first model and better gauge what limited users' conversations. From 2020 to 2022, the experience changed from a simple conversation with CHOM5KY to including in 2023 more creative and collaborative play—helping rethink the values we give to our shared futures with AI.

A third part of the chapter insists on a last and important key finding—and that is the transformative capacity of art to shift collective conversations, and create spaces where citizens are free to rethink preconceptions, challenge, them, and question them. Of course, let us be realistic: research-creations and artistic projects are strange. Weird. Even if VR helps lower the costs of building robots, creative VR experiences can also be tedious, and costly. The entire process of this research-creation took six years. But we believe it helped gauge important opportunities and limitations of human/AI interactions.

By choosing to address important issues through storytelling and play, we crafted a space where visitors could *wonder*—about machines, about how AI really differed in capacities to human collaboration and creativity, and in particular, where they could rethink what they hoped for the future of AI. CHOM5KY vs. CHOMSKY is not a didactic experience. We wanted people to question, enquire how AI works, and learn about machines and about human minds. The VR worldview is itself filled with metaphors where machine programming is compared to the flight patterns of bees, where the unknown of neural connections is described like the bacteria in our guts. As we look back on the goals of the experience, we consider public participation endeavors should learn from playful explorations and experiences, as they create a unique opportunity to let preconceptions aside, and create a space for citizens to feel and test their capacity, interests, and the limits to their agency. Art, we believe, has helped reveal sense-making processes necessary for social change. But our most important lesson from the experience is that art can also be sense-creating.

Emergent, unexpected storytelling experiences, such as CHOM5KY vs. CHOMSKY, attract visitors because they are unexpected and weird. Researchers, artists, and policymakers are not often striving for weirdness when hoping to foment public participation, citizen engagement within AI debates, or even citizen AI literacy. Yet there is much to learn from what art can bring to the table. At a time where ChatGPT keeps making headlines and is constantly presented as creative,[18] flirty, or even having deep feelings, CHOM5KY vs. CHOMSKY decided to create a space where visitors and citizens alike can debunk and demystify how such systems function—helping rethink complex notions, helping question their sense of self, of machines, and of each other. As such, an overarching lesson from this research-creation is that art can be used to nurture the specificity of human agency in current technological AI advancements. Why? Because our everyday practices drive meaning.Because art, in its many forms, helps us pause, reflect and question such practices and meanings. And because our meanings drive future shared concepts, agency, and values. Let us not forget that we humans must remain the architects of our own future. Artists, policymakers, and scientists should have a role to play in creating alternative imaginaries and alternative paths to human-centered AI. By highlighting the power of collective choices, we simultaneously make salient the beauty and interdependence of our organic and artificial worlds. We have so much to wonder about; the conversation is just starting.

## NOTES

1 See Adams (2003).
2 The phrase was popularized by American futurist Ray Kurzweil (2005).
3 In March 2023, Yann Lecun shared on Twitter: "The *only* reason people are hyperventilating about AI risk is the myth […] that the minute you turn on a super-intelligent system, humanity is doomed. This is preposterously stupid and based on a *complete* misunderstanding of how everything works."
4 The image was created by Pablo Xavier (last name unrevealed), a 31-year-old construction worker from Chicago. See https://techcrunch.com/2023/03/29/ai-pope-midjourney-goncharov/ or https://www.forbes.com/sites/danidiplacido/2023/03/27/why-did-balenciaga-pope-go-viral/?sh=2ef6e38b4972
5 One could question if these warning cries are part of marketing strategies.
6 Co-produced by the National Film Board of Canada (Montreal) and Schnelle Bunte Bilder studio (Berlin), directed and authored by Sandra Rodriguez. It received support from Medienboard, Sundance Institute, John T. and Catherine D. MacArthur Grant for Non-Fiction Storytelling, MIT Libraries and MIT Open Documentary Lab. See Rodriguez (2022).
7 2022 has seen the emergence of artificial intelligence programs that create images from textual prompts, scrape the Internet and find patterns, and replicate and imitate.
8 Recent reports flagged this phenomenon under the term "pseudo-AI" Solon (2018).
9 The metaverse has been described in science fiction as a hypothetical iteration of the Internet away from screens and into an immersive world. Recent developments in VR and AR, WebVR, and Meta (previously Facebook), have seen huge investments in VR platforms

10 R&D, testing, and development of VR still take time and money.
11 MetaHuman, by Unreal Engine, promises to create high-fidelity digital humans
12 See early responses to the release of OpenAI's ChatGPT and Microsoft's Bing chatbot.
13 Deepfake technology was here used to adapt a voice that resembles that of Professor Chomsky's, yet is left fake or robotic enough, so that one could never confuse the bot for the real person.
14 See "Noam Chomsky: The False Promise of ChatGPT" New York Times, March 8, 2023
15 https://en.wikipedia.org/wiki/Stephanie_Dinkins
16 See referred work Do Not Track 2016 - https://donottrack-doc.com/en/episode/5; and Manic VR 2018: https://ars.electronica.art/outofthebox/en/manic-vr/
17 One can here think of the invention of the World Wide Web and what social media, the Internet, and meme culture have rather become.
18 In an interview for a recent The New York Times article, "8 Big Questions about AI", Sebastian Thrun is quoted saying either AI is becoming creative, or creativity is overrated. Yet the definition he gives of creativity goes against learnings in cognitive science, and against any definition an artist would give on creative work.

## REFERENCES


Adams, Tim. 2003. "Question Time," *The Guardian* (30 November). Online. https://www.theguardian.com/books/2003/nov/30/highereducation.internationaleducationnews

"Ai-Da." (n.d.) Online. https://www.ai-darobot.com/

Audry, Sofian, 2021. *Art in the Age of Machine Learning*. Cambridge, MIT Press.

Bender, Emily M. and Koller, A. 2020. Climbing towards NLU: On Meaning, Form, and Understanding in the Age of Data. In *Proceedings of the 58th Annual Meeting of the Association for Computational Linguistics*, Online. Association for Computational Linguistics.

Bhava Nair, Rukmini. 2008. "Noam Chomsky's metaphors as a dialogue across disciplines" Rukmini Bhaya Nair|Indian Institute of Technology Delhi, India.

Bonarini, A. 2020. "Communication in Human-Robot Interaction". *Current Robotics Reports*, 1, 279–285. https://doi.org/10.1007/s43154-020-00026-1

Chomsky, Noam, Roberts, Ian, and Watumull, Jeffrey. 2023. "Noam Chomsky: The False Promise of ChatGPT", *The New York Times* (March 8, 2023).

"Conversations with Bina48." (n.d.) Online. https://www.stephaniedinkins.com/conversations-with-bina48.html

Dautenhahn, Kerstin et al. 'KASPAR – a Minimally Expressive Humanoid Robot for Human–robot Interaction Research'. 1 Jan. 2009: 369–397.

Dervin, Brenda and Micheline Frenette. 2001. « Sense-making Methodology: Communicating Communicatively with Campaign Audiences », in: R.E. Rice et C. K. Atkin (dir. publ), *Public Communication Campaigns*, Thousand Oaks, CA: Sage, pp. 69–87.

Gentelet, Karine and Mizrahi, Sarit K. 2023. A Human-Centered Approach to AI Governance: Operationalizing Human Rights Through Civil Participation. in: C. Régis, J.-L. Denis, M.L. Axente, & A. Kishimoto (Eds.). *Human-Centered AI: A Multidisciplinary Perspective for Policy-Makers, Auditors, and Users* (1st ed.). Chapman and Hall/CRC. https://doi.org/10.1201/9781003320791

Häring, Markus, Kuchenbrandt, Dieta, André, Elisabeth. 2014. "Would you like to play with me?: how robots' group membership and task features influence human-robot interaction" *HRI '14: Proceedings of the 2014 ACM/IEEE international conference on Human-robot interaction* March 2014, pp. 9–16.

Jackendoff, R. 2002. *Foundations of language: Brain, Meaning, Grammar, Evolution*. Oxford University Press, USA.

Katz, Yarden. 2012. "Noam Chomsky on Where Artificial Intelligence Went Wrong: An Extended Conversation with the Legendary Linguist," *The Atlantic*. Online. https://www.theatlantic.com/technology/archive/2012/11/noam-chomsky-on-where-artificial-intelligence-went-wrong/261637

Katz, Yarden. 2017. "Manufacturing an Artificial Intelligence Revolution," *SSRN*. 27 Nov. Online. http://dx.doi.org/10.2139/ssrn.3078224

Koay, Kheng Lee, Syrdal, Dag Sverre, Dautenhahn, Kerstin and Walters, Michael L. 2020. "A narrative approach to human-robot interaction prototyping for companion robots," *Paladyn, Journal of Behavioral Robotics*, 11(1), 66–85. https://doi.org/10.1515/pjbr-2020-0003

Kurzweil, Ray. 2005. *The Singularity is Near: When Humans Transcend Biology*. London: Penguin.

Lafontaine, Céline. 2004. *L'empire cybernétique. Des machines à penser à la pensée machine*. Paris: Le Seuil, 238 pages.

Ma, Lin and Aihua, Liu. 2008. "A Universal Approach to Metaphors", *Intercultural Communication Studies*, XVII(1), 260–268. Harbin Institute of Technology.

Marcus, Gary and Luccioni, Sasha. 2023. "Stop treating AI models like people: No, they haven't decided to teach themselves anything, they don't love you back, and they still aren't even a little bit sentient," https://garymarcus.substack.com/p/stop-treating-ai-models-like-people

Melucci. 1996. *Challenging Codes: Collective Action in the Information Age*. Cambridge University Press.

Pietroski, Paul, 2021. "Chomsky on Meaning and Reference", in *A Companion to Chomsky*, Nicholas Allott, Terje Lohndal, Georges Rey (Eds.), John Wiley & Sons, Inc.

Rodriguez, Sandra, 2013. "Making Sense of Social Change: Observing Collective Action in Networked Cultures", *Sociology Compass*, December. 02 2013, Available online: https://compass.onlinelibrary.wiley.com/doi/abs/10.1111/soc4.12088

Rodriguez, Sandra, 2022. "Chom5ky vs Chomsky: A Playful Conversation on AI", Art work produced by the National Film Board of Canadan and Schnellebuntebilder: https://www.nfb.ca/interactive/chomsky/

Roose, Kevin. 2002. "We Need to Talk About how Good A.I. is Getting", *The New York Times*, August 24, 2022.

Solon, Olivia. 2018. "The rise of 'pseudo-AI': how tech firms quietly use humans to do bots' work," *The Guardian* (6 July). Online. https://www.theguardian.com/technology/2018/jul/06/artificial-intelligence-ai-humans-bots-tech-companies

"The best AI chatbots seem like real humans, and that's scaring people," *South China Morning Post*, 6 December 2022. Online. https://www.scmp.com/lifestyle/gadgets/article/3202134/best-ai-chatbots-seem-real-humans-and-thats-scaring-people?module=perpetual_scroll_0&pgtype=article&campaign=3202134

# 18 Christina Colclough's Commentary

*Christina Colclough*
The Why Not Lab

Christina Colclough is the founder of The Why Not Lab, an organization whose mission is to ensure that the digital world of work is empowering rather than exploitative. She holds a Ph.D. from the University of Copenhagen and has extensive global labor movement experience. That makes her unusually well positioned to bridge the gap between theory and practice and lead the conversation on workers' digital rights. Colclough is included in the all-time Hall of Fame of the World's most brilliant women in AI Ethics. She is also a Fellow of the Royal Society of Arts in the UK, an Advisory Board member of Carnegie Council's new AI and Equality Initiative, and a member of the UNESCO Women4EthicalAI Platform and the OECD One AI Expert Group. She is affiliated with FAOS, the Employment Relations Research Center at Copenhagen University, and in 2021, she was a member of the Steering Committee of the Global Partnership on AI (GPAI).

**Please tell us more about your work.**

My main occupation is capacity building, that is to say, I give workshops and courses for unions and governments around the world. It can be courses for trade union leaders, negotiators, or organizers, or it can be for wider audiences, but each course is tailor-made to the specific organization I am working with.

These courses and workshops cover a significant amount of content. We explain what digital technologies are, what data is, what AI is, and so on and so forth. We also discuss how these new technologies affect the world of work. And we give actionable examples. For instance, we explain that in certain cases, it is possible to deduce what kind of instructions were given to the algorithm that runs behind an automated hiring tool. In addition to that, I sit on various boards around the world advising on issues related to the labor market. I am also part of a team that develops tech for good, that is, technologies that try to avoid the pitfalls of current digital technologies and, instead, empower workers. Lastly, I do a fair amount of public speaking.

**How do you define human-centered AI, and why do you think it's important, especially for workers, which is your main preoccupation?**

DOI: 10.1201/9781003320791-21

I have mixed feelings about that expression. Ensuring that digital technologies benefit those they are intended to serve is very important to me. But the concept of "human centeredness" is quite broad. Not specifying what people we are recentering AI around or why we are recentering AI around those people could lead to undesirable effects. We must not forget that what is good for one group of humans could be absolutely horrible for another. Moreover, human-centered AI could be used as a concept to perpetuate the status quo. "Capitalist-human-centered AI," so to speak, would be human-centric, but it would not be AI for the benefit of all.

Both of the papers I have read—and we will talk about them specifically in a minute—seem to suggest that "human-centered" means that no AI systems should be brought into the world without consideration for the people who will be affected by them. But as a standalone concept, I am not sure human-centered AI will be able to achieve that.

Nevertheless, I do believe that we should not introduce technology for the sake of technology and that we should always consider the consequences of a given technology on those it is going to affect. That way we can ensure that technology benefits those it is intended to serve. That is a large part of what I do with my work on the governance of digital technology.

**Could you give me examples of how, in your opinion, human-centered AI was implemented in some contexts and not implemented in others?**

We are still at the beginning of this journey. But one example that comes to mind when I think about human-centered AI is the Digital Bargaining Hub, a tool I have developed with Public Services International, the global union for public-service workers, and Professor Hannah Johnston at York University. The Hub is an online database of collective bargaining clauses related to the digitalization of work.

One thing the database shows is that workers are starting to negotiate around data and the introduction of digital technologies in the workplace. I am glad to see this is happening. It is a way for the Union movement to ensure that the introduction of AI technologies does not come at the expense of providing decent working conditions to workers.

But there are also gaps, especially with respect to certain uses that are made of AI. For instance, AI not only has immediate impacts, as in, your productivity goes down, your employer fires you; it could also have long-term repercussions. For example, if an AI system determines that a worker's productivity is declining and that that worker is a 52-year-old woman with a BMI of 30, and a long tenure in the company, the system might infer that this is a generalizable "fact" against which future applicants with similar profiles will be measured. This kind of future projection is an issue that unions are not grappling with at the moment. If unions do not understand algorithmic inferences, of course, they will not be able to put adequate guardrails in place. To prevent that, we are aiming at helping unions by writing model clauses they should include in collective bargaining agreements.

This is a good segue to a very important point that was mentioned in Lévesque et al.'s article, and which I want to emphasize: it takes two to tango. For meaningful negotiations to take place, both parties must "know what they need to know." And what the unions have found—especially those that I have done a lot of in-depth training with—is that when they go to management and ask for guardrails around the use of algorithmic inferences or the right to opt out of the selling of data, they are met with a certain kind of bewilderment. We tend to assume—and this goes for all of the literature on impact assessment and audits—that management understands AI and other technologies rather well. But it is not necessarily the case. In fact, one of the biggest mistakes many people make is assuming that managers really understand the technologies they are adopting or using.

**You read two articles from the book: "Crafting human-centered AI in workspaces for better work," by Lévesque and his colleagues, and "Human-centered AI for industry 5.0," by Passalacqua and his colleagues. Could you tell me a bit more about these papers and how they changed your point of view, the way you see things?**

Reading these papers confirmed several things that I think all politicians need to be reminded of. First of all, the framework presented in "Crafting human-centered AI" is very interesting. I like that it proposes a way to measure the various dimensions of the quality of work—or "better work," as the authors call it. I think that framework would be very helpful for politicians, as I mentioned, but also for unions. However, the authors could have taken their argument about inclusive governance a bit further. The framework works well to measure how AI systems can impact good work "here-and-now," but I would have liked to read more about how they think it ties into inclusive governance.

And this leads to another limit I see with this paper. It covers today's impacts quite well, but it doesn't really address the long-term consequences of AI, even though, as I mentioned earlier, they are, in my opinion, one of the most dangerous and threatening aspects of the digitalization of work. Take another example, that of an automated scheduling tool. It might benefit workers in the short term, as it could become easier for them to change their shifts or to ask for schedules that better suit their needs. However, these systems are collecting a lot of data, and there could be long-term negative consequences if this data were to be used to infer new information about employees. The system might conclude, for instance, that clients prefer workers who speak English with a British accent. This inference could then become a "generalizable fact" that would end up penalizing workers with foreign accents—they could be either fired or given the worst shift. Given the magnitude of these future harms, it would have been interesting to discuss them in the paper.

The second article I read, "Human-centered AI for Industry 5.0," was quite dense, and keeping track of all the different models presented in it was complex. But this chapter confirms that co-deployment and co-design really do pay off. And it clearly shows that organizations should not simply impose technology on workers. Instead, they should actually involve them in technology projects.

That said, I think something important is missing from the article. The authors conclude that AI will change the tasks, jobs, and skills needed, but they do not mention who should be responsible for reskilling and upskilling workers. For example, should employers who introduce disruptive technologies be obliged to continuously train disrupted workers, to help them onward in their careers? Right now, the situation mostly looks like a free-for-all: everybody can disrupt work, and nobody is taking any responsibility for the aggregated effects. Given how Passalacqua and his colleagues discuss how disruptive technologies should be introduced in cooperation with affected people, it would have been interesting to come up with policy recommendations or reflections on who's responsible for what when work and workers are impacted by AI.

**Speaking of that, what are possible recommendations you would make to policymakers and users, the 2–3 things that they really should pay attention to when we are talking about human-centered AI and its implementation in work settings?**

My first recommendation is directly related to what we just discussed. Employers who are introducing powerful technologies like AI should have "disruption obligations," that is, they should be forced to provide lifelong learning and career development opportunities to the workers affected by these tools.

My second recommendation is for mandatory inclusive governance. Right now, governments are leaning towards a market-access approach to regulating AI. This type of regulation heavily relies on certifications and standards. The EU–US Trade and Technology Council, the G7, the G20, and the European Commission all seem to agree to use the OECD AI Principles as a starting point, and they're trying to agree on what AI certifications and standards should include. However, none of them are discussing the stringent requirements of inclusive governance.

This is truly disappointing. First, the certification bodies (ISO, IEEE, and others), are 99% industry-dominated. Second, how do you certify something that, by nature, is fluid and changeable? In my view, the only way to justify doing so is to establish strong requirements regarding the elaboration and implementation of inclusive and transparent governance mechanisms. If you do not, how are you going to ensure that all other laws—antidiscrimination laws, working time laws, etc.—are respected? Take, for example, the situation of a biased automated hiring system. A company finds out the system it uses discriminates against young women. They ask the system's developer to fix this issue. The developer comes back and says: "We can do that with 89% accuracy." Is that OK? What about the remaining 11%? Inclusive governance mechanisms should be put in place to make that kind of decision: workers should be involved in it, the processes applied to correct the problem should be transparent to the authorities, etc. After all, at the end of the day, discrimination is illegal and must be fought.

I have one final and, I think, important comment to make. Both of the articles I read correctly expose that AI could free us of routine tasks. However, neither recognizes that these routine tasks that we will be liberated from might play a useful role in our lives. I met an Italian professor at the G7 some years ago, and he said something that made a strong impression on me: "We might all hate the routine tasks that are part of our jobs, but they make it possible for us to pause our brain." If we optimize our work so much that those brain pauses disappear, how will that play out in relation to burnout and stress? I think that this scholar had a very insightful point. We spend a lot of time talking about the benefits of automating routine tasks, but is that really going to be beneficial? Think about what we are experiencing now. The exponential growth of digital technologies goes almost hand in hand with the exponential growth of burnout and stress-related illnesses.

So I think that when we talk about the disruption obligations of organizations, we should indeed take time to reflect on what "good work" should be considered to be in a digital world. Should we have a four-day working week? Should we have a redistribution of working time and tasks? Should we have the right to disconnect? Personally, I would say yes to all of those things.

**This makes me think that working on an article's references or transcribing conversations is routine work that indeed helps me to recharge my batteries.**

For me, it's doing my accounting or travel expenses. These are tasks I can put some good music on and do, and just have that needed pause. French composer Debussy is famous for saying that "music is what is made between the notes." We should never forget how "our" music is made in the pause.

# 19 Joseph Nsengimana's Commentary

*Joseph Nsengimana*

Mastercard Foundation Center for Innovative Teaching and Learning

Joseph Nsengimana is Director of the Mastercard Foundation Center for Innovative Teaching and Learning. The Mastercard Foundation is a private Canadian foundation—one of the largest in the world—with a mission to advance access to education and deepen financial inclusion. The foundation's work has been focused on Africa since 2009, reaching hundreds of millions of people. In 2018, the foundation launched a strategy called Young Africa Works, laying out a bold plan to enable 30 million young people in Africa, particularly young women and groups facing the highest barriers to opportunity such as displaced or disabled youth, to access dignified and fulfilling work by 2030. The Mastercard Foundation Center for Innovative Teaching and Learning helps to drive this goal by advancing the use of technology to deliver at-scale access to relevant education and skills training, a precondition for dignified work. The education sector is likely to be significantly impacted by the mainstreaming of AI. This drives Mr. Nsengimana's interest in human-centered AI.

**AI can be used to train young people and fight unemployment, but at the same time, it could also make jobs disappear. I guess that you're interested in both sides of the equation?**

Absolutely. I think that AI has the potential to do a lot of good and create educational and economic opportunities, but only if we make the right investments now. So, at the Center, we're trying to figure out how AI will impact training, education, and job creation. And once we understand what the impact is going to be, we ask ourselves, "What do we need to do? How can we best address some of those challenges?"

Now, technology affecting the job market is not unprecedented. From automation to industrialization, every time there is a new revolution, every time a new technology is introduced, it has an impact on jobs. The difference now is the speed at which this is happening. The changes are happening so fast it's hard to get ahead of the transformation. That's the challenge: "How can we take advantage of this rapid transformation to get young people into employment?"

 DOI: 10.1201/9781003320791-22

**The book's focus is really on human-centered AI—a concept that is not always as well defined as it should be. How do you define it?**

To me, human-centered AI is all about the direction we give to the undergoing transformation. We know that transformation is going to happen, but how do we make sure it happens in a way that is beneficial to humanity, and to society at large? That's how I see human-centered AI: systems informed by the needs, cultures, and experiences of those it seeks to support or help; systems that are designed and implemented in a way that is transparent, accountable, and aims at ensuring that the positive outcomes outweigh the negatives.

**We don't talk much about what's going on with AI in Africa. I guess, in a way, because we don't know much about it. Could you tell me a bit about one or two projects in which you would say AI is used in a human-centered fashion?**

You're right. The lack of information is one of the main challenges for human-centered AI on the continent. AI relies on available information, and unfortunately, at this moment in Africa, there isn't much information that can be used to build AI systems that are truly human-centered. It would be great if more research on AI was done on the continent—and if more data was produced on the continent.

But to your point, there are some interesting African initiatives in the field of education. To employ the terminology used in Poellhuber et al.'s article ("Artificial Intelligence in Higher Education: Opportunities, Issues and Challenges"), they focus more on building adaptive learning tools than state-of-the-art AI, but still, they're related. For instance, I worked with a company in South Africa called Siyavula Education. They created this adaptive platform that teaches math, physics, and chemistry. The way it works is that the system guides the students, records how students are responding, and then, adapts the pedagogical content based on their responses. So Siyavula's platform has some elements of AI in its adaptive nature. That's an application of AI that seems quite beneficial to me. We've seen huge improvements, for example, in students' performances in math. And I think that as technology matures, we'll see more applications of this kind of technology, and its impact will be even more meaningful.

**You just mentioned one of the articles you read, Poellhuber et al.'s. What did you think of this article and the others you read? How did they change the way you think about AI?**

The articles I read are very informative. And I like the way they are written. They are accessible. Nonacademics can read them and understand them, and I think that's great. Especially given how mainstream AI has become, I think it's important to have these types of articles out there in the public. A lot of people fear AI. And often, it's because they don't understand it. But that's a problem because when people are scared of AI, there is a risk that their fears become self-fulfilling prophecies: they think it's dangerous, so they don't engage with the technology, so they don't generate data, and AI systems end up being dangerous.

That's something that came to my mind as I read Poellhuber et al.'s article and, also, Mele et al's paper, "AI-based service provision for an engaged university." I thought, let's hope this gets out to a broader audience so people can read it, be inspired by it, and actually contribute to the development and deployment of human-centered AI.

As to how the papers impacted my view of AI, let's say they reinforced the perception I already had, which is that there are a lot of opportunities that can come out of AI in education. But the articles also did a good job highlighting the dangers. For example, Poellhuber et al.'s article talks about ChatGPT and the fact that it relies on information available on the Internet up to the year 2021, and when I look back and see just how much misinformation was out there during that period of time … It's scary … It's scary to think that this is what is feeding any AI system.

**Is there something you didn't see in these articles that you would have liked to see? For example, things they did not address and that you're really worried about …**

There are a couple of things that caught my attention. First, in his conclusion, Poellhuber et al. ask if AI will eventually be able to replace teachers. And though they immediately follow-up by answering that question with a clear "No," I feel the question maybe should have been framed differently. Many people will start reading the paper with this preconception that, somehow, AI will diminish the role of teachers. But when you ask a question like that, even if you only do it rhetorically, you reinforce people's preconceptions. If they had rephrased the question and asked something like this instead, "How can AI enable teachers to better personalize instruction and therefore drive learning?" then I think they would have avoided reinforcing people's bias.

Second, I liked Mele et al.'s article, which talks about universities' role in society, but I was puzzled by something. The authors argue that a new role is emerging for universities—that they now have to engage with society. I think it might be interesting to take a moment to ask: ok, but is that really new? Universities have always done research that somehow impacts society. I have to admit that I struggled to understand why this is new now. The mission is not that new. It's just the tools that are different. The only new thing I see is the application of the technologies. We're just talking about the application of a new technology.

**I'd like to go back to something you mentioned earlier. You said that building large language models such as ChatGPT requires a lot of information available online. We know that ChatGPT is fed with vast quantities of English and French material, of course, but I guess not so much with content in African languages. To what extent is this an issue and how could it be corrected?**

I think you nailed it on the head. There just isn't enough quality information available. If you look at the data, what really is out there *from* the continent? The data that covers the continent … pay attention to where it's from. It's just as important. Who wrote what's out there? And how well did they actually understand the continent and its complexity?

When you go on ChatGPT and ask questions, I think you can see that the answers are not entirely correct. You can see that there are gaps. Worst, as Poellhuber et al. mention, it's not only that the information isn't available, it is also that these AI models actually make up answers when the information isn't available.

Mind you, it's not necessarily a problem with the technology itself. The problem is in the data we're using. So, what's the solution? More research, more data production and publication should be done by experts—*local* experts who understand the context in which the information and AI will be used. I think we need to generate more local data to counter this problem.

Again, that's why I said that articles like the ones in your book are so important, because the more people read them, the more they understand their role in generating data that will feed AI models/platforms, and in turn, they will provide more accurate responses.

**What recommendations would you make to policy-makers, auditors, users, etc., to strengthen the development and implementation of human-centered AI in our societies—in African societies, especially?**

Policy-makers or auditors should require more transparency of the developers, they should disclose the source of their data, what's in the datasets, and what are the key elements informing some recommendations/decisions that the AI systems are making. I think that knowing where the data is coming from would allow people to decide whether or not they can trust what the AI systems are proposing. So more transparency would be at the top of my recommendations.

Like it or not, AI is going to impact every aspect of our lives. So, let's make sure that people understand what AI is and how it works. Take ChatGPT, it can generate false information, but the way the information is presented is so convincing that you don't even question what you are reading. It will throw in quotes, mention a couple of experts you have already heard of, and if you're not aware of how it works, you'll believe it. But if you understand where it's getting the data, if you know the pitfalls of the technology, then, you won't be as likely to fall for it. So, in addition to transparency, you need education so people can understand these systems: what to expect, what to watch out for, and so on. So, yes, education is another thing that I would recommend.

My third recommendation is more global, not necessarily focused on the continent. It's about ethical guidelines. There are a lot of people developing AI systems, and there will be even more. What rules are they subjected to? Right now, there aren't any. Everybody is doing whatever they want. We're hearing this from folks who were at the forefront of AI's development about its dangers. So I think there's a need to agree on ethical guidelines regarding these systems.

# *Section III*

## *Lessons Learned and Promising Practices*

# 20 A Human-Centered Approach to AI Governance

## *Operationalizing Human Rights through Citizen Participation**

*Karine Gentelet*
Université du Québec en Outaouais, Gatineau, Québec, Canada

*Sarit K. Mizrahi*
University of Ottawa, Ottawa, Ontario, Canada

## INTRODUCTION

Advances in artificial intelligence (AI) have wrought a surge in human rights violations that are increasingly preoccupying the minds of citizens and policymakers worldwide (Raso et al., 2018). Algorithms created to regulate online speech instead censor content related to religion, sexual diversity, and far beyond (Penney et al., 2018). AI systems developed to detect cancers (Lashbrook, 2018; Madhusoodanan, 2021) or assess the flight risk of criminal defendants are prejudiced against Black people (Angwin et al., 2016). Facial recognition technologies deployed in high schools misidentify students of color as security threats (Simonite et Barber, 2019). The examples abound. Marginalized communities are recurrently and disproportionately impacted by human rights violations at the proverbial hands of AI technologies.

With a growing body of research illustrating the urgent need to devote increased attention to respecting human rights in the AI context, numerous initiatives are underway to define governance tools for the use of these technologies. Some advance ethical frameworks or principles for responsible AI (Council of Europe; World Economic Forum; Germany, 2019; Australian Government, 2019). Others attempt to develop tools that seek to assess the human rights impacts of these technologies

* This research was conducted as part of the Chair in Social Justice and Artificial Intelligence - Abeona-ENS-Obvia (2020–2022)

DOI: 10.1201/9781003320791-24

throughout their conception, development, and deployment (Government of Canada; Metcalf et al., 2021). All these approaches, however, tackle AI systems in and of themselves. They neither reinforce human rights in the use of these technologies nor reflect on the needs of the individuals affected and their potential recourses. In other words, these initiatives address technological challenges rather than societal ones.

The primary reason for this shortcoming is that the source of algorithmic discrimination was, until very recently, thought to be technical in nature (Crawford, 2021). Efforts to debias algorithms therefore relied on solutions that were also technological: neutralizing training data (Chander, 2017), developing ethical AI (Council of Europe), or controlling algorithmic risks.[1] But as the literature has come to demonstrate, the bias propagated by AI is—at its very core—a *systemic issue* rooted in inequitable social and power relations (Boyd et Crawford, 2012; Birhane, 2021: 2; Lupton, 2018: 6; Noble, 2018: 6; Benjamin, 2019; Broussard, 2018; Eubanks, 2018). In this sense, any mode of AI governance that focuses on tackling bias through strictly technical means will likely do little more than give the illusion of legal compliance without addressing the foundational elements that give rise to these algorithmic injustices (Zimmermann et al., 2020).

To truly seize AI's discriminatory effects, we must prioritize a system of AI governance that goes beyond assessing the impacts of these technologies, instead positioning human rights as an overarching legitimizing framework. It's precisely such a system that we propose in our chapter—one that empowers the people affected by the use of AI systems by focusing on human rights as mobilized *by them for them*. We begin by situating algorithmic bias in its preexisting societal context, illustrating how these biases are socially (rather than technologically) produced and merely amplified by AI. We proceed by demonstrating how the normative solutions currently favored to address this issue fall short of attacking the root of the problem. Because they protect entitlements to outcomes defined through top-down processes that don't respond to citizens' lived experiences, these solutions remain unable to provide them with any concrete protection; to permit them to leverage their human rights in ways that respond to their needs. Achieving this feat, we advance, requires us to instead reshape the AI governance narrative by positioning human rights as an *operative* principle—one that includes the social, economic, and cultural dimensions of human beings and their livelihoods, while simultaneously advancing an approach to AI governance that's grounded in citizen action. It's a system that supports the active, voluntary, and deliberate participation of citizens in the governance of technological tools; that allows people whose rights are infringed by AI systems to mobilize their lived experiences through concrete processes of participation and legal recourse founded on the agency and autonomy of each and every member of society (Chateauraynaud, 2021).

## INVISIBLE SOCIAL CONTEXTS

Oftentimes developed in environments that favor dominant hegemonic structures at the detriment of other narratives, AI is notorious for being biased (Crawford, 2021). This issue is due in part to the data used to train them, which often misrepresent groups that have been socially, culturally, and historically marginalized (Zimmermann et al., 2020; Richardson et al., 2019; Harrington, 2020). While many have suggested "neutralizing" the data on which algorithms are trained to overcome their obvious propensity for discrimination (Chander, 2017), we cannot—in many situations—*debias* data because they are drawn from preexisting social structures in which discrimination still lives strong. And AI quite simply amplifies this bias due to its overvaluation of data.

In some circumstances, minorities are *underrepresented* in AI datasets in ways that contribute to their *invisibility* in society. It's recently been highlighted, for instance, that healthcare applications designed to diagnose melanoma are incapable of detecting it in people with darker skin tones (Lashbrook, 2018; Madhusoodanan, 2021). While trained predominantly on images of skin cancer in White people, the underrepresentation of people of color isn't the result of some oversight. Rather, it's because similar images of melanoma in people with darker skin tones are lacking, primarily due to their inequitable access to healthcare (Morosini, 2021; O'Donnell et al., 2016). The bias propagated by skin cancer detection apps therefore reflects preexisting societal structures that place less emphasis on the needs and health of minorities in ways that contribute to their *invisibilization*.

In other cases, minorities are *overrepresented* in AI datasets in ways that contribute to their *hypervisibility* in society. Such is the case, for example, with the risk-assessment algorithm used by several American courts (Equivant). But an investigation by ProPublica journalists led to the discovery that this algorithm is biased against Black people, more likely to classify a Black person as high risk than it would a White person with a much more violent criminal history (Angwin et al., 2016). And this miscalculation is a direct result of Black peoples' *overrepresentation* in its training data, which is inextricably linked to their overrepresentation in the criminal justice system (Raso et al., 2018).

And contributing to marginalized groups' misrepresentation in machine learning datasets is the fact that the data used to represent them are drawn from Big Data, collected across multiple levels without any consideration to the social context surrounding their production (individual, local, institutional, and national) (Benjamin, 2019; Noble, 2018). Moreover, when being prepared for machine learning processes, these data are labeled and categorized based on a variety of classifications (such as gender, socioeconomic status, or ethnicity) but rarely from an intersectional perspective that could actually help connect the various underlying dimensions of peoples' lived experiences.

Take, for instance, the social interactions generated by a woman from a racialized group, who is also a single mother and a medical caregiver during the COVID-19 pandemic. Use of her data by algorithms won't be able to capture their social dimensions because they're disconnected from the context of a worldwide health crisis and therefore unable to capture the fact that these experiences have a significant impact on her own path, as well as the paths of women sharing similar characteristics. This issue, therefore, doesn't stem from the impacts surrounding AI's use (Andrus et al., 2021). Rather, it arises from the classification of the data and the choice of indicators that are based on collective representations (Zimmermann et al., 2020).

While it's true that data are collected on an individual basis, the choice of creating links between particular socioeconomic indices is based on common characteristics that remain specific to societal modes of organization (Delacroix, 2022: 12) and to indicators used to define the paths of individuals, such as level of education or criminal record. Linking these indicators can have contradictory effects when operationalized by machine learning algorithms (Ibid) and can have discriminatory impacts on people sorted on the basis of these indicators. Moreover, when these datasets are used for secondary purposes, they become disconnected from the context in which they were produced. In the case of collective impacts on large groups of people sharing common characteristics, it becomes important to document these systemic issues on an institutional level and to determine the root of these societal biases to remedy them.

Once we become aware of the impacts of AI tools on entire populations, it becomes impracticable to perpetuate a mode of responsibility that exists at the individual level alone. Rather, responsibility becomes collective since the impacts are societal and, through automation, become systemic (Zimmermann et al., 2020). Grasping this collective dimension is therefore fundamental. To do so, however, we must begin by seizing the human person in an integral and holistic fashion by considering her social identity above all; by contextualizing her identity based on her position within society (Birhane, 2017). It's crucial to consider the societal ecosystem that contributes to defining this social identity (based on a multitude of socioeconomic or racialized factors, for instance) to determine to which degree these groups will be impacted by the use of algorithms. Unfortunately, however, this collective dimension doesn't feature explicitly enough in the regulatory tools currently developed to address the human rights impacts of AI systems.

## THE DISCONNECT BETWEEN HUMAN RIGHTS PROTECTION AND NORMATIVE APPROACHES TO AI

With society's increasing awareness of AI systems' propensity for discrimination (Shull, 2018; Follows, 2021: 19), industry and government alike have created normative frameworks geared toward ensuring that they're designed, developed, and deployed with an eye to human rights. But there's an important distinction to be made between governance frameworks elaborated with an eye to human rights and protecting rights *as* rights (Kennedy, 2022: 178). And in neglecting to make this distinction, the normative frameworks currently favored to address AI's human rights impacts remain unable to offer citizens any concrete protection to this effect.

These frameworks generally come in one of three forms. The first is ethical codes, which advance a certain set of values to be respected in the creation of AI systems (Yeung et al., 2019; Council of Europe, n.d.; World Economic Forum, n.d.; Germany, 2019; Australian Government, 2019). In view of promoting responsible AI, most ethical codes emphasize that AI systems should be transparent, explainable, and contestable (Germany, 2019; Australian government, 2019; OECD; Malta Digital Innovation Authority; the Netherlands; Switzerland, the Federal Council; European Commission). Equity, nondiscrimination, and justice are also commonly favored values (Germany, 2019; Australian Government, 2019; OECD, 2019; The Netherlands, n.d.; Switzerland, The Federal Council, n.d.; European Commission, 2019; China Daily, 2019). While specifying that AI systems must be inclusive and accessible, most ethical codes recognize that algorithmic bias is difficult to eliminate entirely, only requiring that it be reduced as much as possible.

While all these values are laudable, ethical codes remain difficult to implement. And the reasons are threefold. First, ethical principles vary across the spectrum.

While some do recur, context and culture play a large role in defining ethical standards and current approaches fail to account for such elements (Yeung et al., 2019). Second, these codes seek to advance certain outcomes deemed desirable but do little to permit individuals to *concretely* enjoy their human rights (Gentelet et Mizrahi, 2021). Third, ethical frameworks are voluntary codes of conduct that lack the necessary enforcement mechanisms to ensure accountability for human rights violations (Yeung et al., 2019). Often referred to by critics as ethics washing, this approach does little more than delay the development of concrete human rights protections in the AI context (Waldman, 2021). Ethical frameworks, in other words, make it *appear* as if the values they advance are at the forefront of AI development, but it's little more than a smokescreen.

The second normative approach to regulating AI is algorithmic impact assessments ("AIAs"), which is a governance practice that seeks to render visible the harms propagated by AI systems and to ensure that practical steps are taken to rectify those harms prior to deployment (Metcalf et al., 2021). They consist of a questionnaire that allows developers to test the impact level of a particular algorithm. Considered to address the issue of algorithmic bias, it is increasingly being promoted as a primary AI governance tool (Karlin et Corriveau, 2018; Kaminski et Malgieri, 2020); Mantelero, 2018; Raji et al., 2020; Reisman et al., 2018; Selbst, 2018; Government of Canada, n.d.).

Although AIAs certainly seem more proactive than ethical frameworks in that they seek to assess and address algorithmic harms prior to deployment, they're similarly ineffective at advancing human rights. The reason is that AIAs "are not neutral measuring instruments. [Rather,] the *impacts* at the center of AIAs are *constructs* that act as *proxies* for the often conceptually distinct sociomaterial *harms* algorithmic systems may produce" (Metcalf et al., 2021: 735). These tools define harms in terms of impacts, but oftentimes the impacts identified don't correlate to the lived experiences of those affected.

AIAs used to assess facial recognition algorithms, for example, will measure their impacts on people's ability to get equal service. So long as there's not more than a 10% difference in the service received by a young Black woman as opposed to an older White male, the AI system will be considered low risk (Kozlowska, 2021). But this metric doesn't measure the underlying emotional, psychological, social, or economic harms that marginalized people often suffer due to being consistently misidentified by facial recognition algorithms (Johnson, 2022). In this sense, AIAs only serve to protect individuals primarily from *indirect* impacts as opposed to *actual* harms (Metcalf et al., 2021: 735).

The third normative approach generally favored to regulate AI is legislation. Seeking to create a reasonable balance between protecting individual rights and corporate interests,[2] regulatory frameworks tend to come in either one of two forms: risk-based regulations and rights-based ones. Risk-based regulations assess the scope of risk associated with various uses of AI systems and then subject developers to obligations that are proportional to their potential threats. A prime example of such a regulation is the *AI Act* proposed by the European Union in 2021.[3] It's a fairly comprehensive bill that provides a framework for the ethical development and deployment of AI systems, prohibiting the use of certain AI systems in situations deemed undesirable while heavily regulating those considered high risk.

Although the *AI Act* does create some promising initiatives, its approach remains anchored in inequality. It permits AI systems to be deployed as long as their operational risks to human rights are low. But low risk doesn't mean *no risk.* In neglecting to account for situations where risks are low in the aggregate but are disproportionately felt by particular groups and communities (Leprince-Ringuet, 2021), this approach characterizes a certain measure of human rights violations as acceptable and conceives of fundamental freedoms as negotiable (Access Now, 2021; Hidvegi et al., 2021). As such, the very fact that a human right has been infringed doesn't necessarily entitle citizens to redress (Liu, 2019: 77). Rather, it's only when a high-risk materializes that accountability becomes more likely under this regulatory mechanism (Ibid).

For their part, rights-based regulations create specific rights for citizens and impose solutions that organizations must comply with to ensure that those rights are respected. But they often advance a one-size-fits-all approach that doesn't adequately account for societal challenges, making it difficult for many citizens to mobilize the solutions favored. Take, for instance, the EU's *General Data Protection Regulation* ("GDPR").[4] Considered the gold standard of digital privacy protection, it grants citizens rights to manage their personal data. We're probably all familiar with the pop-ups appearing on websites that request permission to use our personal information—this is a direct result of the GDPR. But most people quite simply click "accept" without taking any additional steps to manage their data, such that this aspect of the GDPR hasn't had its desired effect (Friedman, 2019; Modoono, 2021).

While they do cover a lot of ground, none of these solutions are truly effective at empowering citizens. Although seemingly guided by the moral imperatives we hold dear, each of these frameworks is, in its own way, laden with value judgments and normative claims that particular impacts of AI systems are wrong and must be addressed for the benefit of society or that specific solutions will meet everyone's needs (Kennedy, 2022: 178). But there's a certain subjectivity to this kind of reasoning that remains distinct from rights arguments. "The point of an appeal to a right, the reason for making it" notes legal philosopher Duncan Kennedy, "is that it can't be reduced to a mere 'value judgment' that one outcome is better than another" (Ibid: 184).

Protecting human rights, in other words, doesn't translate into safeguarding citizens' entitlements to certain outcomes. Rather, it means protecting citizens' entitlements to *capabilities*; providing them with *actual* possibilities to pursue *their* vision of what it looks like to *concretely* enjoy their fundamental freedoms in ways that are relevant to them (Nussbaum, 2002; Sen, 1984). It's at precisely this juncture that current approaches to AI regulation are left wanting. While they may be *informed* by human rights, they're not quite *grounded* in them. By neglecting to position human rights as an overarching legitimizing framework, they fall short of creating entitlements to capabilities that endow citizens with the power to *concretely* enjoy their human rights in ways that are relevant to them.

## CITIZEN PARTICIPATION AS A TOOL FOR ACHIEVING SOCIAL JUSTICE

Many of the tools developed to address human rights infringements by AI systems seek to control the *impacts* of those systems. This approach places human rights on the periphery of the AI ecosystem, with the rights being viewed as "potentially" affected or at risk. The solutions developed from this perspective focus primarily on outcomes deemed desirable; outcomes often defined through top-down approaches that exclude citizens, shadowing the viewpoints of their lived experiences as well as the intersectionality of their profiles. This angle doesn't allow us to think about participation in any adequate sense, nor does it support processes that permit citizens' engagement in order to obtain justice.

By justice, we refer to justice in its holistic sense, *qua* social justice—a global principle that seeks to promote equity, fairness, and equality, providing them with the leverage to act and demand redress where their rights are violated. The goal, then, of our approach is to shift the narrative of current approaches to AI governance; to move away from a disembodiment of the risks focused on protecting citizens' entitlements to outcomes, and to concentrate instead on ensuring that the *processes*, through which these outcomes are defined will better place us to identify concrete solutions that advance the principles of social justice.

To do so, these processes must meet three conditions for inclusive and active participation. First, they must take steps to ***deconstruct*** universalist representations of AI's effects. Given their great potential, AI technologies are wrongly presented as possessing universal benefits (Kalluri, 2020). To successfully conceptualize inclusive and representative participation processes, it's crucial to deconstruct both these universalist representations and their ability to hinder inclusivity. In other words, to consider inclusion and participation, we must first comprehend *exclusion*. Until these mechanisms of exclusion are acknowledged for what they are, it will be difficult to propose effective processes of participation that avoid reproducing similar outcomes.

Approaching participation processes from the lens of exclusion makes it possible to account for social context and overcome the tendency to treat the impacts of AI systems as universal and undifferentiated. In the words of computer scientist Abeba Birhane, understanding must take precedence over prediction (Birhane, 2017: 6). Groups living in societies that use AI systems consistently experience significant social, political, cultural, or religious discrimination. This systemic discrimination greatly impacts how they participate in society, limiting their freedom in the spaces that should be encouraging their participation. These spaces, in other words, don't respond to these groups' needs and often neglect to integrate objectives that will achieve this feat.

But supporting participation and fostering inclusion doesn't rely on understanding exclusion alone. It also necessitates the creation of tools enabling the consideration of the various contexts within which people and groups evolve (Zimmermann et al., 2020). Defining context can be a major challenge at the technological level as it requires understanding the complexity of these contexts and their underlying dimensions (Suchman, 2007). Requiring a deep knowledge of societies and their histories, this exercise could be simplified by consulting with those impacted rather than taking an external and conceptual approach. This methodology would make it possible to understand these contexts from the viewpoints of those concerned, while simultaneously accounting for the intersectionality of their profiles. This *ex ante*, as opposed to *ex post*, strategy provides the dimension necessary to determine the extent of AI's impacts on human rights (Zimmermann et al., 2020), while simultaneously making it possible to create a human-centered process that integrates the collective issues underlying individuals' identities as social actors (Birhane, 2021).

Second, to be inclusive and active, participatory processes must ***reinforce*** citizens' capabilities. In current processes, little room is left for the voices and experiences of citizens. But given that *they* are the ones affected, it remains crucial to consider societal impacts for what they actually are, rather than for what we might anticipate them to be. A key factor in understanding these impacts from a real-life/ground perspective is to extend deference to the expertise of citizens as authorities over their own lives. It's only through these citizens' agency—through their capability to identify AI's impacts on their livelihood—that it will be possible for society to tackle some of its most socially damaging impacts.

And reinforcing their capabilities in this sense relies also on empowering citizens through rights education in the AI context, providing them with the tools to voice their lived experiences. Rights education effectively provides groups or individuals with the knowledge base necessary to define injustices in terms of a body of fundamental rights. To know what rights are being impacted, they must be aware of what their rights are. In the AI context, which involves a loss of control and human agency (Delacroix, 2022), rights education fulfills the objective of empowering actors. It supports inclusion and encourages participation by extending citizens the *capability* to control and act on their needs and to project themselves as actors and stakeholders regarding the issues that affect them.

The condition of their participation therefore rests on their mastery and knowledge of the tools necessary to identify and address infringements to their rights. And this education must equally extend to *digital* rights to provide individuals with the foundation necessary to grasp the fundamental freedoms at play in the AI context. Unfortunately, however, the current discourse surrounding digital rights and AI is very technical, making it difficult for everyday citizens to comprehend their correlation to social and cultural rights. This lack of awareness causes them to forego certain digital rights that have important impacts on their lives.

From this perspective, reinforcing their capabilities also involves talking to potentially impacted groups in terms to which they can relate; in terms that place *human rights*—rather than the technology itself—as the central point of reference. Human rights risks in the AI context are currently predicted based on the state of knowledge, technologies, and other foreseeable characteristics such as misuse of AI systems or noncompliance with regulatory mechanisms, for instance. But this approach doesn't provide citizens with the opportunity to define risks based on injustices they experience at the hands of AI systems[5]—injustices that may not necessarily qualify as human rights violations, but that nevertheless interfere with citizens' ability to be recognized within society.

Moreover, because everyone's rights aren't impacted in the same way, it's important to permit citizens to define impacted rights without systematically linking them to the AI context. One of the main challenges that individuals meet concerns lived experiences that lead to feelings of injustice, which they're initially unable to link to any human rights risks associated with AI systems because they're not sufficiently familiar with the technicalities involved to pinpoint the rights that are interacting with these systems.

Allowing for the identification of human rights impacts from the perspective of justice/injustice could avoid this pitfall, while also promoting inclusion in two other important ways. First, it includes groups and individuals in defining solutions that best address the injustices they face in the AI context, which could be structural or epistemic and have deep impacts on individuals' paths (Rafanelli, 2022). Indigenous organizations, for instance, have identified digital sovereignty (rather than personal data protection) as a more adequate solution for addressing impacts arising from their misrepresentation in datasets (Gentelet et Bahary-Dionne, 2022). Second, it could help identify societal structures that play a role in these injustices; that are wrongly labeled as neutral or invisible, when they in fact contribute to the exclusion of minorities from societal processes. It therefore also makes it possible to identify nontechnical systemic factors that predate the use of digital technologies but are amplified by AI's overvaluation of data.

The third and final condition necessary for participation to be effective, active, and deliberate, is that it must respond to the challenge of mobilization. To ***mobilize*** citizens, it's crucial to provide access to effective tools that enable their participation. To achieve this feat, these tools must make it possible to reverse any process rooted in inequity or discrimination and support what Chateauraynaud defines as *réversabilité* (Chateauraynaud, 2021: 10). This reversibility renders citizen action effective and thus enables tangible results to arise from this participation.

Among the elements necessary for moving toward reversibility is acknowledging the important role of civil society in documenting human rights abuses by integrating them into AI governance. Because of their close field contacts, civil society organizations[6] are in the position to document the most severe human rights abuses, as well as those that most often go unrecorded (Andrus et al., 2021: 250). Unfortunately, however, because AI governance mechanisms aren't accessible to their organizations, civil society is deprived of taking legal action for want of financial and administrative force.

And if civil society lacks the resources necessary to act, it's little wonder that citizens from marginalized communities find themselves similarly positioned. To involve citizens in governance and regulatory processes, it's therefore essential to develop structures that are both accessible and easily mobilizable in the event that they wish to denounce infringements to their rights. Given that there are currently few direct (if any) recourses for citizens to challenge AI's human rights impacts, it remains crucial to ensure the availability of accountability mechanisms to citizens, be they democratic or legal.

At the moment, however, most AI governance tools are designed unilaterally from the top-down. They concentrate mostly on pressuring developers—who are not legally equipped to address the issues—to respect ethical and socially acceptable practices, particularly those involving equity (Andrus et al., 2021). Although both national and international institutional actors are working together to define AI regulations, accountability mechanisms are mostly concentrated and operable at a macro level alone.

But the tools and access to be prioritized must exist on several scales, creating a model of inclusion that relies on a variety of spaces that enable both participation and accountability. Current models are limited to intervention by state institutions and high-level authorities. Creating processes that can be easily mobilized by noninstitutional actors—initiated by the people or groups affected—would facilitate a dynamic interpretation of data in ways that give them meaning and limit their use to the specific social context surrounding their production (Ibid).

We could, for instance, rely on preexisting tools for citizen participation and consultation that are integrated into regular modes of state or municipal governance without having to create structures dedicated to AI issues. We could therefore employ these existing low-scale democratic forums and assemblies to include and involve groups and individuals impacted by AI uses. The decision-making processes would then be in formats that citizens are accustomed to, and the issues would be framed on a smaller scale at a neighborhood level.

In terms of access to courts, it might be relevant to extend the jurisdiction of certain low-intensity courts, such as small-claims courts or administrative tribunals, to permit them to preside over cases related to discrimination and violations of social, cultural, and economic rights arising from uses of AI tools. Allowing the people primarily affected by these issues to address courts by bringing forth the injustices they face would certainly contribute to capturing the interconnectedness between the different rights impacted by AI at a macro and state level.

But to make this possible, it would also be necessary to put an end to the current legal estrangement between *de facto* social, economic, and cultural rights and digital rights, allowing citizens to pursue on these bases. In so doing, infringements of rights that are linked to societal issues could be integrated into legal procedures that are directly mobilizable by citizens, grassroots organizations, or civil society. These procedures, pursued by citizens in low-intensity courts, could then be the first step in a much larger process, induced at national and international levels and conducted by civil liberties ombudsmen, human rights commissions, or specialized tribunals, for instance (Table 20.1).

**TABLE 20.1**
**Steps toward Citizen Participation**

| Deconstructing | Reinforcing | Mobilizing |
|---|---|---|
| - Structural mechanisms of exclusion;<br>- Social and political contexts;<br>- Socioeconomic determinants;<br>- The intersection between profiles and rights;<br>- Technological universalism;<br>- The primacy of the expert word;<br>- The relationship between the individual and the collective. | - Human rights education;<br>- The link between human rights and technological rights;<br>- The capabilities of those impacted to describe the harms;<br>- Cooperation across several levels by creating collaborative spaces. | - By working with civil society;<br>- By concentrating on human rights rather than technology;<br>- By facilitating the connection between rights and violations;<br>- By speaking in terms of justice;<br>- Lived experiences as part of the process;<br>- Existing democratic mechanisms of governance and accountability;<br>- Realistic approaches for access to courts and other accountability mechanisms. |

## CONCLUSION

The integration of human rights principles into AI governance is, above all, a matter of social justice. And for AI governance to respond to principles of social justice—for it to *truly* be human-centered—it must look beyond the technical aspects of this technology and respond to the preexisting societal structures that have bred algorithmic bias. Normative solutions developed to address algorithmic injustices must integrate the perspectives of those citizens whose lives, paths, and human rights are impacted by this technology. While reflecting on equitable design or more socially sensitive deployment practices is certainly a good place to start, it is far from sufficient. And the reason is that this approach concentrates strictly on developing solutions to address AI's impacts—to protect citizens' entitlements to certain outcomes—all the while overlooking the importance of the *process* through which they're defined.

To pinpoint effective solutions, we must shift our current focus toward identifying processes that extend citizens the capabilities to participate in defining how algorithmic tools are deployed. Citizens must be provided with the democratic spaces allowing them to participate in the governance of these tools—to define solutions and make decisions about how these technologies will shape their narratives. And including them in this process necessarily involves making it accessible to them in terms that are meaningful to them, based on a human rights language that they can master in light of their lived experiences: the language of injustice, whose essence is grounded in notions of human dignity and serves to encompass the social, economic, and cultural dimensions of human identity while also grasping the discriminatory societal interactions experienced by citizens.

It's only by involving citizens in governance and decision-making processes that we'll be able to define outcomes that are better adapted to societal realities. It's therefore crucial to shift the focus of our approach to AI governance away from risks, results, and entitlements to specific outcomes, and to focus our energies on developing inclusive processes through which we can collectively define our algorithmic future. While doing so might increase the complexity of AI governance, and might even lead to some uncertainties, maintaining the current state of affairs is socially unacceptable. Permitting citizens to participate in defining how AI shapes their narratives on the basis of a human rights framework remains the most effective way to foster social justice. And if we focus on the process, the appropriate solutions will reveal themselves in due course.

## NOTES

1 *Artificial Intelligence Act*, EU COM/2021/206.
2 *Artificial Intelligence Act*; *General Data Protection Regulation.*
3 *Artificial Intelligence Act.*
4 *General Data Protection Regulation.*
5 Injustices that could be defined as unfair, inequitable, or discriminatory practices (Rafanelli, 2022).
6 By civil society, we include NGOs, parapublic institutions but also community, grassroots, and field organizations.

## BIBLIOGRAPHY


### Legislation

*Artificial Intelligence Act, EU COM/2021/206, online: https://eur-lex.europa.eu/legal-content/EN/TXT/?qid=1623335154975&uri=CELEX%3A52021PC0206.*

*Canadian Charter of Rights and Freedoms, Part 1 of the Constitution Act, 1982, being Schedule B to the Canada Act 1982 (UK), 1982, c 11.*

*Charter of Human Rights and Freedoms, CQLR c C-12.*

*Civil Code of Quebec, CQLR c CCQ-1991, art 1457.*

*Competition Act, RSC 1985 c C-34.*

*Consumer Protection Act, CQLR c P-40.1.*

*Digital Markets Act, Regulation (EU) 2022/1925.*

*General Data Protection Regulation, EU 2016/679.*

### Jurisprudence

*Jane Doe 464533 v DN, 2016 ONSC 4920.*

*Jane Doe 72511 v Morgan, 2018 ONSC 660.*

*Jones v Tsige, 2012 ONCA 32.*

### Doctrine

Access Now. (2021, April 21). EU Takes Minimal Steps to Regulate Harmful AI systems, must go further to protect fundamental rights. Retrieved from https://www.accessnow.org/eu-minimal-steps-to-regulate-harmful-ai-systems/

Andrus, McKane, Spitzer, E., Brown, J., & Xiang, A. (2021). What We Can't Measure, We Can't Understand: Challenges to Demographic Data Procurement in the Pursuit of Fairness. In *Proceedings of the 2021 ACM Conference on Fairness, Accountability, and Transparency* (p. 249).

Angwin, J., Larson, J., Mattu, S., & Kirchner, L. (2016, May 23). Machine Bias. Retrieved from ProPublica: https://www.propublica.org/article/machine-bias-risk-assessments-in-criminal-sentencing

Australian Government, Department of Industry, Science and Resources. (2019, November 7). Australia's Artificial Intelligence Ethics Framework. Retrieved from https://www.industry.gov.au/publications/australias-artificial-intelligence-ethics-framework

Bedoya, A. M. (2016, January 18). The Color of Surveillance: What an Infamous Abuse of Power Teaches Us About the Modern Spy Era. Retrieved from Slate: https://slate.com/technology/2016/01/what-the-fbis-surveillance-of-martin-luther-king-says-about-modern-spying.html

Benjamin, R. (2019). *Race After Technology: Abolitionist Tools for the New Jim Code*. Cambridge: Polity Press.

Birhane, A. (2017). Descartes was wrong: 'a person is a person through other persons'. Retrieved from Aeon: https://aeon.co/ideas/descartes-was-wrong-aperson-is-a-person-through-other-persons

Birhane, A. (2021). Algorithmic Injustice: A Relational Ethics Approach. *Patterns*, *2*(2), 100205.

Boyd, D., & Crawford, K. (2012). Critical Questions for Big Data: Provocations for a cultural, technological, and scholarly phenomenon. *Information, Communication & Society*, *15*(5), 662.

Broussard, M. (2018). *Artificial Unintelligence: How Computers Misunderstand the World*. Cambridge: MIT Press.

Chander, A. (2017). The Racist Algorithm. 115 *Mich L Rev*, 1023.

Chateauraynaud, F. (2021). Des expérimentations démocratiques en tension. L'œuvre des citoyens dans le travail politique des bifurcations. *Cahiers du GRM, 18*. Retrieved from https://doi.org/10.4000/grm.3238

China Daily. (2019, June 17). Governance Principles for the New Generation Artificial Intelligence – Developing Responsible Artificial Intelligence. Retrieved from http://www.chinadaily.com.cn/a/201906/17/WS5d07486ba3103dbf14328ab7.html

Costanza-Chock, S. (2020). *Design Justice: Community-Led Practices to Build the Worlds We Need*. Cambridge: MIT Press.

Council of Europe. (n.d.). Ethical Frameworks. Retrieved from https://www.coe.int/en/web/artificial-intelligence/ethical-frameworks

Crawford, K. (2021). *Atlas of AI*. New Haven: Yale University Press.

Delacroix, S. (2022). Diachronic Interpretability & Machine Learning Systems. *Journal of Cross-disciplinary Research in Computational Law*.

Equivant. (n.d.). Northpointe Suite Risk Needs Assessments. Retrieved from https://www.equivant.com/northpointe-risk-need-assessments/

Eubanks, V. (2018). *Automating Inequality: How High-Tech Tools Profile, Police, and Punish the Poor*. New York: St Martin's Press.

European Commission, High-Level Expert Group on AI. (2019, April 8). Ethics Guidelines for Trustworthy AI. Retrieved from https://ec.europa.eu/digital-single-market/en/news/ethics-guidelines-trustworthy-ai

Follows, T. (2021). *The Future of You: Can Your Identity Survive 21st-Century Technology?* London: Elliot and Thompson.

Foucault, M. (1995). *Discipline and Punish*. New York: Vintage Books.

Friedman, V. (2019, April 10). Privacy UX: Better Cookie Consent Experiences. *Smashing Magazine*. Retrieved from https://www.smashingmagazine.com/2019/04/privacy-ux-better-cookie-consent-experiences/

Gentelet, K., & Bahary-Dionne, A. (2022). Les angles morts des réponses technologiques à la Covid-19: des populations marginalisées invisibles. *Éthique publique*, *23*(2). Retrieved from https://doi.org/10.4000/ethiquepublique.6441

Gentelet, K., & Mizrahi, S. K. (2021, September 26). We Need Concrete Protections From Artificial Intelligence Threatening Human Rights. *The Conversation*. Retrieved from https://theconversation.com/we-need-concrete-protections-from-artificial-intelligence-threatening-human-rights-168174

Germany, Daten Ethik Kommission. (October 2019). Opinion of the Data Ethics Commission. Retrieved from https://www.iicom.org/feature/germany-opinion-of-the-data-ethics-commission/

Google Images. (2022a, December 6). Nurse. Retrieved from https://www.google.com/search?q=nurse&rlz=1C1VDKB_enCA1015CA1015&sxsrf=ALiCzsZbocwC1J5Hn1WIE_O6aLIN1jPyMA:1670340387335&source=lnms&tbm=isch&sa=X&ved=2ahUKEwjlm8PrpuX7AhUXk4kEHagsAg8Q_AUoAXoECAEQAw&biw=954&bih=944&dpr=1

Google Images. (2022b, December 6). Professor. Retrieved from https://www.google.com/search?q=nurse&rlz=1C1VDKB_enCA1015CA1015&sxsrf=ALiCzsZbocwC1J5Hn1WIE_O6aLIN1jPyMA:1670340387335&source=lnms&tbm=isch&sa=X&ved=2ahUKEwjlm8PrpuX7AhUXk4kEHagsAg8Q_AUoAXoECAEQAw&biw=954&bih=944&dpr=1

Government of Canada. Algorithmic Impact Assessment Tool. (n.d.). Retrieved from https://www.canada.ca/en/government/system/digital-government/digital-government-innovations/responsible-use-ai/algorithmic-impact-assessment.html

Harrington, C. N. (2020). The Forgotten Margins: What is Community-Based Participatory Health Design Telling Us? *Interactions*, *27*(3), 24.

Hidvegi, F., Leufer, D., & Massé, E. (2021, February 17). The EU Should Regulate AI on the Basis of Rights, Not Risks. *Access Now*. Retrieved from https://www.accessnow.org/eu-regulation-ai-risk-based-approach/

India. (2019, June 13). *National Strategy for AI*. Retrieved from https://indiaai.gov.in/research-reports/national-strategy-for-artificial-intelligence

Infocomm Media Development Authority & Personal Data Protection Commission Singapore. (2020). *Model Artificial Intelligence Governance Framework Second Edition*.

Italy. *National Strategy on Artificial Intelligence*. Retrieved from https://knowledge4policy.ec.europa.eu/ai-watch/italy-ai-strategy-report_en

Johnson, K. (2022, March 7). How Wrongful Arrests Based on AI Derailed 3 Men's Lives. *Wired*. Retrieved from https://www.wired.com/story/wrongful-arrests-ai-derailed-3-mens-lives/

Kalluri, P. (2020). Don't ask if artificial intelligence is good or fair, ask how it shifts power. *Nature*, *583*, 169.

Kaminski, M. E., & Malgieri, G. (2020, October 12). Algorithmic Impact Assessments under the GDPR: Producing Multi-layered Explanations. Retrieved from https://papers.ssrn.com/sol3/papers.cfm?abstract_id=3456224

Karlin, M. (2018, March 18). A Canadian Algorithmic Impact Assessment. Retrieved from https://medium.com/@supergovernance/a-canadian-algorithmic-impact-assessment-128a2b2e7f85

Karlin, M., & Corriveau, N. (2018, August 7). The Government of Canada's Algorithmic Impact Assessment: Take Two. Retrieved from https://medium.com/@supergovernance/the-government-of-canadas-algorithmic-impact-assessment-take-two-8a22a87acf6f

Kennedy, D. (2022). The Critique of Rights in Critical Legal Studies. In W. Brown & J. Halley (Eds.), *Left Legalism/Left Critique* (pp. 178–228). Durham: Duke University Press.

Kim, P. T. (2020). Manipulating Opportunity. *Virginia Law Review*, *106*(4), 867.

Kozlowska, I. (2021, April 4). Algorithmic Impact Assessments – What Impact Do They Have? *Montreal AI Ethics Institute*. Retrieved from https://montrealethics.ai/algorithmic-impact-assessments-what-impact-do-they-have/

Lashbrook, A. (2018, August 16). AI-Driven Dermatology Could Leave Dark-Skinned Patients Behind. *The Atlantic*. Retrieved from https://www.theatlantic.com/health/archive/2018/08/machine-learning-dermatology-skin-color/567619/

Leprince-Ringuet, D. (2021, April 6). Facial Recognition Tech is Supporting Mass Surveillance. It's Time for a Ban, Say Privacy Campaigners. *ZDNet*. Retrieved from https://www.zdnet.com/article/facial-recognition-tech-is-supporting-mass-surveillance-its-time-for-a-ban-say-privacy-campaigners/

Liu, H.-Y. (2019). The Digital Disruption of Human Rights Foundations. In S. Mart (Ed.), *Human Rights, Digital Society and the Law* (pp. 75). London: Routledge.

Lupton, D. (2018). How do Data Come to Matter? Living and Becoming With Personal Data. *Big Data & Society*, *5*(2), 205395171878631.

Madhusoodanan, J. (2021, August 28). These Apps Say They Can Detect Cancer. But Are They Only For White People? *The Guardian*. Retrieved from https://www.theguardian.com/us-news/2021/aug/28/ai-apps-skin-cancer-algorithms-darker/

Malta Digital Innovation Authority. (2019). *AI ITA Guidelines*. Retrieved from https://mdia.gov.mt/wp-content/uploads/2019/10/AI-ITA-Guidelines-03OCT19.pdf

Mantelero, A. (2018). AI and Big Data: A blueprint for a human rights, social and ethical impact assessment. *Computer Law & Security Review*, *34*(4), 754.

Metcalf, J., Moss, E., Watkins, E. A., Singh, R., & Elish, M. C. (2021). Algorithmic Impact Assessments and Accountability: The Co-construction of Impacts. *ACM*. Retrieved from https://dl.acm.org/doi/10.1145/3442188.3445935

Mizrahi, S. K. (2018). Ontario's New Invasion of Privacy Torts: Do They Offer Monetary Redress for Violations Suffered via the Internet of Things? *UWO J Leg Stud*, *8*(1), 3. Retrieved from https://www.canlii.org/en/commentary/doc/2018CanLIIDocs65#!fragment//BQCwhgziBcwMYgK4DsDWszIQewE4BUBTADwBdoByCgSgBpltTCIBFRQ3AT0otokLC4EbDtyp8BQkAGU8pAELcASgFEAMioBqAQQByAYRW1SYAEbRS2ONWpA

Modoono, M. (2021, April 22). How Can We Protect Our Privacy in the Era of Facial Recognition? *Northeastern*. Retrieved from https://news.northeastern.edu/2021/04/22/how-can-we-protect-our-privacy-in-the-era-of-facial-recognition/

Morosini, D. (2021, June 11). Why Doctors Still Struggle to Detect Skin Cancer in Black People. Retrieved from https://www.thecut.com/2021/06/why-doctors-struggle-to-detect-skin-cancer-in-black-people.html

Nadar, S. (2014). "Stories are Data with Soul" – Lessons from Black Feminist Epistemology. *Agenda*, *28*(1), 18.

Noble, S. U. (2018). *Algorithms of Oppression: How Search Engines Reinforce Racism*. New York: New York University Press.

Norwegian Ministry of Local Government and Modernisation. (n.d.) *National Strategy for Artificial Intelligence*. Retrieved from https://www.regjeringen.no/contentassets/1febbbb2c4fd4b7d92c67ddd353b6ae8/en-gb/pdfs/ki-strategi_en.pdf

Nussbaum, M. (2002). Capabilities and Social Justice. *International Relations and the New Inequality*, *4*(2), 123.

O'Donnell, P., Tierney, E., O'Carroll, A., Nurse, D., & MacFarlane, A. (2016). Exploring Lever and Barriers to Accessing Primary Care for Marginalised Groups and Identifying Their Priorities for Primary Care Provision: A Participatory Learning and Action Research Study. *International Journal for Equity in Health*, *15*, 197.

O'Neil, C. (2016). *Weapons of Math Destruction: How Big Data Increases Inequality and Threatens Democracy*. New York: Crown.

OECD. (2019). *Recommendation of the Council on Artificial Intelligence*. Retrieved from https://legalinstruments.oecd.org/en/instruments/OECD-LEGAL-0449

Pabarcus, A. (2011). Are 'Private' Spaces on Social Networking Websites Truly Private? The Extension of Intrusion Upon Seclusion. *William Mitchell Law Review*, *38*, 397.

Parikh, N. (2021, October 14). Understanding Bias In AI-Enabled Hiring. *Forbes*. Retrieved from https://www.forbes.com/sites/forbeshumanresourcescouncil/2021/10/14/understanding-bias-in-ai-enabled-hiring/?sh=621536c87b96

Penney, J., McKune, S., Gill, L., & Deibert, R. J. (2018). Advancing Human-Rights-by-Design in the Dual-Use Technology Industry. *Journal of International Affairs*, *71*(2), 103.

Rafanelli, L. M. (2022). Justice, Injustice, and Artificial Intelligence: Lessons from Political Theory and Philosophy. *Big Data & Society*, *9*(1). Retrieved from https://doi.org/10.1177/20539517221080676

Raji, I. D., Smart, A., White, R. N., Mitchell, M., Gebru, T., Hutchinson, B., Smith-Loud, J., Theron, D., & Barnes, P. (2020). Closing the AI Accountability Gap: Defining an End-to-End Framework for Internal Algorithmic Auditing. In *Conference on Fairness, Accountability, and Transparency (FAT* '20)*, Barcelona, ES.

Raso, F. A., Hilligoss, H., Krishnamurthy, V., Bavitz, C., & Kim, L. (2018, September 25). Artificial Intelligence & Human Rights: Opportunities and Risks. *Berkman Klein Center for Internet & Society*. Retrieved from https://cyber.harvard.edu/publication/2018/artificial-intelligence-human-rights

Reisman, D., Schultz, J., Crawford, K., & Whittaker, M. (2018). Algorithmic Impact Assessments: A Practical Framework for Public Agency Accountability. *AI Now*. Retrieved from https://ainowinstitute.org/aiareport2018.pdf

Richardson, R., Schultz, J., & Crawford, K. (2019). Dirty Data, Bad Predictions: How Civil Rights Violations Impact Police Data, Predictive Policing Systems, and Justice. *New York University Law Review*, *94*. Retrieved from https://ssrn.com/abstract=3333423

Schneier, B. (2015). *Data and Goliath: The Hidden Battles to Collect Your Data and Control Your World*. New York: W. Norton & Company.

Sedler, R. A. (1984). Constitutional Protection of Individual Rights in Canada: The Impact of the New Canadian Charter of Rights and Freedoms. *Notre Dame Law Review*, *59*, 1191.

Selbst, A. D. (2018). Disparate Impact in Big Data Policing. *Georgia Law Review*, *52*, 109.

Selingo, J. (2017, April 11). How Colleges Use Big Data to Target the Students They Want. *The Atlantic*. Retrieved from https://www.theatlantic.com/education/archive/2017/04/how-colleges-find-their-students/522516/

Sen, A. (1984). Rights and capabilities. In Sen, A. (ed.), *Resources, Values and Development*. Cambridge: Harvard University Press.

Shull, A. (2018, August 16). The Charter and Human Rights in the Digital Age. *CIGI*. Retrieved from https://www.cigionline.org/articles/charter-and-human-rights-digital-age

Simonite, T., & Barber, G. (2019, October 17). The Delicate Ethics of Using Facial Recognition in Schools. *Wired*. Retrieved from https://www.wired.com/story/delicate-ethics-facial-recognition-schools/

Solove, D. J. (2011). *Nothing to Hide: The False Tradeoff between Privacy and Security*. New Haven: Yale University Press.

Suchman, L. (2007). *Human-Machine Reconfigurations: Plans and Situated Actions*. Cambridge: Cambridge University Press.

Switzerland, The Federal Council. *Digital Switzerland Strategy*. Retrieved from https://www.digitalerdialog.ch/en/

The Government of the Republic of Serbia. (n.d.). *Strategy for the Development of Artificial Intelligence in the Republic of Serbia for the Period of 2020-2025*. Retrieved from https://www.srbija.gov.rs/tekst/en/149169/strategy-for-the-development-of-artificial-intelligence-in-the-republic-of-serbia-for-the-period-2020-2025.php

The Netherlands. (n.d.). *Strategic Action Plan for Artificial Intelligence*. Retrieved from https://www.government.nl/documents/reports/2019/10/09/strategic-action-plan-for-artificial-intelligence

Waldman, A. E. (2021). *Industry Unbound: The Inside Story of Privacy, Data, and Corporate Power*. Cambridge: Cambridge University Press.

Winner, L. (1986). *The Whale and the Reactor: A Search for Limits in an Age of High Technology*. Chicago: University of Chicago Press.

World Economic Forum. *AI Ethics Frameworks*. Retrieved from https://www.weforum.org/projects/ai-ethics-framework

Yeung, K., Howes, A., & Pogrebna, G. (2019). AI Governance by Human Rights-Centred Design, Deliberation and Oversight: An End to Ethics Washing. In M. Dubber & F. Pasquale (Eds.), *The Oxford Handbook of AI Ethics*. Oxford: Oxford University Press.

Zimmermann, A., Di Rosa, E., & Kim, H. (2020, January 9). Technology Can't Fix Algorithmic Injustice. *Boston Review*. Retrieved from https://www.bostonreview.net/articles/annette-zimmermann-algorithmic-political/

# 21 Operationalizing AI Regulatory Sandboxes for Children's Rights and Well-Being

*Vicky Charisi**
European Commission, Joint Research Centre, Seville, Spain

*Virginia Dignum*
Umeå University, Umeå, Sweden

## INTRODUCTION

### MOTIVATION AND GENERAL SCOPE OF THE CHAPTER

Children and youth are surrounded by AI in many of the products they use in their daily lives, from social media to education technology, video games, smart toys, and voice assistants. AI can affect the videos children watch online, their curriculum as they learn, and the way they play, and interact with others. For the purposes of this chapter, we adopt UNICEF's definition of AI. AI can be used to facilitate children's development and empower them in their activities, but children are also vulnerable to potential risks posed by AI, including bias, cybersecurity, data protection and privacy, and lack of accessibility. AI must be designed to respect the rights of the child user and to provide equal opportunities for all children. Several organizations, including UNICEF and the World Economic Forum (WEF), have developed guidelines for child-centric AI, but the challenge of designing and implementing responsible and trusted child-centered AI remains complex to address. In this chapter, we describe challenges and propose a regulatory sandbox environment as a means to address children, businesses, regulations, and society as a whole. For the rest of the chapter, whenever we use the term "sandboxes," we refer to "regulatory sandboxes."

The proposed approach to the operationalization of sandboxes for age-appropriate AI systems aims to bring competent authorities closer to companies that develop AI

* European Commission, Joint Research Centre. The views expressed are purely those of the author and may not in any circumstances be regarded as stating an official position of the European Commission.

DOI: 10.1201/9781003320791-25

and to define best practices that will guide the implementation of child-centered AI systems. Such a sandbox provides a controlled environment to connect innovators, regulators, child experts, and users to cooperate in order to facilitate the development, testing, and validation of innovative AI systems with a view to ensuring alignment with existing guidelines on AI for children.

## Review of Relevant Work on Sandboxes in General and in AI Systems

Regulatory sandboxes emerged in the last decade in the context of FinTech (Allen 2019), aiming to gather data about a novel technology and to promote evidence-based regulatory reforms. Using sandboxes, regulators can assess the effectiveness of laws and policies, in particular how these affect specific groups, such as children. At the same time, regulatory sandboxes also allow regulators to assert how well new AI applications fit within existing legal frameworks (Ranchordas, 2021).

The European Commission's proposal for a Regulation on AI,[1] the AI Act, refers to sandboxes as a means for providers and prospective providers to lower uncertainty, innovate, experiment with AI technology, and contribute to evidence-based regulatory learning. The supervision of the AI systems in the AI regulatory sandbox should therefore cover their development, training, testing, and validation before the systems are placed on the market or put into service as well as any occurrence of substantial modification that may require a new conformity assessment procedure (Council of the European Union, 2022). The European Commission (EC) is following up on this work and preparing a Staff Working Document to be published in summer 2023, with the working title "Guidance on relevant use cases of regulatory sandboxes, test beds and living labs in the EU.[2] This work is in line with the science for policy brief published by the that analyses the typical features of regulatory sandboxes (Kert et al., 2022).

The basic idea of the regulatory sandbox is to provide exploratory, dialogue-based guidance to selected projects in exchange for full openness about the assessments that are made. It provides a safe environment for the testing of innovations and regulations in market conditions to improve legal certainty (Kert et al. 2022). In this way, the sandbox builds up a base of practical examples in a field where both the technology and the law are complicated and relatively new.

Sandboxes have the potential to be beneficial in driving forward the improvement of products and services. For example, in the context of healthcare, regulatory sandboxes aim to improve healthcare experience and outcomes, but also the experiences of healthcare providers, commissioners, and regulators, supporting stakeholders to collaborate in determining "what good looks like," and to design and drive innovation without some of the associated risks (Leckenby et al. 2021). Commonly used approaches to sandboxes include assessment lists and repositories. Repositories support knowledge management (e.g., Gutierrez's soft law database[3] lists soft laws related to AI procurement), or document information about AI in public administrations (e.g., the AI registry of the City of Amsterdam[4]).

Several countries are currently using the method of sandboxes in the context of Artificial Intelligence. For example, in 2018, Japan introduced a sandbox regime open to organizations and companies both in and outside Japan willing to experiment with new technologies, including blockchain, AI, and the Internet of Things (IoT) in fields such as financial services, healthcare, and transportation (IFLR, 2019). In Europe, both Norway and the United Kingdom (UK) have developed AI sandboxes. Norway established a regulatory sandbox as part of its national AI strategy, to provide guidance on personal data protection for private companies and the public sector (Datatilsynet, 2022). In the UK, a sandbox has been set up to explore new technologies, such as voice biometrics and facial recognition technology, and related data protection issues.[5] In 2022, the European Commission launched the first pilot on regulatory sandboxes on AI with the Spanish government (European Commission, 2022). As such, it seems that experimenting within the safe environment sandboxes provide is a technique that has started already to be used in the context of AI.

However, regulatory sandboxes also come with a risk of being misused or abused and need the appropriate frameworks to succeed. For example, sandboxes could open the door to favoritism in the regulatory process, potentially creating a quid pro quo relationship between companies and regulators (Knight & Mitchell, 2020), or to regulatory capture. This is particularly important when experimenting with services and products that address vulnerable populations such as children. In the following paragraph, we summarize the characteristics that are child-specific and require special attention when designing sandboxes for AI-based products and services for children or that have the potential to be used by children.

## What Is Special about Children?

While the use of sandboxes for the experimentation of policies and technology applications for adults is a method that has been used in other sectors, we propose that in the case of users under the age of 18, there are special characteristics that need to be considered.

The first years of every child's life are a period of great opportunity. The child's brain is filled with curiosity, imagination, wonder, and determination, which develop through interaction with the physical and social environment. Curiosity-driven learning together with intrinsic motivation have been argued to be fundamental ingredients for efficient education as key cognitive processes in fostering active learning and spontaneous exploration (Oudeyer & Smith, 2016). It is well-known that environmental factors during childhood affect the pace of brain development and the development of cognitive and social-emotional skills, which has long-lasting impacts on children's current and future well-being and the outcomes throughout schooling and later in adulthood (Tooley et al., 2021; Bolton et al. 2022). But how can we support the development of those faculties and mitigate the emerging risks when children act within the digital world and interact with systems that intervene in their decision-making process and their development?

With the recent technical advances in machine learning and, more recently, in foundation models, the landscape of the environmental factors that affect children's everyday lives has dramatically changed. We observe that what we consider a "children's environment" includes virtual or physical AI-based systems (Hupont et al., 2023) that are not specially designed for them. In such contexts, we identify specific elements that require special attention when designing methods for policies for children's protection, AI-based products and services that might be accessed by children, and techniques to evaluate the alignment of AI-based systems with current regulatory frameworks, such as in the case of sandboxes.

### The Importance of Variability of Contexts for the Developing Brain

Brain sciences and computational approaches in developmental psychology have shed new light on our understanding of children's cognition and how critical individual variability is during children's rapid development. The dynamic nature of children's behavior is highly affected by the opportunities for exploratory activities that foster curiosity and critical and creative thinking. Pelz and Kidd (2020), for example, emphasize the importance of free play and indicate the developmental aspects of children's exploratory behaviors which appear more elaborate throughout human childhood. A diverse environment fosters children's problem-solving, language learning, inductive reasoning, visual perception, etc. (Raviv et al. 2022; Charisi et al., 2020). Individual variability, especially for neurodiverse children, such as children with autism, is of particular importance. Various contextual factors are catalytic when designing experimentation and assessment for age-appropriate AI and children such as situational contexts, individual, spatial, and social contexts. For example, a child can use an AI system during free or structured play (Situational context), while being at home or school (Spatial context), either alone or with parental supervision (Social context). The variability of contexts plays an important role in a child's development and should be considered when designing sandboxes for age-appropriate AI for the effectiveness of a policy or an innovative solution. This information can be used to develop the indicators and metrics for evaluation of the system in an effective way in different contexts (see also Figure 21.3).

### The Challenge of Protecting Children's Privacy and Safety at the Same Time

Research shows that children tend to define privacy as being alone, managing information, being unbothered, and controlling access to places (Wolfe, 1978). However, while research shows that children recognize that privacy involves autonomy, autonomy does not necessarily define privacy for them, since a lack of autonomy at a younger age is what they experience over much of their life (ibid.). These insights were developed in contexts related to the physical world, while children's understanding of privacy and safety in the context of AI is still to be understood.

More recent research indicates that there are developmental differences in terms of an individual's level of maturity among children of different ages when it comes to their understanding of personal data and privacy in digital spaces (Stoilova et al., 2021). Similarly, Kumar et al. (2017) found that while individuals recognized certain privacy and security components, children aged 5-7 had gaps in their knowledge regarding these notions. As such, while certain children can develop strategies to manage concerns about privacy, younger children largely rely on parents for support (ibid.).

Anecdotal evidence, however, indicates that while parental supervision is essential for young children's involvement with AI systems, overreliance on parental intervention is problematic. In addition to the sense of lack of privacy, parental intervention might create, large discrepancies in terms of children's protection, due to differences in parents' knowledge and availability.

A responsible design of an AI-based system takes into account how to scaffold children's understanding of privacy and safety and it supports their capacity to consent. As such, children can take an increasingly active role in terms of their privacy and safety during their interaction with AI-based systems. A responsible and developmentally appropriate approach to the system design in terms of child's scaffolding allows children to exercise their right to privacy in a safer way.

### Agentic Capacity in Childhood

Agency has been defined as "the capacity to act" (Manyukhina et al., 2019) and entails the element of intentionality. Albert Bandura takes an agentic perspective (1989) and explains that agency refers to the human capacity to influence one's functioning and the course of events through one's actions. This is part of a causal structure that is socially situated and considers that people are intentional contributors to their life circumstances. Synofzik et al. (2008) draw an important distinction between the "feeling of agency" and the "judgment of agency." The feeling of agency is a lower-level nonconceptual feeling of being an agent; it is the background buzz of control we feel for our voluntary actions when not explicitly thinking about them. The judgment of the agency, on the other hand, is a higher-level concept and arises in situations where we make explicit attributions of agency to the self or others.

In this context, Greene and Nixon (2020) take a developmental psychology perspective and reject the view of the child as fully autonomous agent, rational, and essentially competing for a vision of the child as a person with agentic capacity over time, and at times capable of forming and asserting their choices and their opinions but not always heard, and not always effective. They explain their view based on power relations as a factor in how the agency is expressed or not. Both perspectives, social cognition theory, and developmental psychology theories, highlight the importance of context for children's experience for the manifestation of their agency, such as the context of family, school, or society (Sorbring & Kuczynski, 2018). From an educational point of view, Baker et al. (2021) propose that developmentally appropriate and playful activities create the space for children to develop their agency.

The abovementioned considerations indicate the complexity that needs to be considered when designing policies and innovative solutions for children as well as methods for experimentation and metrics for evaluation in the context of AI, while oversimplification of such factors might have consequences on children's current and future well-being.

## METHODOLOGY

As discussed in the previous section, children have special characteristics and needs that should be considered when designing and implementing AI systems that might be used by them and when considering experimentation methods for the evaluation of such systems.

We take into consideration the current urgency for legislation regarding AI-based systems and the increasing activity from national and international organizations regarding policy recommendations for AI, specifically for users under the age of 18.

The majority of these initiatives propose that in addition to the general policy frameworks and legislations for adult citizens, special considerations are needed regarding users under the age of 18.

For our work, we first consider the existing policy initiatives that have been developed specifically for children and we highlight the points that might need to be reconsidered, in light of children's rights. Then, we elaborate on the existing techniques for the design of sandboxes in different fields and we propose a framework for the design of a sandbox for AI and children, which we then combine with the canvas proposed by UNICEF for the operationalization of the policy guidance for AI and children. Finally, we present a proof of concept of the proposed framework with indicative examples for elaboration.

## OVERVIEW OF POLICY INITIATIVES FOR CHILDREN AND IDENTIFICATION OF COMMON CHARACTERISTICS

An increasing number of national and international organizations recognize the need for child-specific approaches to policy recommendations for the development of AI systems.

We take as a starting point the United Nations Convention on the Rights of the Child (UNCRC),[6] which is a legally binding international agreement setting out the civil, political, economic, social, and cultural rights of every child, regardless of their race, religion, or abilities. The convention has 54 articles, and it has become the most widely ratified human rights treaty in history, helping transform children's lives around the world. The UNCRC has been ratified by all EU Member States and neighboring countries.

Given the evolving and expanding digital environment and the opportunities and risks that the digital world brings for children, the UN adopted Comment 25[7] of the rights of the child in the digital environment after a large consultation with children and all interested parties. This Comment elaborates on the four basic principles of the Rights of the child, namely:

1. Nondiscrimination;
2. The best interest of the child;
3. Right to life, survival, and development; and
4. Respect for the views of the child.

In 2019, UNICEF and the Government of Finland launched a two-year project to focus specifically on children's rights in the context of artificial intelligence. The organization of a series of global and regional workshops with experts and children was instrumental to the development of those guidelines. UNICEF involved 245 children from different countries with an emphasis on underrepresented geographical areas, ran a series of workshops with experts, and reviewed 20 national strategies on AI to conclude the following requirements[8]:

1. Support children's development and well-being;
2. Ensure inclusion of and for children;

3. Prioritize fairness and nondiscrimination for children;
4. Protect children's data and privacy;
5. Ensure safety for children;
6. Provide transparency, explainability, and accountability for children;
7. Empower governments and businesses with knowledge of AI and children's rights;
8. Prepare children for present and future developments in AI;
9. Create an enabling environment for child-centered AI.

In May 2021, given the technological, legal, and policy advances, the OECD published a revision of the "Recommendation on the Protection of Children Online"[9] which was renamed "Recommendation on Children in the Digital Environment." Regarding AI-related activities for education, UNESCO adopted the UN General Comment 25 on Children's Rights in Relation to the Digital Environment, and following a similar approach to OECD, UNESCO categorized AI-based applications for education into two large categories that contribute to the improvement of learning and equity for all children,[10] namely:

1. AI to promote personalization and better learning outcomes; and
2. Data Analytics in Education Management Information Systems (EMIS) and the Evolution to Learning Management Systems (LMS).

In 2022, the Joint Research Centre (JRC) of the European Commission, considering the existing work on AI and children's rights conducted a series of workshops with experts, children, and policymakers and proposed an integrated agenda for research and policy.[11] This multistakeholder approach proposes the following requirements:

1. AI minimization, valuable purpose, and sustainability;
2. Transparency, explainability, communication, and accountability;
3. Inclusion and nondiscrimination;
4. Privacy, data protection, and safety;
5. Integration and respect of children's agency.

Overall, the abovementioned initiatives are aligned to a certain extent regarding the risks and opportunities of AI for children, but they differ in terms of goals and priorities. All initiatives recognize the need for practical, actionable methodologies that would pilot policy guidance for AI and children's rights and eventually support AI development and use.

A recent review of the existing major policy initiatives that have been designed especially for children, identifies ten principles that they seem to have in common, while UNICEF's framework includes all of them (Wang et al., 2022). For this reason, for the purposes of this work, we adopt UNICEF's framework, which we combine with a set of principles for the design of sandboxes specifically for AI and children.

It should be noted that the first regulatory actions, especially for age-appropriate AI, have started to appear, such as the recently signed the California Age-Appropriate Design Code Act,[12] commencing July 1, 2024. This will require a business that provides an online service, product, or feature likely to be accessed by children to comply with specified requirements. However, the examination of specific legal frameworks and Acts is beyond the scope of this chapter, and it will be considered in a later stage of our work.

## SANDBOXES FOR AGE-APPROPRIATE AI AND CHILDREN

UNICEF has developed a number of concrete steps to support the development and use of AI in a child-centered manner. This requires firstly a thorough knowledge of how children and AI systems intersect. However, according to UNICEF, "it is not adequate to simply mention human or child rights in the ethics chapters of AI documents (a common occurrence in national AI strategies)" (UNICEF, 2021).

Policymakers, leaders, and AI system developers should be aware and have sufficient knowledge of child rights, AI-related opportunities for children's development, and, where appropriate, the use of AI for the achievement of the SDGs, either for their policies or their products or services.

A starting point for any initiative should always be the question of whether an AI-based solution is the best approach for the issue at hand. Answering this question should force us to look at the AI system or use it from all perspectives. These considerations guide how AI should be developed and deployed. Note that AI technology is increasingly built-in into many products and services. Moreover, AI systems can be developed from scratch, bought off the shelf, or added to systems already in use.

The use of sandboxes that specifically support experimentation on children-related AI applications contributes to trusted and transparent AI solutions for children. Using sandboxes enables providers and developers to test and evaluate AI applications for safety and alignment with responsible and ethical principles. At the same time, a sandbox approach can support consumers and the wider public to understand the effects of AI technology and, where appropriate, raise demands for technology services to have the right safeguards in place.

Based on current best practices (German Federal Ministry for Economic Affairs and Energy, 2019), we propose a basic structure for a children-centered AI development sandbox (See also Figure 21.1):

1. **Define the purpose and scope of the application**:
   - What are the goals for the sandbox?
   - Who can participate?
   - What type of products/services are within the scope of the sandbox?
   - How will the findings be used?
   - How is it ensured that the regulators and practitioners can learn from the regulatory sandbox?
2. **Map the AI regulatory ecosystem, including policies/legal frameworks, and current industry standards**
   - What are relevant regulations and standards to be considered?
3. **Define responsibilities**
   - Who is responsible for supervision and evaluation?

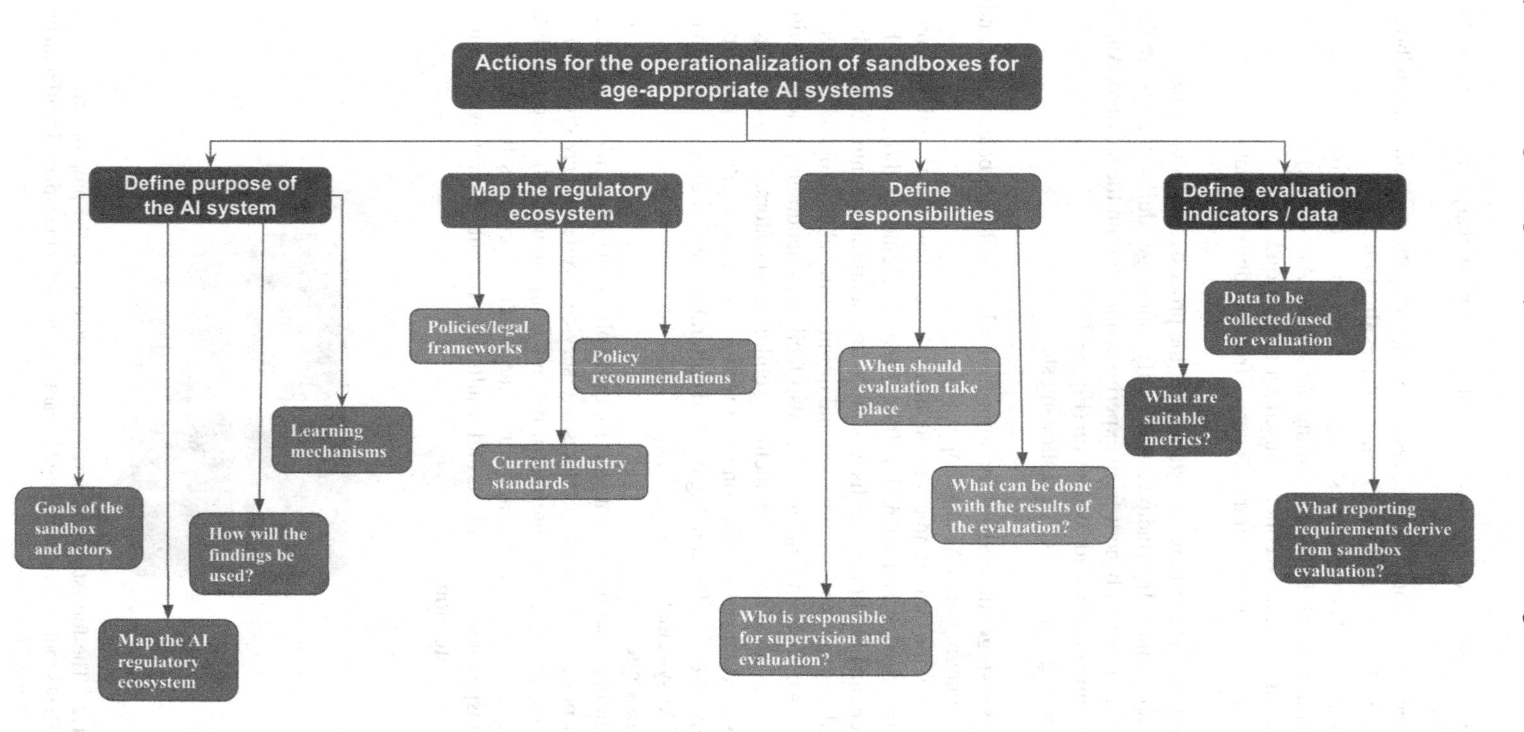


**FIGURE 21.1** Proposed framework for the development of sandboxes for age-appropriate AI and children and the required actions as discussed in Section "Sandboxes for Age-Appropriate AI and Children."

- When should evaluation take place (continuously, fixed moments, long-term)?
- What can be done with the results of the evaluation within the sandbox?

4. **Define indicators and data sources for evaluation**
   - What are suitable metrics? (e.g., objective and subjective measures of children's well-being concerning the specific AI system.)
   - Which data can be collected/used for evaluation?
   - What reporting requirements derive from sandbox evaluation?

Once the aims of the sandbox are established, the processes on how to use the sandbox can be determined. In principle, we identify four stages during the system development process that can benefit from experimentation within the sandbox Analysis, Design, Implementation, and Evaluation (Figure 21.2).

More specifically we define the following steps:

A. **Analysis stage**: understand the impact of AI on children, bearing in mind the national/regional AI landscape.
   - The sandbox environment implements the regulations and policy recommendations based on national/local inputs, including those from children, and facilitates the discussion of how a children-centered AI policy and requirements support national or international development plans.
   - The sandbox environment can also facilitate/encourage co-design with all relevant stakeholders, including children, educators, NGOs, researchers, and others working on child well-being in the digital world to perceive the social, technological, and cultural context of the use and impact of AI systems;

B. **Design stage**:
   - Within the sandbox, strategies for the application of the children-centered AI policy to concrete products and services are tested, creating a space to experiment with alternative options to implement child-centered AI requirements. At the same time, the sandbox provides the means for transparency, with projects and results open for public scrutiny and third-party evaluation.

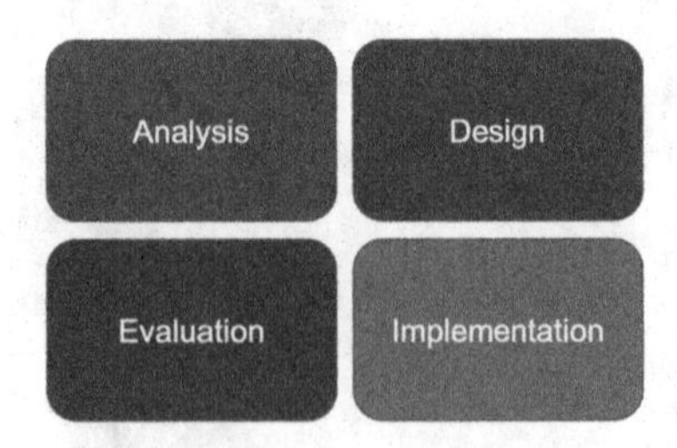


**FIGURE 21.2** The four stages of development according to the proposed framework for sandboxes. They should not be considered linear but as modules that appear in different phases as shown in Figure 21.3.

**TABLE 21.1**
**UNICEF's Canvas Proposed by UNICEF for the Operationalization of the Policy Guidelines for AI and Children**

| **Project / Application Title:** | | |
|---|---|---|
| **Date:** | **Version:** | **Authors:** |
| Purpose / Motivation: | | |
| Short description of the project: | | |
| Impact on children | Values/Principles | Regulation |
| How are children impacted directly or indirectly?<br>Who is responsible for the impact of the AI system on children? | Which child rights, human values, and ethical principles are upheld, and which are put at risk? | Which law/regulations apply to the AI system? |
| Design Requirements | | |
| Positive effects on children | Mitigating negative effects on children | Child development and well-being |
| Data protection and privacy | Fairness/nondiscrimination | Transparency |
| Accountability and redress | Safety | Inclusion |
| Explainability | Conscious users | Prepare for the future |

C. **Implementation stage**:
- Support to develop and execute different implementation strategies, including careful steps to test, redesign, and adapt design according to concrete use and context, while ensuring that implementation choices and decisions are grounded on the relevant legislation and existing guidelines

D. **Evaluation stage**:
- By verifying the adherence to the child-centered AI requirements, sandboxes provide support to the monitoring of the impact of different strategies, applying impact metrics, and leading to the update of design plans and the policies and strategies based on data-driven evidence.

To support these steps, we consider UNICEF's proposed canvas to analyze AI applications in a structured way[13]: Table 21.1 shows the main areas of consideration for the operationalization of the policy guidance, which includes the examination of existing regulations that would apply to the specific AI system and the evaluation of the system regarding its impact on children.

## PROOF OF CONCEPT: A 4-STEP, PROCESS-ORIENTED APPROACH

We consider UNICEF's model as appeared in Figure 21.1 for the operationalization of the policy guidance and we combine it with the structure we proposed in the section "Sandboxes for Age-Appropriate AI and Children" (Figures 21.1 and 21.2) to elaborate on the following framework for the design of a sandbox for AI and Children.

It should be noted that this framework should not be seen as a checklist; rather we encourage its elaboration as a process-based approach with the participation of all the involved stakeholders, including children.

In addition, the proposed framework is a modular tool, the elements of which can be used at the different stages of the development of the policy. Figure 21.3 presents a proof of concept of our proposed approach for the design of a sandbox for AI and children as appears in Figure 21.1 and the four stages of "Analysis, Design, Implementation, and Evaluation" with indicative examples. In each step of our framework, we provide potential answers, which are not exclusive. We highlight the elements that relate to the UNICEF guidelines as appear in Figure 21.2.

Figure 21.3 presents a proof of concept of the proposed framework with indicative examples of potential answers for the proposed Actions. We included all the elements that appear in the canvas proposed by UNICEF for the operationalization of the policy guidelines for AI and children (UNICEF, 2021). In addition, in the first column of Figure 21.3, we indicate at which stage of development each of the Actions should be considered, emphasizing the process-oriented approach and the iterative nature of our framework.

As seen in Figure 21.3, the proposed Actions and the questions to be answered per action are interrelated. However, we would like to highlight the importance of **transparency** throughout the process of experimentation. Transparent documentation of the whole process can contribute to the development of trust among the different stakeholders especially among the policymakers, developers, and users. This requires a special effort from all the involved stakeholders and constant dialogue among them. Resolving possible conflicts should be based on values that serve the best interest of the child.

In addition, since the specific framework is designed for the experimentation of policies and AI applications for children, an in-depth knowledge of **children's characteristics, development, and needs** is necessary. Action 1a of the framework (What are the goals of the sandbox) should be considered in relation to Action 4a (What are suitable metrics), which include aspects of the effectiveness of the policies, their technical implementation into a system, and the acceptability by and the impact (positive and negative) on children. As mentioned in the introduction of this chapter, the variability of the context in which children develop, as well as the large variability among children's cognitive and socio-emotional maturity and abilities, and their cultural differences should be taken into account for the design of systems that are inclusive and accessible to all (Charisi et al. 2020, Newbutt et al. 2022). As such, our framework for the development of sandboxes for age-appropriate AI and children proposes that testing should be made by taking into consideration the variety of contextual elements.

In a similar line, Action 1b of the proposed framework highlights the importance of accessibility of the process to underrepresented populations. This does not only include children as core stakeholders for the development of age-appropriate AI but also people who might require special support to participate in decision-making throughout the process.

Currently, we observe an increase in awareness of the need for policies and assessment tools for AI applications to be specifically designed for children, and many organizations are working toward this direction in addition to the scientific work

| Stage of development | Actions based on the proposed sandbox for age-appropriate AI and children | |
|---|---|---|
| Analysis<br>Design<br>Implementation | **1. Define the purpose and scope of the application** | |
| | **Question** | **Potential Answers** |
| | a. What are the goals for the sandbox? | Experimentation with policies and AI system implementations regarding children's rights, such as:<br>- support children's development and well-being<br>- ensure inclusion of and for children,<br>- prioritize fairness and non-discrimination,<br>- protect children's data and privacy,<br>- ensure safety for children<br>- provide transparency, explainability and accountability for children<br>Empower governments and businesses with knowledge of AI and children's rights<br>Prepare children for present and future developments in AI<br>Create an enabling environment for child-centered AI<br><br>This should be done in a safe environment for evidence-based regulation and development through dialogue. |
| | b. Who can participate?<br>- Accountability and redress<br>Inclusion and diversity of the team<br>Empowering children | Policymakers, developers, parents, educators, children, designers, and experts from various sectors, such as governmental and non-governmental institutions, industry, academia, schools, etc.<br>Design of sandboxes with accessibility in mind for inclusion of under-represented population |
| | c. What type of products/services are within the scope of the sandbox? | Smart toys, recommender systems, intelligent tutoring systems, chatbots, decision-making support applications, social robots, voice assistance, video |

**FIGURE 21.3** A proof of concept with indicative examples of the Actions included in the proposed framework for the design of sandboxes to experiment with policies for age-appropriate AI and children.

| | | |
|---|---|---|
| | | games, behavior-change applications, etc. |
| | d. How will the findings be used? | Evidence will be used to inform policymakers, developers, industry, practitioners, etc. |
| | e. How is it ensured that the regulators and practitioners can learn from the regulatory sandbox? | Documentation, transparency, dialogue, and exchanges throughout the whole process of the sandboxing. |

| Analysis<br>Design | 2. Map the relevant AI regulatory ecosystem | |
|---|---|---|
| | **Question** | **Potential Answers** |
| | a. Which policies/legal frameworks (child-specific but also non-specific) apply to the specific AI system? | Relevant policies on AI and children include California Age-Appropriate Design Code Act, which, commencing July 1, 2024<br>GDPR, etc. |
| | b. Which current industry standards apply to the specific AI system? | Standards published by IEEE, CEN-CENEL, and other relevant standards organizations. |
| | c. Which policy recommendations apply to the specific AI system? | Include existing policies by design<br>Perform an ethical assessment based on the current regulations and policy recommendations<br>Recommendations that need to be considered (See section 3 of this paper for a short overview). |

**FIGURE 21.3** (Continued)

| | 3. Define responsibilities | |
|---|---|---|
| Design | **Question** | **Potential Answers** |
| Evaluation | a. Who is responsible for supervision and evaluation of the product?<br><br>How can accountability be ensured? | Responsible parties should be identified in advance (e.g. the policymaker, the educator, the developer, etc). |
| | b. When should the evaluation take place? | At the end of every stage, at fixed moments or continuously, after the implementation |
| | c. What can be done with the results of the evaluation within the sandbox? | Redesign of products/services, withdrawal of permission to operate |

| | 4. Define indicators and data sources for evaluation | |
|---|---|---|
| Design | **Question** | **Potential Answers** |
| Implementation<br>Evaluation | a. What are suitable metrics? | Effectiveness of the policies-under-experimentation in different contexts (see Table 1)<br>Acceptability of the policies<br>- by the policy institution<br>- by the users and the citizens<br>- by experts<br>- by the company<br><br>Effectiveness of the product<br>- Positive effect on children<br>- Negative effect on children<br>Inclusion of children's needs<br>- Children's agentic capacity<br>- Is the product playful by design |

**FIGURE 21.3** (Continued)

| | | |
|---|---|---|
| | b. What data can be collected/used for evaluation? | - Certify that suppliers have the needed knowledge of AI methods and techniques<br>- Request information about data governance mechanisms that are in place, including information about training data<br>- Address any bias or flaws of the (training) data before using the AI applications.<br>- Revisit the AI impact assessment at key decision points during the procurement process.<br>- Redesign of products/services<br>- withdrawal of permission to operate |
| | c. What reporting requirements derive from sandbox evaluation? | Possible ways to address socio-ethical requirements<br>Transparency:<br>- include different stakeholders<br>- make available reports on choices and decisions, risks assessment, and on usage incidents<br>Robustness:<br>- check if the system meets performance requirements and behaves as intended<br>Accountability:<br>- Possible ways to address: external 'AI Ethics' board, internal 'AI Ethics Officer'<br>Human and environment well being:<br>- eg. demand use of green energy in any data centres or high-performance computing needed for the project |

**FIGURE 21.3** (Continued)

made by academic institutions. As indicated in Action 2 of our framework, existing initiatives can provide valuable guidance when experimenting with sandboxes for age-appropriate AI and children.

Lastly, Action 4 indicates the need for valid and scientifically sound metrics to measure not only the effectiveness of policies and innovative solutions but also their acceptability and ecological validity in generalizable terms, treating all children as having the same opportunities for protection and participation.

## CONCLUSIONS

Among the great challenges posed to democracy today is the use of technology, data, and automated systems in ways that threaten the rights of children. Although these systems have brought about extraordinary benefits, the use of algorithms has put children's well-being at risk: their privacy violations, the excessive use of online services and exposure to content that is particularly harmful to children raise many questions with respect to safeguarding the rights of the child. According to the UN Convention on the Rights of the Child and the EU Fundamental Rights Charter children have a right to protection and participation.

The analysis presented in this chapter suggests that the use of regulatory sandboxes could be a positive means to experiment and test regulatory frameworks and AI applications for children, enabling policymakers, providers, and developers to evaluate AI applications for their impact on children's rights, safety, and well-being.

This will provide means to operationalize child-centered approaches to policy recommendations for the development of AI systems, such as those put forward by UNICEF, the UN, the OECD, UNESCO, and the JRC of the European Commission. To realize these objectives, it is important to highlight the need for policymakers, leaders, and AI system developers to have sufficient knowledge of child rights and AI-related opportunities for children's development.

Ultimately, building responsible AI requires a focus on these three interconnected core principles: accountability, responsibility, and transparency [Dignum 2019, Charisi et al. 2022]. While the first two, must be solved through socio-legal means and intervention by policymakers, the last requires novel engineering solutions (Charisi et al. 2017) and the establishment of good-design practices, based on our knowledge about child development, while legislation can enforce the adoption of such practices. Yet, to ensure the long-term adherence of our socio-technical systems to such good practices, all major stakeholders must stop considering ethics and the upcoming AI-related standards as an afterthought or "obstacles to be dealt with in time." Instead, policy considerations and governance mechanisms to satisfy said considerations should become integral parts of any project implemented and evaluated throughout the AI system's lifecycle.

## NOTES

1 https://eur-lex.europa.eu/legal-content/EN/TXT/?uri=celex:52021PC0206
2 https://eur-lex.europa.eu/legal-content/EN/TXT/HTML/?uri=CELEX:52022DC0332
3 https://docs.google.com/spreadsheets/d/1YIKOCzVNmM9eM7y4gmq7XIxJ38-orXCp8nSp0DEIeIc/edit#gid=566110847
4 https://algoritmeregister.amsterdam.nl/en/ai-register/
5 https://www.europarl.europa.eu/RegData/etudes/BRIE/2022/733544/EPRS_BRI(2022)733544_EN.pdf
6 https://www.ohchr.org/en/professionalinterest/pages/crc.aspx
7 https://www.ohchr.org/EN/HRBodies/CRC/Pages/GCChildrensRightsRelationDigitalEnvironment.aspx
8 https://www.unicef.org/globalinsight/media/2356/file/UNICEF-Global-Insight-policy-guidance-AI-children-2.0-2021.pdf
9 https://legalinstruments.oecd.org/en/instruments/OECD-LEGAL-0389

10 https://unesdoc.unesco.org/ark:/48223/pf0000377897
11 https://publications.jrc.ec.europa.eu/repository/handle/JRC127564
12 https://leginfo.legislature.ca.gov/faces/billCompareClient.xhtml?bill_id=202120220AB2273&showamends=false
13 https://www.unicef.org/globalinsight/media/1166/file/UNICEF-Global-Insight-tools-to-operationalize-AI-policy-guidance-2020.pdf

## REFERENCES

Allen, Hillary J. (2019). Regulatory Sandboxes. *George Washington Law Review*, vol. 87, 579–645.

Baker, S. T., Le Courtois, S., & Eberhart, J. (2021). Making space for children's agency with playful learning. *International Journal of Early Years Education*, 1–13.

Bandura, A. (2006). Adolescent development from an agentic perspective. *Self-efficacy Beliefs of Adolescents*, *5*(1–43).

Bolton, J. L., Short, A. K., Othy, S., Kooiker, C. L., Shao, M., Gunn, B. G., ... & Baram, T. Z. (2022). Early stress-induced impaired microglial pruning of excitatory synapses on immature CRH-expressing neurons provokes aberrant adult stress responses. *Cell Reports*, 38(13), 110600.

Charisi, V., Chaudron, S., Di Gioia, R., Vuorikari, R., Escobar Planas, M., Sanchez, M. J. I., & Gomez Gutierrez, E. (2022). *Artificial intelligence and the rights of the child: Towards an integrated agenda for research and policy* (No. JRC127564). Joint Research Centre.

Charisi, V., Dennis, L., Fisher, M., Lieck, R., Matthias, A., Slavkovik, M., ... & Yampolskiy, R. (2017). Towards moral autonomous systems. arXiv preprint arXiv:1703.04741.

Charisi, V., Gomez, E., Mier, G., Merino, L., & Gomez, R. (2020a). Child-robot collaborative problem-solving and the importance of child's voluntary interaction: a developmental perspective. *Frontiers in Robotics and AI*, *7*, 15.

Charisi, V., Malinverni, L., Schaper, M. M., & Rubegni, E. (2020b). Creating opportunities for children's critical reflections on AI, robotics and other intelligent technologies. In *Proceedings of the 2020 ACM interaction design and children conference: extended abstracts* (pp. 89–95).

Council of the European Union (2022) https://data.consilium.europa.eu/doc/document/ST-14954-2022-INIT/en/pdf

Datatilsynet (2022). Sandbox for responsible artificial intelligence https://www.datatilsynet.no/en/regulations-and-tools/sandbox-for-artificial-intelligence/

Dignum, Virginia. *Responsible artificial intelligence: how to develop and use AI in a responsible way*. Springer, 2019.

European Commission (2022). Launch event for the Spanish Regulatory Sandbox on Artificial Intelligence https://digital-strategy.ec.europa.eu/en/events/launch-event-spanish-regulatory-sandbox-artificial-intelligence

German Federal Ministry for Economic Affairs and Energy, 2019). https://www.bmwk.de/Redaktion/EN/Publikationen/Digitale-Welt/handbook-regulatory-sandboxes.pdf?__blob=publicationFile&v=2

Greene, S., & Nixon, E. (2020). *Children as Agents in Their Worlds: A Psychological–Relational Perspective*. Routledge.

Hupont Torres I, Charisi V, De Prato G, Pogorzelska K, Schade S, Kotsev A, Sobolewski M, Duch Brown N, Dunker C and Vespe M. 2023. *Next Generation Virtual Worlds: Societal, Technological, Economic and Policy Challenges for the EU*. Publications Office of the European Union, Luxembourg (Luxembourg). https://publications.jrc.ec.europa.eu/repository/handle/JRC133757

IFLR (2019). Japan: sandbox reform https://www.iflr.com/article/2a63afeimueh7mpx1lb0g/japan-sandbox-reform

Kert, K., Vebrova, M. and Schade, S. (2022). Regulatory learning in experimentation spaces, European Commission, JRC130458.

Knight, B. R., & Mitchell, T. E. (2020). The sandbox paradox: Balancing the need to facilitate innovation with the risk of regulatory privilege. *The South Carolina Law Review*, *72*, 445.

Kumar, P., Naik, S. M., Devkar, U. R., Chetty, M., Clegg, T. L., & Vitak, J. (2017). 'No Telling Passcodes Out Because They're Private' Understanding Children's Mental Models of Privacy and Security Online. *Proceedings of the ACM on Human-Computer Interaction*, *1*(CSCW), 1–21.

Leckenby, E., Dawoud, D., Bouvy, J. et al. The Sandbox Approach and its Potential for Use in Health Technology Assessment: A Literature Review. *Applied Health Economics and Health Policy* 19, 857–869 (2021). https://doi.org/10.1007/s40258-021-00665-1

Manyukhina, Y., & Wyse, D. (2019). Learner agency and the curriculum: A critical realist perspective. *The Curriculum Journal*, *30*(3), 223–243.

Newbutt, N., Rice, L., Lemaignan, S., Daly, J., Charisi, V., & Conley, I. (2022). Co-designing a social robot in a special educational needs school: Listening to the ambitions of autistic children and their teachers. *Interaction Studies*, *23*(2), 204–242.

Oudeyer, P. Y., & Smith, L. B. (2016). How evolution may work through curiosity-driven developmental processes. *Topics in Cognitive Science*, *8*(2), 492–502.

Pelz, M., & Kidd, C. (2020). The elaboration of exploratory play. *Philosophical Transactions of the Royal Society B*, *375*(1803), 20190503. https://doi.org/10.1098/rstb.2019.0503

Ranchordás, S. (2021). Experimental Regulations for AI: Sandboxes for Morals and Mores. *Morals & Machines*, *1*(1), 86–100. https://doi.org/10.5771/2747-2021-1-86

Raviv, L., Lupyan, G., & Green, S. C. (2022). How variability shapes learning and generalisation. Trends in cognitive sciences.

Sorbring, E., & Kuczynski, L. (2018). Children's agency in the family, in school and in society: implications for health and well-being. *International Journal of Qualitative Studies on Health and Well-Being*, *13*(sup1), 1634414.

Stoilova, M., Nandagiri, R., & Livingstone, S. (2021). Children's understanding of personal data and privacy online–a systematic evidence mapping. *Information, Communication & Society*, *24*(4), 557–575.

Synofzik, M., Vosgerau, G., & Newen, A. (2008). Beyond the comparator model: a multifactorial two-step account of agency. *Consciousness and Cognition*, *17*(1), 219–239.

Tooley, U. A., Bassett, D. S., & Mackey, A. P. (2021). Environmental influences on the pace of brain development. *Nature Reviews Neuroscience*, *22*(6), 372–384.

UNICEF (2021a). Policy Guidance on AI for Children https://www.unicef.org/globalinsight/reports/policy-guidance-ai-children

UNICEF (2021b). Tools to operationalize the UNICEF policy guidance on AI for children https://www.unicef.org/globalinsight/media/1166/file/UNICEF-Global-Insight-tools-to-operationalize-AI-policy-guidance-2020.pdf

Wang, G., Zhao, J., Van Kleek, M., & Shadbolt, N. (2022). Informing Age-Appropriate AI: Examining Principles and Practices of AI for Children. In *Proceedings of the 2022 CHI Conference on Human Factors in Computing Systems* (pp. 1–29).

Wolfe, M. (1978). Childhood and privacy. *Children and the Environment*, 175–222.

# 22 Towards the Social Acceptability of Algorithms

*Marina Teller**
Université Côte d'Azur—GREDEG, Nice, France

## INTRODUCTION: TOWARDS A WORLD GOVERNED BY ALGORITHMS

Law and computer science are two very close disciplines. First, law is a language that is embodied in lines of code, just like computer science. Second, the law expresses values as well as an underlying logic. Legal reasoning is the result of a syllogism that confronts factual situations with rules. This process is typically algorithmic. Law and algorithms are therefore closely related, both in form and substance. This proximity is such that it may have led to fears of a subversion of the legal field by algorithmic logic. Indeed, the introduction of algorithmic devices coupled with artificial intelligence systems leads to a form of digitization of law, whose deeper ambition would be to substitute a technological standard for a legal norm (Teller, 2020a).

These debates are not just technical. They reveal another, even more profound confrontation that is ideological, even political. Indeed, the legal rule has an anthropological function: it enshrines values, reserving a privileged place for notions, such as fundamental rights, freedom, justice, free will, and consent. The law is also the mirror of a society, and it embodies its essential values.

What is the effect of technology on the law? It is difficult to summarize these effects because they are numerous and diffuse. We can say, in summary, that the law is reduced to a rational approach, which reduces intrinsically human experiences to a processing of data. The law is assessed in the light of data processing: indeed, the person is assimilated to a *data subject*, according to the terminology of Article 4 of the GDPR, that is to say, a data provider. The algorithmization of all our lived experiences gives rise to a particular regulation that academics have called "algorithmic governance" (Rouvroy et al., 2013).

This governance model is based on technology and statistical power. It has the effect of disqualifying the subject of law as a person to see in him only a transmitter of data, a unit of account soluble in *big data* (Rouvroy, 2018). The management of

* Professor, Université Côte d'Azur – GREDEG, CNRS; 3IA Côte d'Azur, Chaire Droit économique et Intelligence Artificielle. This work benefited from State aid managed by the National Research Agency under the Investissements d'Avenir 3IA Côte d'Azur project with the reference n° ANR-19-P3IA-0002.


DOI: 10.1201/9781003320791-26

people by data processing then transforms the regulation of social interactions and this is the whole anthropological dimension of the topic. Let us not be fooled by the technological aspect of algorithms: under the code, there are truly and essentially political questions (Boddington, 2020) that do not appear clearly. It is an overlooked aspect in academic literature dedicated to the societal aspects of AI: in order to truly conceptualize the idea of human-centered AI, it is necessary to consider the subject from a political standpoint of governance. Governance, in fact, serves as the gateway to essential concepts for life in society, such as consensus, debate, voting, and collective deliberation. If human-centered AI is defined as systems that are designed by people, for people, and with people, in such a way that the ultimate design aim is the promotion of human flourishing, then it is imperative to contemplate the governance of AI, particularly for the last aspect of the definition: "with people."

**What is "algocracy"?**—The power of algorithms competes with that of traditional regulatory forces such as the market and bureaucracy (Aneesh, n.d.). Academics have focused on the institutional dimension of algorithmic power, which is analyzed as an unofficial counter-power guided by computer code (Chrisley, 2020).

This new power is called "algocracy" (Danaher, 2016) and it transforms the relationship between law, enforcement, and free will. Indeed, rules are embedded within computer code, leading to a loss of understanding about what is at stake for citizens or subjects of rights (Kitchin, 2017: 14). What happens to the freedom and autonomy of individuals in ecosystems governed by algorithmic programming? What about the intention and willingness to comply—or not—with the rule? This "algocracy," seen as the omnipresence and omnipotence of algorithms, transforms the very idea of decision-making, which is based on conscious and voluntary deliberation. Free and informed decision-making would risk being called into question in its essence by these technological rules that have given rise to algorithmic law (G'Sell, 2020: 86). The fear is that artificial intelligence tools go beyond their role as mere decision-making aids. They have the potential to gradually but most certainly become substitutes for decision-making (Teller, 2020b: 461).

Finally, the highly technical nature of the algorithmic rule makes it easier to escape the control of the rule of law. The legal system is undergoing an "algorithmic turn" (Restrepo Amariles, 2020: 133) which therefore raises essential questions: is it compatible with the democratic model of the rule of law? Is algocracy part of the logical continuation of democracy and technocracy? Is it quite naturally the "*next step or the next move*" (Bersini, 2018)? Should we fear the slide from a "coup d'état" to a "*coup data*" (Basdevant, 2020: 46)?

This chapter aims to explore the interactions between artificial intelligence (AI) and law: more specifically, it focuses on the consequences of AI in the legal and social fields. As Marshall McLuhan said, "We shape our tools, and then our tools shape us." How can new technologies affect our legal and political systems? Is democracy in danger? Indeed, the introduction of new technologies is not neutral for a given society and this can affect several different categories of people at the same time: consumers, users, but also citizens. However, individuals are never consulted, and their opinion is not considered to assess the societal balance related to new technologies. As the effects of new technologies are increasingly impactful, it becomes necessary to ensure the consensus of citizens on a large scale. Technology is now becoming a political issue. If we want to establish the principle of human-centered AI as an effective principle with tangible effects for society, it is necessary to reflect on the conditions for the acceptance of AI systems by individuals. Indeed, individuals are more than just users of AI systems; they are providers of data. They are therefore at the core of AI at all levels. The main challenge lies in establishing a framework that ensures the will of individuals by granting them the opportunity to express themselves and consent to the use of algorithmic systems. In a nutshell, AI must be conceived "with people" and not solely within the laboratories of data scientists or big tech companies. That is why, to provide a comprehensive understanding of the concept of human-centered AI, it is necessary to also focus on the conditions for its acceptance by society.

This chapter calls this issue "the social acceptability of algorithms." The aim of our proposal is to rethink the processes of promulgation of our legal rules to strengthen the conditions for the democratic debate that must preside over the development of any norm (Teller, 2022a). The ambition is therefore to move from algocracy to algorithmic democracy (section "From Algocracy to Algorithmic Democracy"), which implies rethinking the conditions to guarantee real social acceptability of algorithms (section "Toward the Social Acceptability of Algorithms").

## FROM ALGOCRACY TO ALGORITHMIC DEMOCRACY

Algorithms have colonized every space of our daily lives: advertising, IoT, health, transport, education, public administrations, "smart cities," LegalTech, cryptoassets, etc. Many professions as well as legal specialties are now confronted with algorithmic systems, which profoundly modifies the balance in terms of governance.

The legal framework has a particular role to play here: indeed, it is necessary to rethink the interactions between humans and algorithms from their design phase. This is an essential condition for ensuring a particular form of explainability, to open the "black box." Mandatory disclosure of computer code, as may be required in the context of algorithmic transparency, is not sufficient to account for algorithmic logic. Indeed, to understand the effects of an algorithmic system, it is necessary to have access to its internal logic (code and database), in order for example to detect possible biases. The legal framework is therefore essential because it also makes it possible to ensure upstream the possibility of human intervention. This is what gave birth to the concept of "human guarantee" or human in the loop.

Algorithmic governance thus requires a specific legal framework (section "A Specific Legal Framework"). Indeed, if the algorithm can indisputably claim efficiency, the debate must first focus on its legitimacy (section "Effectiveness *versus* Legitimacy").

### A Specific Legal Framework

**The challenge of complexity**—In his *Traité d'algocracy*, computer scientist Hugues Bersini (2023) announces the increasing steering of public policies by algorithmic systems due to the complexity and multiplicity of crises to be managed: global warming, resource scarcity, migratory movements, economic inequalities, energy transition, distribution of commons, global pandemics, etc. (Bersini, n.d.).

Computer science would be the only truly effective response because there is a common point to all these crises: they make inevitable trade-offs in terms of selection and preferences (thus, selecting energy-intensive or frugal practices, virtuous or prohibited behaviors from an ecological point of view, setting up the categories of population eligible for vaccination and according to what priority, etc.). These trade-offs are complex, and policymakers will be tempted to find help through a variety of algorithmic tools: algorithms that recommend, advise, select, or constrain, with a view to improve or optimize. These tools will select or decide based on pre-established criteria and instantaneously. Most certainly, this fully automated handling of public decisions will arouse resistance (which we want) and debates (which we also hope).

**Thinking algorithmic governance**—We note that in some legal fields, supervisory or regulatory authorities have already validated this "algorithmization" and are in favor of the use of algorithmic tools and artificial intelligence systems. However, the missing links in AI governance have been highlighted by academics (UNESCO-Mila, 2023) and the topic remains complicated or even subversive. Regulators have even begun to consider reforming some governance rules.

This is particularly visible in the banking and financial sector. Thus, public recommendations and white papers directly target the governance of algorithms (ACPR, 2020) so that actors integrate new issues: new risks (cyber risks and risks related to the outsourcing of models, hosting or technical skills), new audit missions and revision of validation functions (to integrate compliance concerns "by design," i.e. during the design of an algorithm, but also throughout the life cycle of the algorithmic system, in order to ensure continuous compliance with evaluation principles, such as adequate data processing, absence of instability, and validity of explanations of system decisions).

More fundamentally, academics have clearly identified that the fundamental issues of algorithmic governance tackle a new way of thinking about law with the appearance of a new writing that feeds a new myth, that of an organization of social relations without third parties and without law, by the sole set of algorithmic writings (Garapon & Lassègue, 2018). Let's say even more clearly the stakes: the algorithm is the modeling of a choice that preexists computer coding. This choice, which is a matter of political decision-making, must remain part of the public debate. A rule must be effective, but it must first be legitimate.

## Effectiveness *versus* Legitimacy

**A legal framework focused on data and market**—Algorithmic decisions are already well framed by law (CNIL, 2017; European Commission, 2021a, 2021b; European Commission, 2020). Important work has been carried out nationally and internationally to regulate artificial intelligence systems. Today, the risks associated with the "black box effect" are well identified and denounced: risk of bias (Bertail et al., 2019) and discrimination (Crawford, 2013), the opacity of decision-making processes (Burrell, 2016), difficulty in preventing the risk of collusion (Marty, 2017), and risk of technological capture and infringement of sovereignty (European Commission, 2022). In European law, legislators have already built a strong legal framework around data, including the General Data Protection Regulation (European Parliament and Council, 2016), the Regulation on the free flow of nonpersonal data

(European Parliament and Council, 2018), the Cybersecurity Regulation (European Parliament and Council, 2019a), and the Open Data Directive (European Parliament and Council, 2019b). The construction of the European Digital *Single Market* (DSM) is carried out through a succession of texts or initiatives in progress: *The Digital Markets* Act (European Parliament and Council, 2020), the *Digital Service* Act (European Commission, 2020), and the *Data Governance Act* (act (European Parliament and Council, 2020).

Undoubtedly, the legislator has understood the need to regulate the effects of algorithms, more specifically regarding their impact on markets and personal data. Nevertheless, it seems to us that one dimension is still relatively little present in the European regulatory framework: that of fundamental rights and the procedural legitimacy of decisions based on algorithmic artificial intelligence systems.

**The debate around processual legitimacy**—The legitimacy of the choices that govern algorithmic systems is a very important issue that underpins the engagement and trust of the public in technology. The acceptance of those who must apply it is part of the legitimacy of the rule. As Habermas recalled, "the more law is solicited as a means of political regulation and social structuring, the greater is the burden of legitimation that the democratic genesis of law must bear" (Habermas, 1997).

The rule must circulate between different places of dialogue and discussion forums. This will make it possible to postulate the rational and consented nature of the content of the norm (Chevallier, 2005). The legitimacy of the norm therefore remains conditioned by a discussion process that guarantees its social acceptability. It is necessary that "citizens can conceive themselves at any time as the authors of the law to which they are subject" (Habermas, 1997). The challenge is therefore to build the formal processes to achieve a real social acceptability of algorithms. It is here that the law finds its rightful place.

## TOWARDS THE SOCIAL ACCEPTABILITY OF ALGORITHMS

**The role of law**—Many ethical charters, declarations, and other "hard law" texts have already been put in place to regulate algorithms. Our objective is not to make a review (CAHAI Secretariat, 2020) but to question the place that is reserved, within these various texts, to the question of algorithmic legitimacy. As mentioned earlier, this question is crucial to fully grasp the concept of human-centered AI. However, this consideration is often neglected: the debates rather crystallize questions of trust, ethics, or responsibility. Legitimacy is rarely mentioned. Yet, it is essential that the public debate also addresses AI systems from the perspective of their legitimacy because algocracy is a very real source of destabilization: European authorities consider the risk that the technology overrides fundamental rights because "*given the intrusive nature of some applications or uses* of *AI, it could happen that the current framework on human rights, democracy and the rule of law does not protect us enough, or not in time*" (Council of Europe Study, 2020). It is therefore necessary to open the public debate on algorithmic regulation if we want to place the citizen at the heart of the issues.

The social acceptability of algorithms can first be achieved by the rules of law. Several approaches can be proposed: conventionally, we can ensure rights downstream of the creation of algorithmic systems (section "From a Risk-based Approach to a Rights-based Approach"). We can also propose more innovative procedures to involve the citizen upstream, at the algorithmic design stage: a bottom-up approach based on citizen participation appears to be a very inspiring proposition (section "Procedures for Citizen Involvement").

## From a Risk-Based Approach to a Rights-Based Approach

**The limits of a "risk-based approach"**—The regulation of artificial intelligence is being built before our eyes. The proposal for a Regulation of the European Parliament and of the Council laying down harmonized rules on AI (European Commission, 2021b) aims to regulate AI systems, based on a risk-based approach. This proposal is seen as a legal and symbolic advance to get algorithms out of "lawlessness." As innovative as it is, the text is far from perfect and raises concerns: indeed, the starting point of the proposal is AI systems, classified according to their risks and not people. The personal dimension, in terms of rights and freedoms, is not the anchor of the text: whether it concerns end users, mere data subjects, or other individuals affected by the AI system, the absence of any reference to the person affected by the AI system appears as a blind spot (Teller, 2022b) in the proposal. The European Data Protection Board (2021) regretted this technological prism to the detriment of individuals.[1]

**Another way: the "rights-based approach"**—A "rights-based" regulation (Castets-Renard, 2020) makes it possible to put the law back at the center of algocracy's issues. The aim is to highlight the risks that AI systems pose to fundamental rights, well beyond personal data breaches (Kaminski and G. Malgieri, 2019). This topic is serious enough to justify the proposal of new rights and new methods of regulation, to guarantee the "contestability" of algorithmic systems. Several proposals have been made and we can mention some of them: set up national audit platforms to test the code, under the supervision of an independent public authority; use statistical tools that can allow counterfactual assessment; and finally, generalize the mechanism of impact studies before any algorithmic process is put into circulation.[2] Other texts have enshrined specific rights, particularly regarding the processing of personal data, *via* the GDPR.

**The proposal for digital principles**—More recently, the European Commission has proposed a framework of digital principles (European Commission, 2022). This approach is interesting and reveals a change of direction: it is no longer the market or data that are targeted by the legislator, but the individuals. This would give citizens new rights, such as access to high-quality connectivity, sufficient digital skills, and fair and nondiscriminatory online services. These principles would be discussed in the context of a broad societal debate and could be enshrined in a solemn interinstitutional declaration by the European Parliament, the Council, and the Commission.[3] The purpose is clear: in the digital world, citizens and companies should have no fewer rights or protection than in the offline world. These new digital principles are aimed directly at citizens, businesses, administrations, and legislators: they are the starting point for any discussion of algocracy.

**Algorithmic explainability**—The enforcement of rights requires an upstream understanding of algorithmic processes. That is why it seems to us that the rights-and-principles approach should be accompanied by a general principle of algorithmic explainability that integrates several dimensions, including interpretability and auditability. These concepts come from engineering (Desmoulin-Canselier & Le Métayer, 2020) and they are not to be confused with transparency which is only a (very imperfect) way to understand algorithmic results by giving access to the source code of algorithms. Auditability characterizes the practical feasibility of an analytical and empirical evaluation of the algorithm and aims more broadly to obtain not only explanations for its predictions but also to evaluate it according to other criteria (performance, stability, data processing).

Experts still debate the distinction between "explainability" and "interpretability:" the concept of "explainability" is often associated with a technical and objective understanding of how an algorithm works (and would therefore be appropriate for the perspective of an audit engagement), whereas the notion of "interpretability" seems more linked to a less technical discourse (and would therefore rather be addressed to a consumer or an individual). The quality of algorithmic explainability also depends on the context. Thus, to explain *how* an algorithm works and *why* it makes this or that decision, several levels of explanation can be considered to take into account the nature of the recipient (professional, customer, general public).[4]

## Procedures for Citizen Involvement

**Proceduralize public debate—**Can we go even further and go beyond the logic of subjective rights (conferring rights on) to involve citizens in algorithmic governance? The proposal may come as a surprise, but American academics (Desai & Kroll, 2018) reminds us that this approach can be fruitful: in the field of financial security, the legislator has put in place procedures governing whistleblowers by the Sarbanes–Oxley Act in 2002. The objective was to encourage private actors to ensure the proper application of the law, by providing support to the Public Prosecutor's Office by denouncing practices deemed illegal (Desai & Kroll, 2018). In the field of artificial intelligence, the proposal would consist of relying on algorithm users, not to denounce shortcomings, but to establish discussion procedures in which they would be involved (Castelluccia & Le Métayer, 2019).

Given their potentially major effect on society, algorithmic decisions must be part of the public debate. Academics initiate a reflection on the principles of good governance ensuring a quality debate: these discussion procedures must involve all stakeholders, including experts from all disciplines, policymakers, professionals, NGOs, and the general public. They must be conducted rigorously by asking the preliminary question of the legitimacy of the use of an algorithmic solution.

In certain situations, a prohibition in principle could be envisaged, in cases of manifest infringement of fundamental rights (Hannah-Moffat, 2013) (such as fair trial and the presumption of innocence). These proposals sound interesting, but would they stand the test of practice? In this perspective, field observation is decisive and two ongoing projects deserve to be mentioned: the CITICODE project and the FARI institute, both laboratories for experimenting citizen involvement, on the initiative of Brussels computer scientist Hugues Bersini.

**CITICODE and FARI, the laboratories of citizen involvement—***"Algorithms must remain open, entirely in the public domain such that ideally, everyone can keep their say and their line of code, even clumsy, to write"* (Bersini, 2023). Combining theory with practice, Professor Bersini launched the CITICOD project, which concretizes the massive participation of citizens in the development of software devices.

This experiment is interesting because it has made it possible to identify the governance rules applicable to citizen participation in the writing of algorithms. Three successive processes are considered: an elective process (who could contribute to the writing of this software and how?); a pedagogical process (how to train citizens both in the good mastery and detailed understanding of the issues and in the algorithmic approach?); a deliberative process (how to proceed with the development and maintenance of these codes when their updating and continuous improvement are essential?).

The second experiment is the FARI Institute,[5] funded by the European recovery plans in response to the Covid crisis. Its aim is the development of AI algorithms for the management of Brussels public goods (such as access to employment, mobility, public health, energy transition, animal welfare, and administrative simplification).

These projects have as a common denominator a reinvention of the workings of representative democracy where the three powers that define it must be rethought in the light of algocracy. This finally presages "a new form of legislative power, by which ordinary citizens, accompanied by experts (including lawyers) and some elected officials, will question until their software incarnation the new coercive mechanisms allowing society to function more harmoniously." It will be necessary to carefully follow the results of these ongoing experiments that open the prospect of a real transition from algocracy to algorithmic democracy.[6]

In conclusion, we believe that the future of living together will closely depend on the conditions surrounding the introduction of new technologies, particularly AI. Consensus and collective deliberation must remain strong values that should not dissolve in technological design. Serious consideration should be given to better accommodating the will and consent of AI users. This aspect appears crucial to define the concept of human-centered AI, which aims to involve individuals in the decision-making process of algorithms, giving them the opportunity to engage in debate, make choices, and express themselves.

## NOTES

1 "Indeed, the obligations imposed on actors vis-à-vis affected persons should emanate more concretely from the protection of the individual and his rights. Thus, the EDPS and the EDPS urge legislators to explicitly address in the proposal the rights and remedies available to persons subject to AI systems."
2 For a more detailed analysis of these proposals, see Teller (2020b).
3 Four proposals are made as follows: (1) citizens with digital skills and highly qualified digital professionals, (2) secure, efficient, and sustainable digital infrastructures; (3) the digital transformation of companies; and (4) the digitalization of public services
4 This approach is proposed by the ACPR (2020). Depending on the audience, the explanation will be simple, functional, or technical.
5 See https://fari.brussels/ FARI will soon elicit citizens' participation through Citizen Panels, a platform where citizens will not only be asked for their opinions but also their active and regular involvement in different AI projects.
6 "It is up to us to decide, think and write the lines of code known to all that will help us live better. This is why I propose and defend a new form of governance in which citizens must be involved from the outset and throughout the writing of codes that are supposed to

constrain and circumscribe their behaviour. This is undoubtedly the only way to guarantee the legitimacy and acceptance of this increasingly invasive software in the organization and security of our mobility, the choice of our children's school, tax redistribution and crime prevention" (Bersini 2023).

## REFERENCES

ACPR (2020). Governance of artificial intelligence algorithms in the financial sector.

Aneesh, A. (n.d.) *Technologically coded authority: The post-industrial decline in bureaucratic hierarchies*. Stanford University. Retrieved from https://web.stanford.edu/class/sts175/NewFiles/Algocratic%20Governance.pdf

Basdevant, A. (2020). Le « coup data » numérique. *L'ENA hors les murs*, 2020(3), 46.

Benabou, V.-L. (2020). *Un droit vivant. Manifeste pour des juristes incarnés et sensibles à l'heure de l'intelligence artificielle. In Penser le droit de la pensée, Mélanges en l'honneur de Michel Vivant*. Dalloz, p. 715.

Bersini, H. (2018). Big Brother is driving you. Brèves réflexions d'un informaticien obtus sur la société à venir. Académie Royale de Belgique, coll. L'académie en poche (2e édition).

Bersini, H. (2023). *Algocratie*. Deboeck.

Bersini, H. (n.d.). Big Brother is driving you. Brief reflections of an obtuse computer scientist on the society to come.

Bertail, P., Boune, D., Clemençon, S., & Waelbroeck, P. (2019). *Algorithms: bias, discrimination and equity*. Télécom ParisTech.

Boddington, P. (2020). Normative Modes: Codes and Standards. In M. D. Dubber, F. Pasquale, & S. Das (Eds.), *The Oxford Handbook of Ethics of AI* (p. 125).

Burrell, J. (2016). *How the machine 'thinks': Understanding opacity in machine learning algorithms*. Big Data & Society.

CAHAI Secretariat. (2020). Towards a regulation of AI systems: International perspectives on the development of a legal framework for Artificial Intelligence (AI) systems based on Council of Europe standards in the field of human rights, democracy and the rule of law (December 2020). Compilation of contributions.

Castelluccia, C., & Le Métayer, D. (2019). Understanding algorithmic decision-making: Opportunities and challenges. Report for the European Parliament, 72.

Castets-Renard, C. (2020). Le Livre blanc de la Commission européenne sur l'intelligence artificielle: vers la confiance? *Dalloz*, 15, 837.

Chevallier, J. (2005). La gouvernance et le droit. In *Mélanges Paul Amselek*. Bruylant, 189.

Chrisley, R. (2020). A human-centered approach to AI ethics: a perspective from cognitive science. In M. D. Dubber, F. Pasquale, & S. Das (Eds.), *The Oxford Handbook of Ethics of AI* (p. 463).

CNIL. (2017). How to allow Man to keep the hand? Report on the ethical issues of algorithms and artificial intelligence.

Council of Europe Study DGI. (2020). Towards a regulation of AI systems. International perspectives on the development of a legal framework based on Council of Europe standards in the field of human rights, democracy and the rule of law (DGI(2020)16).

Crawford, K. (2013). The hidden biases in big data. *Harvard Business Review*.

Danaher, J. (2016). The threat of algocracy: Reality, resistance and accommodation. *Philos. Technol.*, 29, 245.

Desai, D. R., & Kroll, J. A. (2018). Trust but verify: A guide to algorithms and the law. *Harvard Journal of Law and Technology*, 31, 1.

Desmoulin-Canselier, S., & Le Métayer, D. (2020). Deciding with algorithms – What place for Man, what place for law? Dalloz, Les Sens du Droit.

European Commission. (2020). White Paper on Artificial Intelligence: a European approach to excellence and trust.

European Commission. (2021a). Communication from the Commission. Fostering a European approach to Artificial Intelligence. COM(2021)205 final.
European Commission. (2021b). Proposal for a Regulation laying down harmonised rules on artificial intelligence (Artificial Intelligence Act) and amending certain Union legislative acts, COM(2021) 206 final, 21 April.
European Commission. (2022). Communication from the Commission establishing a European Declaration on Digital Rights and Principles for the Digital Decade. COM(2022)27 final.
European Data Protection Board, EDPB-EDPS. (2021). Joint Opinion 5/2021 on the proposal for a Regulation of the European Parliament and of the Council laying down harmonised rules on artificial intelligence (Artificial Intelligence Act).
European Parliament and the Council. (2016). Regulation (EU) 2016/679 on the protection of natural persons with regard to the processing of personal data and on the free movement of such data.
European Parliament and the Council. (2018). Regulation (EU) 2018/1807 establishing a framework for the free flow of non-personal data within the European Union.
European Parliament and the Council. (2019a). Regulation (EU) 2019/881 on ENISA (European Union Agency for Cybersecurity) and on cybersecurity certification for information and communication technologies.
European Parliament and the Council. (2019b). Directive (EU) 2019/1024 on open data and re-use of public sector information.
European Parliament and the Council. (2020). Proposal for a Regulation on contestable and fair markets in the digital sector. COM/2020/842 final.
Garapon, A., & Lassègue, J. (2018). Justice digitale. Révolution graphique et rupture anthropologique. P.U.F.
G'Sell, F. (2020). Les décisions algorithmiques. In *Le Big Data et le Droit. Dalloz, coll. Thèmes et commentaires* (p. 86).
Habermas, J. (1997). *Droit et démocraties, entre faits et normes*. Gallimard.
Hannah-Moffat, K. (2013). Actuarial sentencing: An "unsettled" proposition. *Justice Quarterly*, 30(2), 270–296.
Kaminski, M. E., & Malgieri, G. (2019). Algorithmic Impact Assessments under the GDPR: Producing Multi-layered Explanations. *U of Colorado Law Legal Studies Research Paper*, 19–28. Available on SSRN: https://ssrn.com/abstract=3456224
Kitchin, R. (2017). Thinking critically about and researching algorithms. *Information, Communication & Society*, 20(1), 14–29.
Marty, F. (2017). Algorithmic pricing, artificial intelligence, and collusive equilibria. *Revue internationale de droit économique*, 2, 83–116.
Restrepo Amariles, D. (2020). Le droit algorithmique: sur l'effacement de la distinction entre la règle et sa mise en œuvre. In *Le Big Data et le Droit. Dalloz, coll. Thèmes et commentaires* (p. 133).
Rouvroy, A. (2018). Homo juridicus est-il soluble dans les données? In *Law, norms and freedom in cyberspace: Droit, normes et libertés dans le cybermonde, Liber amicorum Yves Poullet*. Larcier, p. 417.
Rouvroy, A., Berns, T., & Carey-Libbrecht, L. (2013). Algorithmic governmentality and prospects of emancipation. *Réseaux*, 177(1), 163–196.
Teller, M. (2020a). L'avènement de la Deep Law (vers une analyse numérique du droit?). In *Mélanges en l'honneur d'Alain Couret, Un juriste pluriel*. Coédition EFL-Dalloz.
Teller, M. (2020b). Intelligence artificielle. In *Le droit économique au XXIème siècle* (J.-B. Racine, dir.). LGDJ, p. 461.
Teller, M. (2022a). Vers l'acceptabilité sociale des algorithmes ou comment passer de l'«algocratie» à la démocratie algorithmique. *Revue pratique de la prospective et de l'innovation*, (1).
Teller, M. (2022b). Les droits fondamentaux à l'ère des neurosciences.
UNESCO-MILA. (2023). Missing links in AI governance.

# 23 Human-Centered AI for Industry 5.0 (HUMAI5.0)

## *Design Framework and Case Studies*

*Mario Passalacqua, Garrick Cabour, and Robert Pellerin*
Polytechnique Montréal, Montréal, Québec, Canada

*Pierre-Majorique Léger*
HEC Montréal, Montréal, Québec, Canada

*Philippe Doyon-Poulin*
Polytechnique Montréal, Montréal, Québec, Canada

## INTRODUCTION

Integrating artificial intelligence (AI) in the workplace has created many challenges and opportunities for human work. Increased human–automation collaboration is expected on physical or cognitive tasks. Several disciplines have echoed the fact that these new technologies automate part of the work steps in collaboration with human operators rather than replacing entire professions, such as Information Technologies (Seeber et al., 2020), economics (Frey & Osborne, 2017), work psychology (Parker & Grote, 2022) and human factors & ergonomics (Mueller et al., 2021). The area of Industry 4.0 (I4.0) is at the forefront of the digitalization of human work wherein AI plays a central role. I4.0 intends to increase production system capabilities in terms of productivity, repeatability, flexibility, real-time monitoring, and process standardization (Zheng et al., 2021). This is done by integrating a set of digital, robotic, and automated technologies into production (Kadir et al., 2019) and combining different digital solutions together (Zheng et al., 2021). The latest technological advances in I4.0 have increased the capabilities of machines in performing complex, cognitive tasks (Xiong et al., 2022). However, the development of I4.0 technologies follows a

 DOI: 10.1201/9781003320791-27

technocentric approach (Sony & Naik, 2020). Focusing on technology development first (Carayannis et al., 2022). Bibliometric analyses quantified the technocentric directions of I4.0. A recent literature review noted that out of a sample of 4885 studies with a search strategy that included the terms *Industry 4.0* and *Human Factors*, 4849 studies focused on technical factors and 36 on human factors (Passalacqua et al., 2022). This top-down approach often neglects the contextual factors that govern work systems and their potential integration into situated operational practices (Loup-Escande, 2022).

Consequently, increased attention and principles for more effective and safer human–AI collaboration have been proposed by academics, governments, and industrial groups (Kadir et al., 2019; Maddikunta et al., 2022; Neumann et al., 2021). The European Commission launched the Industry 5.0 (I5.0) stream to operationalize human-centricity, sustainability, and resiliency principles to align new technological development toward social values. New subfields emerged in response to the potential harm created by adopting a techno-centered design approach considering only the technical dimension of AI. First, several studies underlined the potential benefits that I5.0 could bring by relying on potential application cases and by formalizing avenues for future research (Jiao et al., 2020; Maddikunta et al., 2022; Seeber et al., 2020); Second a series of frameworks or design principles were proposed to integrate humans and technologies in collaborative work settings (Dubey et al., 2020; Mueller et al., 2021) or to consider human and organizational factors in digital transformation (Liao et al., 2020; Peeters et al., 2020). However, while these efforts have provided initial coarse-grained guidance, these general principles do not consistently translate into everyday design practices or final AI-based products. Design teams require systematic methods that provide fine-grained guidance to design new solutions that promote long-term human well-being, engagement, and system performance within complex socio-technical environments (Roth et al., 2018).

This chapter presents HUMan-centered AI for Industry 5.0 (HUMAI5.0), an innovative six-step framework that supports design teams in adopting a human-centered approach for designing and implementing AI-fused systems in the workplace. We apply this framework to two use cases in the I5.0 context. This chapter adopts a human factors' perspective wherein human-centered AI focuses on the collaboration between humans and AI systems in the workplace. The rest of the chapter is organized as follows: Section 2 delves into the literature from which the framework is derived; Section 3 presents the framework; and Section 4 applies it to two use cases.

## RELATED LITERATURE

### Existing Design Methods and Frameworks

#### Technology-Centered Design for AI and Limits

AI-infused solutions often follow a technocentric roadmap where figures of merit estimate how mature the hardware and software components are (Sony & Naik, 2020). When the solution reaches a satisfying level of technological readiness, development teams seek areas and problems where the device could be applied. While this top-down approach narrows the problem space by focusing only on technical details, it neglects the contextual and human factors that govern work systems (Spies et al., 2020). Moreover, techno-centrism often accompanies a desire to automate as many functions as possible to reduce human intervention in industrial processes, a source of uncertainty and failure in this philosophy (Millot et al., 2015). However, AI cannot, at present, fully acquire and reproduce tacit knowledge, socio-cognitive processes, and "*advanced know-how rooted in decades of institutional experience*" (Miller & Feigh, 2019; p.2). This technology could however support or automate a specific portion of a task in human–technology collaboration scenarios (Xiong et al., 2022). Complementary design methods that pay equal attention to human, organizational, and technical factors in design processes have emerged, which will be discussed in the following subsection.

## Complementary Design and Analysis Methods to Address Technocentrism Shortcomings

Implementing AI solutions in organizations introduces simultaneous social and organizational challenges in addition to technical ones (Kadir et al., 2019; Maddikunta et al., 2022; Neumann et al., 2021). Several authors advocated to adopt human-centered design approaches to overcome these challenges and ensure a successful transition to digitalized work situations (Neumann et al., 2021; Oppl & Stary, 2019). Human-centered design approaches place end-users at the center of product design (Rapp, 2021). Contrary to technocentric approaches, they are part of an inductive philosophy that considers users' characteristics and their working environment to *"inform and guide the product development process"* (Rapp, 2021; p.1). This subsection will explore the most relevant in the context of I5.0.

### *Systemic Methods of Human Work Analysis to Inform Design*

**Cognitive Work Analysis (CWA)** is a five-phase framework that guides the design and evaluation of socio-technical systems (Vicente, 1999). CWA allows us to gain a deep understanding of the constraints of the work domain, the system, the task, and the operator. CWA's first phase, the work domain analysis (WDA), is of particular importance for this chapter. WDA gives task-independent design information, allowing for better flexibility and adaptability to novelty or change. WDA identifies constraints in the work environment throughout five levels of hierarchy (abstraction hierarchy), ranging from most abstract to most concrete: functional purpose (end-goal of the system), abstract function (physics principles), generalized function (processes), physical function (equipment), and physical form (location and appearance) (Rasmussen, 1985). WDA is useful to map the means–end relationship that links the end-goal of the system to its physical implementation.

**Knowledge elicitation** is a family of methods and frameworks aimed at inferring the mental models and tacit knowledge of end-users (Shadbolt & Smart, 2015). Relying on advanced interviews and observation techniques, interviewers rely on timely presented prompts (e.g., pictures, documents, field notes, videotape of expert realizing an action) and systematic questions to guide the subject matter expert in the recall of situation-specific elements. Users can also be asked to think aloud while concurrently performing a task ("please tell me what you are doing and why"). The best-known methods in this category are Knowledge Audit (Gourova, 2009), Critical Decision Method (Klein et al., 1989), and Cognitive Task Analysis (Crandall et al., 2006).

*Integrative Design Frameworks for Industry 5.0*

**Systems Framework and Analysis Methodology**. Neumann and his colleagues (2021) proposed a five-step framework to systematically consider human factors in designing and implementing I4.0 systems. They proposed that this framework paints a comprehensive portrait of the socio-technical changes resulting from I4.0-related technology, which allows for specific design and implementation recommendations to avoid dysfunctional Human–Machine Interactions (HMI).

**Motivational Design for Human–Machine Interaction**. Szalma (2014) mobilized self-determination theory's conceptualization of motivation for the design of anthropocentric interfaces or systems. The premise of motivational design is the support of employee autonomy, competence, and relatedness in order to build socio-technical interactions that favor long-term work engagement, performance, and well-being (Szalma, 2009, 2014). He puts forward the idea that considering only hedonic and utilitarian goals is not sufficient for employee performance and well-being. Rather, motivational requirements must be considered when designing or implementing socio-technical systems (Szalma, 2014). Autonomy refers to the perception of being in control of one's own decisions and actions, competence refers to the perception of being capable and effective in one's work, while relatedness refers to the perception of positive/meaningful connections and relationships in the work environment.

**Interdependent analysis** is a framework developed by Johnson et al. (2014, 2017, 2018) to understand and design how human and automated systems can effectively team up to achieve a joint activity together. Instead of focusing on designing the machine architecture in traditional automation design, the method supports exploring systematically the design space of joint human–AI work. The goal is to identify the *interdependent relationships*—when an entity (human or artificial) needs support to complete the objectives of a given task—to design the joint work. Three distinct phases are required to deploy the method:

1) Model the joint activity: define the tasks, subtasks, and capabilities required to achieve the operational objectives
2) Assess team members' capacity to perform and capacity to assist with each task and subtask in the joint activity model
3) Identify all possible human–automation workflows: establish all potential pathways to complete the set of tasks and the associated teamwork requirements

## Limitations of the Existing Frameworks

Each of the presented frameworks has its strengths and limitations. Neumann et al. (2021) present a systematic methodology to properly address human factors in an I4.0 setting. It explains, in detail, how to apply each step of the framework in an I4.0 context. However, its proposed analysis of the technology and the human does not properly capture the complexity of the work domain in complex socio-technical systems. It also mentions the importance of examining the changing psychosocial environment (perception of the social environment), with no concrete recommendations.

The motivational design framework for HMI (Szalma, 2009, 2014) is meant to be an addition to other human factors frameworks instead of being a standalone framework. It provides a way to gain a deeper understanding of systems' psychosocial dynamics compared to Neumann et al.'s (2021) framework, which will result in design recommendations to enhance the HMI in socio-technical systems. In addition, it analyzes the operator's cognitive and affective traits. This analysis, absent from other HMI frameworks, can provide valuable insight into the operators' cognition at a deeper level of abstraction, ultimately resulting in more specific implementation recommendations. Although thorough, its scope of analysis is mostly limited to psychosocial interactions within a system. Using it in conjunction with other frameworks would thus be ideal.

Work analysis methods, such as CWA or knowledge elicitation frameworks, provide an in-depth understanding of the current state of knowledge and the work domain. However, transitioning from an in-depth analysis of the current system to designing future work situations is not appropriately supported by current standalone methodologies. Additionally, the work domain analysis does not account for the psychosocial characteristics of systems. It should therefore be used with the other frameworks.

Finally, interdependent analysis is the most comprehensive method that addresses the above drawbacks. As a formative tool, the interdependent analysis framework does not include a summative evaluation step for the proposed design. Its analytical lens is too coarse to grasp the specificities of complex socio-technical work systems, which suggests the need to complement it with other frameworks.

## HUMAN-CENTERED AI FOR INDUSTRY 5.0 (HUMAI5.0) FRAMEWORK

The presented frameworks and methods, found through a review of the literature, each have their benefits and limitations. Each seems to address, to varying depth, different components of the human–AI interaction. The proposed framework (see Figure 23.1) aims to mobilize the advantages of each to mitigate the drawbacks of others, creating an integrative framework to comprehensively understand the human–AI interaction.

**Step 1—Analysis**. This step involves analyzing the work domain and the humans interacting with it. For this step, we recommend using CWA's work domain analysis (Vicente, 1999) to understand the system's constraints. Other candidates' methods to complete these steps could be People, Activity, Context, Technology (PACT), or Activity Analysis (St-Vincent et al., 2014). More specifically, the WDA identifies constraints in the work environment throughout the five levels of hierarchy, that is, the abstraction hierarchy (Rasmussen, 1985). We also recommend using step 2 of Neumann et al.'s (2021), which identifies the human roles that interact with the

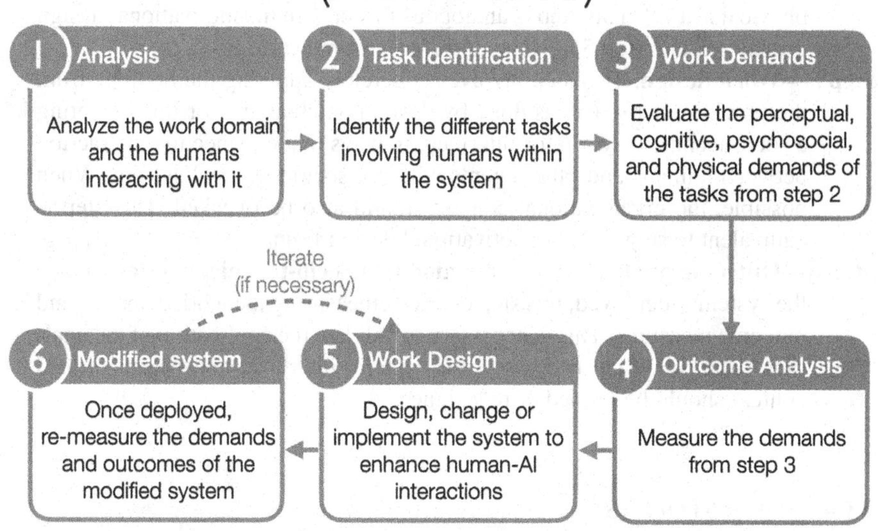


**FIGURE 23.1** HUMAI5.0 six-step framework.

system. Lastly, we recommend using the first step of the motivation design framework, which involves measuring users' cognitive and affective traits. Using these methods offers a more complete understanding of the system's environment.

**Step 2—Task identification**. This step entails identifying the different tasks involving humans within the system. If applicable, changes in task or job characteristics involving humans within the system should be identified. This refers to examining what tasks or job characteristics are added or removed for all humans involved with the system (e.g., additional subsystem monitoring, reduced physical workload). This is analogous to step 3 in Neumann et al.'s model.

**Step 3—Work demand analysis**. This step necessitates evaluating the perceptual, cognitive, psychosocial, and physical demands of the tasks identified through the previous step (step 4 in Neumann et al.'s framework). Additionally, we recommend identifying the environment and task features that affect (positively or negatively) operators' feelings of autonomy, competence, and relatedness (e.g., monotony, isolation, decision latitude). For example, task repetition and monotony can hinder these needs, while supervisory support and skill-developing programs can facilitate them. This refers to step 2 of the motivational design framework. Lastly, we recommend using the available literature to understand how the identified work demands affect (directly and indirectly) the success criteria.

**Step 4—Outcome analysis**. For this step, the perceptual, cognitive, psychosocial, and physical demands should be quantitatively or qualitatively measured using various tools (e.g., questionnaires, neurophysiology). Additionally, the relevant outcomes should be measured (e.g., performance, mental and physical health). This step is analogous to step 4 in the motivational design framework and step 5 in Neumann et al.'s framework.

**Step 5—Work design**. This step involves concretely applying the findings from the previous steps. This is done by designing, changing, or implementing a socio-technical system or interface, in ways that enhance the interaction between humans and other agents within a socio-technical system. When possible, the organizational context should also be tweaked. This step is equivalent to step 5 of the motivational design framework.

**Step 6—Outcome analysis of new or modified socio-technical system**. Once the system is deployed, this step entails remeasuring the work demands and outcome measures. This is necessary to validate the application of the analyses (step 5). When necessary, the system, interface, and organizational context should be refined or redesigned.

## 23.4 CASE STUDIES

In this section, we illustrate the application of HUMAI5.0 involving the (re)design of human work in collaboration with AI agents in two domains: error detection in manufacturing and quality control in aircraft maintenance.

### Error Detection in Manufacturing

#### Context

Within a snowshoe manufacturing plant, the final operator on the assembly line is tasked with error- and defect detection before the product is sent to the customer. In essence, this final operator is in charge of product quality control. Nondefective snowshoes are put on one side of the workstation, while defective ones are put on the other. Figure 23.2 shows the workstation used by the operators. Currently, the quality control is being done manually, that is, without support from any technological system. To aid the operator, the intention is to implement an artificial intelligence-based error detection system (AIEDS). This AIEDS will use computer vision to detect possible errors and defects within assembled snowshoes and present its results to the operator on a computer screen at their workstation.

Before proceeding with the implementation, HUMAI5.0 was applied to gain a complete understanding of the human–AI interaction. Two versions of the AIEDS were tested and compared to the current method of functioning (No AIEDS). In version A, the AIEDS instructs the operator whether an issue is present. It has perfect reliability (100%) in terms of error/defect detection and the AIEDS decides whether to send or not the snowshoe to the customer. In Version B, AIEDS has an imperfect reliability (83%), and the decision-making authority is shifted toward the operator who decides to send or not the product to the customer.

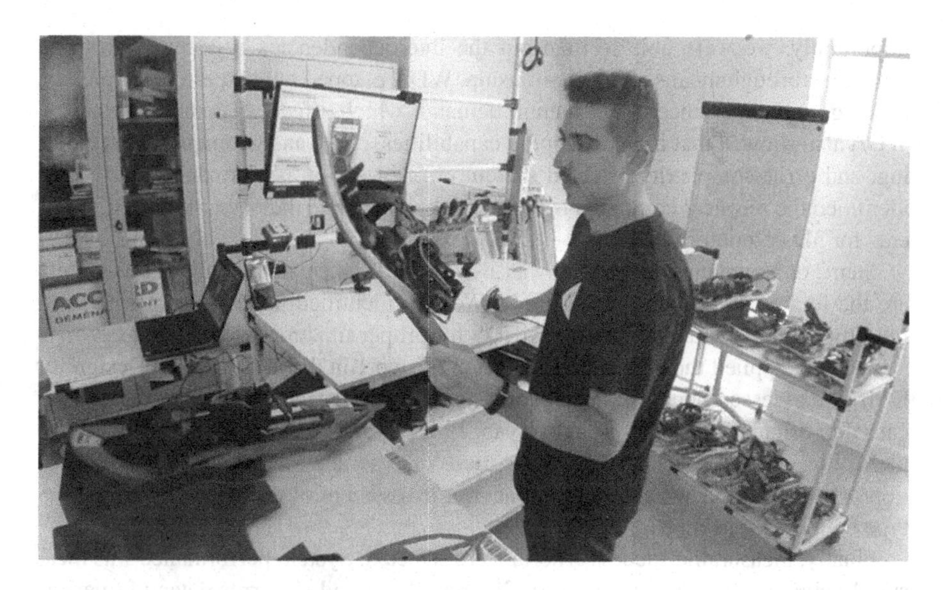

FIGURE 23.2 Workstation used by operators.

## Framework Application

*Step 1—Analysis*

CWA's work domain analysis was applied. It allowed us to first delineate the purpose of the system, that is, customer satisfaction through the delivery of defect-free snowshoes. Then, the domain values were derived (system performance, task engagement, balanced mental workload, balanced stress levels, etc.), followed by the domain functions (mental demand, motivation, image mapping capabilities, depth and geometry understanding, error complexity, training, experience, etc.). Most

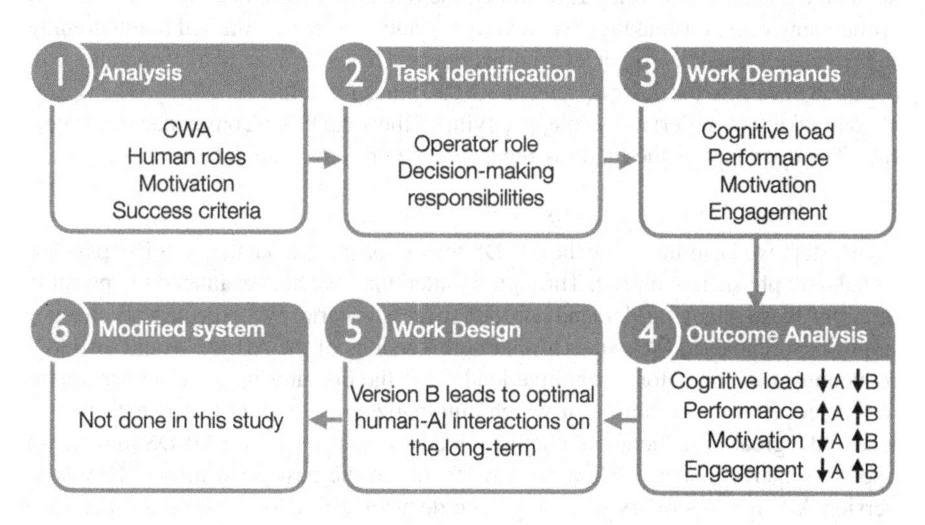


FIGURE 23.3 Framework application to case study #1.

importantly, we were able to map out the interdependencies between values and functions through means–ends relationship. WDA mapped out that error complexity, training, and experience affect mental demand, which affects system performance. WDA also showed that image mapping capabilities, depth and geometry understanding, and error complexity affected system error-detection performance. Lastly, we identified the physical components of the work system and their functions (e.g., cameras for 3D computer vision).

Step 2 of Neumann et al. (2021) was also applied. The main humans in this system are the final operators on the assembly. We also identified the engineers who must initially calibrate and maintain the AIEDS for proper functioning.

We then applied the user analysis portion of step 1 in the Motivational Design for HMI. We assessed operators' personality traits (trait engagement) using the general causality orientations scale (Deci & Ryan, 1985; Ryan & Deci, 2008). This questionnaire allowed us to control for predispositions to experience a certain level of motivation or engagement, which could impact our assessment of the effect of the AIEDS type on the various metrics.

Finally, measurable success criteria were selected. System performance was measured using (1) time taken to verify 30 snowshoes and (2) error-detection performance (percentage of snowshoes correctly categorized). Other important metrics derived from the work domain analysis, such as operator mental demand (workload), motivation, and engagement were measured using questionnaires, and physiological measures when possible. The questionnaires and physiological measures were chosen based on a search of the literature.

### *Step 2—Task Identification*

We identified how the work task is being modified by the introduction of the AIEDS. Using version A, the operator will no longer check for errors/defects in snowshoes. They will simply look at the computer screen to determine whether to put the snowshoe on one side or the other. Essentially, the operator's work becomes automatized without any decision-making responsibilities, and their role is limited to intervening if the system breakdowns or exceptions occur. Using version B, the operator will continue checking for defects, since the AIEDS has imperfect reliability. The operator will take on a supervisory role, verifying if the AIEDS's recommendation is correct. The operator has the final say regarding the presence of a defect.

### *Step 3—Work Demand Analysis*

In this step, we examined how the AIEDS may impact perceptual, cognitive, psychosocial, and physical demands. Through the literature, we also evaluated the possible effect of these changing demands on our success criteria. We expected no changes in physical demands. We expected that both versions of the AIEDS would lead to a reduction in the operator's cognitive load since the operator must no longer manually detect errors. Version A, in which automatization is highest, was expected to lead to the greatest reduction in cognitive load. We expected both AIEDS to increase operators' performance, with version A leading to the best performance. However, version A leaves operators with very little decisional freedom and high monotony. Therefore, we expected version A to lead to lowered perceived autonomy, intrinsic motivation, and task engagement. On the other hand, version B provides operators with full decisional latitude, that is, agreeing or overriding the AIEDS. Therefore, we expected version B to lead to improved perceived autonomy, competence, intrinsic motivation, and engagement.

*Step 4—Outcome Analysis*

To evaluate the impacts AIEDS has on the operator's task, we set up a pilot study in which nine engineering students participated. Participants were randomly assigned to one of three conditions (Manual, AIEDS version A, or AIEDS version B). Participants were trained to complete the task and interact with the AIEDS by detecting errors/defects. After this training task, participants completed the experimental task, which consisted of sorting 30 snowshoes, with six of them having a defect.

Performance time and error detection—Both versions of the AIEDS led to better performance time and error-detection rate than manual execution of the task. However, no differences were observed when comparing versions A and B.

Mental demand (cognitive workload)—Cognitive workload was the lowest for version A of the AIEDS. No differences were detected between version B and manual execution. When interpreting the mean values according to guidelines presented by the authors of the scale (Hart & Staveland, 1988), both versions of the AIEDS resulted in *moderate* mental demand, while manual execution led to *somewhat high* demand.

Perceived autonomy and competence—Version B of the AIEDS led to the greatest perceived autonomy. No differences were found between version A and manual execution. No differences were observed for perceived competence

Intrinsic motivation—We found that version B of the AIEDS led to the greatest intrinsic motivation. No differences were observed between version A and manual execution.

Task engagement—Perceived engagement was higher for version B of the AIEDS than for version A. No differences were observed between version B and manual execution, or between version A and manual execution. For physiological task engagement, version B was the highest, followed by manual execution, followed by version A.

*23.4.1.2.5 Step 5—Work Design*

Results from the outcome analysis indicate important differences between the three conditions. Both versions of the AIEDS led to a significant increase in performance, with no differences being observed when comparing versions A and B. However, version B led to better outcomes in terms of perceived autonomy, intrinsic motivation, and task engagement compared to version A, all while maintaining an acceptable cognitive workload. These outcomes are directly related to long-term performance, well-being, technology acceptance, employee retention, and organizational citizenship behavior (Deci et al., 2017; Van den Broeck et al., 2021). The outcome analysis therefore indicates that version B of the AIEDS will lead to the most optimal human–AI interaction in the long term and should thus be implemented.

**Conclusion**

Applying the steps to this case context allowed us to consider all aspects of the human–AI relationship. We systematically compared three methods of completing the work task to determine which would lead to the best outcomes for both the operator and the organization. While version A of the AIEDS was fully automatized and thus restricted operator autonomy, version B gave operators a greater decisional latitude by allowing them to have the final say over the AIEDS' recommendation. Partially automating the task led operators take on a more supervisory role, which was perceived as being more meaningful and gratifying, leading to more positive short-term outcomes than version A. In turn, these positive short-term outcomes will develop into positive long-term outcomes for both operators and organizations.

## Quality Control of Aircraft Maintenance

### Context

This real-world case study deals with integrating human factors in a research project to design a Cyber-Physical System (CPS) to automate aircraft components' inspection. Industrial inspection is a labor- and knowledge-intensive task carried out by skilled operators. Their objective is to ensure that aircraft components are free of harmful defects. Operators carefully examine each component for potential anomalies, and when one is detected, they evaluate and compare it to several decision criteria. A machine vision-based system was envisioned to support detecting and diagnosing the defects on service-run components such as gas turbines, discs, shafts, and fan blades to aid the inspectors in this tedious task. Starting from a proof of concept (Technology Readiness Level 3), the technology development plan was to reach Technology Readiness Level 6, that is, where the technology functionalities are demonstrated in a relevant (operational) environment. HUMAI5.0 was applied to incorporate human factors input into a project that follows a technology-centric design pathway (see Figure 23.4). In addition, previous implementation failures due to a mismatch between human and automated diagnostics have motivated project leaders to understand how inspectors perform an inspection in detail. The framework

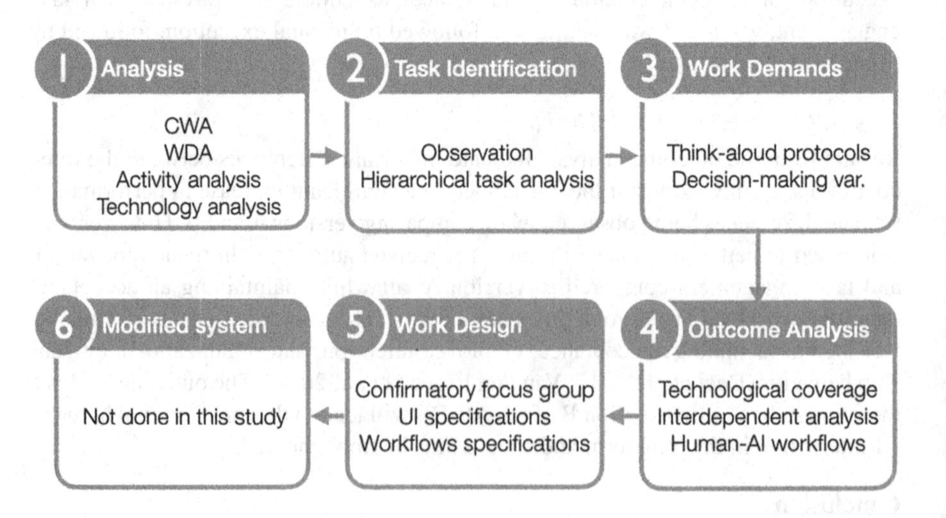


**FIGURE 23.4** Framework application to case study #2.

allowed us to understand, model, and evaluate how an inspection is performed manually. From this understanding, requirements were derived to inform and guide the engineering process while considering future joint human–AI inspection workflows.

Due to space constraints, this case study's details are presented in this book's online supplementary material. Readers are also invited to consult previous publications on this topic (Cabour, 2023; Cabour, Ledoux, et al., 2022a; Cabour, Morales-Forero, et al., 2022b).

## CONCLUDING REMARKS

The introduction of AI technology in complex socio-technical environments changes the operator's role and responsibilities in the workplace. In this chapter, we showed that adopting a human-centric approach to the design of human–AI collaboration in the industry leverages the best abilities of both parties to develop a work environment respectful to the long-term well-being of human workers and beneficial to the organization. To this end, we presented the HUMAI5.0 framework to analyze the work domain systematically, (re)design the human–AI interactions, and evaluate their outcomes. Indeed, the framework emphasizes the need for performing ethnographic studies to grasp the operational realities in which the AI-infused technology will be deployed. This emphasis resonates with other frameworks and methods cited in this chapter that rely on user-centered design for technology development projects, for example, CWA. The strength of HUMAI5.0 is to integrate these existing methods used in silo into an integrative design approach that enables a shared understanding among stakeholders of AI's impacts in their workplace and how best to address it. For example, what tasks are being fully automated or partially automated? How can we suitably redesign work for human–AI collaboration? Since the integration of AI changes work practices, these methods help to envision and design future work practices during technology development cycles (and not post-implementation) and thus avoid falling classical automation issues, such as task-technology mismatches (Loup-Escande, 2022; Spies et al., 2020). Case study 2 illustrates this by methodically designing effective human–AI workflows and user interfaces that align with the work demands. Case study 1 delves into the connection between work performance and motivation, exploring the more profound meaning people ascribe to their work.

The rise of AI has been a topic of much discussion in recent years, with many experts predicting a bleak future of mass unemployment (Employment & Skills, 2014; Forum, 2016). However, this chapter takes a more nuanced approach, acknowledging that while AI will undoubtedly transform the nature of work, it may not necessarily lead to widespread job loss. Instead, AI is more likely to automate specific tasks that humans previously carried out, while new tasks will emerge from the collaboration between humans and AI. For example, AI can monitor the human state, such as detecting signs of fatigue or stress, while humans can validate the output of AI systems for critical decisions. This transformation of work will require a more holistic view of the human–machine system. Rather than viewing humans and machines as separate entities, it's essential to consider them as a whole. HUMAI5.0 can help bridge this gap by enabling a more cohesive approach to design effective and humane joint human–AI work.

The chapter illustrated the successful application of HUMAI5.0 in two cognitively demanding domains, that is, error detection in manufacturing and quality control in aircraft maintenance. This also demonstrates the versatility of the approach by applying it to both experimental and naturalistic contexts. We believe the framework could be applied beyond industrial environments to support the shift from technocentric to more human-centric design in AI projects (e.g., social-environmental projects). Applying the framework to other domains would create opportunities to investigate how it supports decision-makers in considering human and organizational factors in AI projects. Additionally, applying our framework to larger-scale scientific experiments examining human–AI interaction in the workplace could help further refine it.

## REFERENCES

Cabour, G. (2023). *Reconcevoir le travail dans des logiques de collaboration humain-technologie à l'ère du digital: Le cas de l'inspection industrielle* [Thèse de doctorat]. *Polytechnique Montréal.*

Cabour, G., Ledoux, É., & Bassetto, S. (2022a). Aligning work analysis and modeling with the engineering goals of a cyber-physical-social system for industrial inspection. *Applied Ergonomics*. In-Press.

Cabour, G., Morales-Forero, A., Ledoux, É., & Bassetto, S. (2022b). An explanation space to align user studies with the technical development of Explainable AI. *AI & Society*. https://doi.org/10.1007/s00146-022-01536-6

Carayannis, E. G., Dezi, L., Gregori, G., & Calo, E. (2022). Smart environments and techno-centric and human-centric innovations for industry and Society 5.0: A Quintuple Helix innovation system view towards smart, sustainable, and inclusive solutions. *Journal of the Knowledge Economy*, *13*(2), 926–955. https://doi.org/10.1007/s13132-021-00763-4

Crandall, B., Klein, G. A., & Hoffman, R. R. (2006). *Working Minds: A Practitioner's Guide to Cognitive Task Analysis. MIT Press.*

Deci, E. L., Olafsen, A. H., & Ryan, R. M. (2017). Self-determination theory in work organizations: The state of a science. *Annual Review of Organizational Psychology and Organizational Behavior*, *4*, 19–43.

Deci, E. L., & Ryan, R. M. (1985). The general causality orientations scale: Self-determination in personality. *Journal of Research in Personality*, *19*(2), 109–134.

Dubey, A., Abhinav, K., Jain, S., Arora, V., & Puttaveerana, A. (2020). HACO: A Framework for Developing Human-AI Teaming. *Proceedings of the 13th Innovations in Software Engineering Conference on Formerly Known as India Software Engineering Conference*, 1–9. https://doi.org/10.1145/3385032.3385044

Employment & Skills. (2014). The future of work: jobs and skills in 2030. *Evidence Report*, *84*.

Forum, W. E. (2016). The future of jobs: Employment, skills and workforce strategy for the fourth industrial revolution. In: *World Economic Forum Geneva*.

Frey, C. B., & Osborne, M. A. (2017). The future of employment: How susceptible are jobs to computerisation? *Technological Forecasting and Social Change*, *114*(C), 254–280.

Gourova, E. (2009). Knowledge audit concepts, processes and practice. *6*(12).

Hart, S. G., & Staveland, L. E. (1988). Development of NASA-TLX (Task Load Index): Results of empirical and theoretical research. In *Advances in psychology* (Vol. 52, pp. 139–183). Elsevier.

Jiao, J. (Roger), Zhou, F., Gebraeel, N. Z., & Duffy, V. (2020). Towards augmenting cyber-physical-human collaborative cognition for human-automation interaction in complex manufacturing and operational environments. *International Journal of Production Research*, 1–23. https://doi.org/10.1080/00207543.2020.1722324

Johnson, M., Bradshaw, J. M., & Feltovich, P. J. (2017). Tomorrow's human–machine design tools: From levels of automation to interdependencies. *Journal of Cognitive Engineering and Decision Making*. https://doi.org/10.1177/1555343417736462

Johnson, M., Bradshaw, J. M., Feltovich, P. J., Jonker, C. M., van Riemsdijk, M. B., & Sierhuis, M. (2014). Coactive design: Designing support for interdependence in joint activity. *Journal of Human-Robot Interaction*, *3*(1), 43–69. https://doi.org/10.5898/JHRI.3.1.Johnson

Johnson, M., Vignati, M., & Duran, D. (2018). Understanding human-autonomy teaming through interdependence analysis. 1–20.

Kadir, B. A., Broberg, O., & da Conceição, C. S. (2019). Current research and future perspectives on human factors and ergonomics in Industry 4.0. *Computers & Industrial Engineering*, *137*, 106004. https://doi.org/10.1016/j.cie.2019.106004

Klein, G. A., Calderwood, R., & MacGregor, D. (1989). Critical decision method for eliciting knowledge. *IEEE Transactions on Systems, Man, and Cybernetics*, *19*(3), 462–472. https://doi.org/10.1109/21.31053

Liao, Q. V., Gruen, D., & Miller, S. (2020). Questioning the AI: Informing Design Practices for Explainable AI User Experiences. In *Proceedings of the 2020 CHI Conference on Human Factors in Computing Systems* (pp. 1–15). Association for Computing Machinery. https://doi.org/10.1145/3313831.3376590

Loup-Escande, E. (2022). Concevoir des technologies émergentes acceptables: Complémentarité des approches expérimentale, écologique et prospective. *Activités*, *19*(1). https://doi.org/10.4000/activites.7345

Maddikunta, P. K. R., Pham, Q.-V., B. Prabadevi, Deepa, N., Dev, K., Gadekallu, T. R., Ruby, R., & Liyanage, M. (2022). Industry 5.0: A survey on enabling technologies and potential applications. *Journal of Industrial Information Integration*, *26*, 100257. https://doi.org/10.1016/j.jii.2021.100257

Miller, M. J., & Feigh, K. M. (2019). Addressing the envisioned world problem: A case study in human spaceflight operations. *Design Science*, *5*. https://doi.org/10.1017/dsj.2019.2

Millot, P., Pacaux-Lemoine, M.-P., & Trentesaux, D. (2015, October 26). Une conception anthropo-centrée pour casser le mythe de l'"Humain Magique" en Génie Industriel?

Morineau, T., & Flach, J. M. (2019). The heuristic version of cognitive work analysis: a first application to medical emergency situations. *Applied Ergonomics*, *79*, 98–106.

Mueller, S. T., Veinott, E. S., Hoffman, R. R., Klein, G., Alam, L., Mamun, T., & Clancey, W. J. (2021). Principles of Explanation in Human-AI Systems. ArXiv:2102.04972 [Cs]. http://arxiv.org/abs/2102.04972

Neumann, W. P., Winkelhaus, S., Grosse, E. H., & Glock, C. H. (2021). Industry 4.0 and the human factor – A systems framework and analysis methodology for successful development. *International Journal of Production Economics*, *233*, 107992. https://doi.org/10.1016/j.ijpe.2020.107992

Oppl, S., & Stary, C. (2019). *Designing Digital Work: Concepts and Methods for Human-centered Digitization*. Springer Nature. https://doi.org/10.1007/978-3-030-12259-1

Parker, S. K., & Grote, G. (2022). Automation, Algorithms, and Beyond: Why Work Design Matters More Than Ever in a Digital World. *Applied Psychology*, *71*(4), 1171–1204. https://doi.org/10.1111/apps.12241

Passalacqua, M., Pellerin, R., Doyon-Poulin, P., Boasen, J., Léger, P-M. (2022, July). Human-Centred AI in the Age of Industry 5.0: A Systematic Review Protocol. *HCI International 2022 Conference*, Montréal, QC, Canada.

Peeters, M. M. M., van Diggelen, J., van den Bosch, K., Bronkhorst, A., Neerincx, M. A., Schraagen, J. M., & Raaijmakers, S. (2020). Hybrid collective intelligence in a human–AI society. *AI & Society*. https://doi.org/10.1007/s00146-020-01005-y

Rapp, A. (2021). In Search for Design Elements: A New Perspective for Employing Ethnography in Human-Computer Interaction Design Research. *International Journal of Human–Computer Interaction*, *37*(8), 783–802. https://doi.org/10.1080/10447318.2020.1843296

Rasmussen, J. (1985). The role of hierarchical knowledge representation in decision making and system management. *IEEE Transactions On Systems, Man, and Cybernetics*, *2*, 234–243.

Roth, E., Depass, B., Harter, J., Scott, R., & Wampler, J. (2018). Beyond Levels of Automation: Developing More Detailed Guidance for Human Automation Interaction Design: Proceedings of the Human Factors and Ergonomics Society Annual Meeting. https://doi.org/10.1177/1541931218621035

Ryan, R. M., & Deci, E. L. (2008). Self-determination theory and the role of basic psychological needs in personality and the organization of behavior. In *Handbook of personality: Theory and research*, 3rd ed. (pp. 654–678). The Guilford Press.

Seeber, I., Bittner, E., Briggs, R. O., de Vreede, T., de Vreede, G.-J., Elkins, A., Maier, R., Merz, A. B., Oeste-Reiß, S., Randrup, N., Schwabe, G., & Söllner, M. (2020). Machines as teammates: A research agenda on AI in team collaboration. *Information & Management*, *57*(2), 103174. https://doi.org/10.1016/j.im.2019.103174

Shadbolt, N., & Smart, P. R. (2015). Knowledge Elicitation: Methods, Tools and Techniques. In J. R. Wilson & S. Sharples (Eds.), *Evaluation of Human Work* (pp. 163–200). CRC Press. https://eprints.soton.ac.uk/359638/

Spies, R., Grobbelaar, S., & Botha, A. (2020). A Scoping Review of the Application of the Task-Technology Fit Theory. *LNCS-12066 (Part I)*, 397. https://doi.org/10.1007/978-3-030-44999-5_33

Sony, Michael & Naik, Subhash. (2020). Industry 4.0 integration with socio-technical systems theory: A systematic review and proposed theoretical model. *Technology in Society*. https://doi.org/10.1016/j.techsoc.2020.101248

St-Vincent, M., Vézina, N., Bellemare, M., Denis, D., Ledoux, É., & Imbeau, D. (2011). L'intervention en ergonomie. Éditions Multimondes.

St-Vincent, M., Vézina, N., Bellemare, M., Denis, D., Ledoux, E., & Imbeau, D. (2014). Ergonomic Intervention. Institut de recherche Robert-Sauvé en santé et en sécurité du travail.

Szalma, J. L. (2009). Individual differences in human–technology interaction: incorporating variation in human characteristics into human factors and ergonomics research and design. *Theoretical Issues in Ergonomics Science*, *10*(5), 381–397.

Szalma, J. L. (2014). On the application of motivation theory to human factors/ergonomics: Motivational design principles for human–technology interaction. *Human Factors*, *56*(8), 1453–1471.

Van den Broeck, A., Howard, J. L., Van Vaerenbergh, Y., Leroy, H., & Gagné, M. (2021). Beyond intrinsic and extrinsic motivation: A meta-analysis on self-determination theory's multidimensional conceptualization of work motivation. *Organizational Psychology Review*, *11*(3), 240–273.

Vicente, K. J. (1999). *Cognitive work analysis: Toward safe, productive, and healthy computer-based work*. CRC Press.

Xiong, W., Fan, H., Ma, L., & Wang, C. (2022). Challenges of human—Machine collaboration in risky decision-making. *Frontiers of Engineering Management*, *9*(1), 89–103. https://doi.org/10.1007/s42524-021-0182-0

Zheng, T., Ardolino, M., Bacchetti, A., & Perona, M. (2021). The applications of Industry 4.0 technologies in the manufacturing context: A systematic literature review. *International Journal of Production Research*, *59*(6), 1922–1954. https://doi.org/10.1080/00207543.2020.1824085

# 24 Agile Governance as AI Governance

## *A Challenge for Governance Reformation in Japan*

*Kodai Zukeyama*
Kyushu Sangyo University, Fukuoka, Japan

*Tomoumi Nishimura*
Kyushu University, Fukuoka, Japan

*Haluna Kawashima*
Tohoku Fukushi University, Sendai, Japan

*Tatsuhiko Yamamoto**
Keio University, Tokyo, Japan

## INTRODUCTION

In 2019, the Japanese government issued the "Social Principles of Human-Centric AI" (CAO 2019). As a result of investigations by and deliberations with industry, academia, and the government, and their perspectives, this document was created for use in international discussions, including OECD and G7 meetings, regarding the principles of AI.

In the document, the basic philosophy of the social principles of human-centric AI is underpinned with specific principles called the "Social Principles of AI," which the government must consider, as well as "AI R&D and Utilization Principles" that have to be kept in mind by those who develop and operate AI. While the first principles have been elaborated by the government, the latter are treated as items to be

* We gratefully acknowledge the support of Keio University Global Research Institute "Research Project Keio 2040." This chapter partially contains work supported by JSPS KAKENHI Grant Number 20K22059.

DOI: 10.1201/9781003320791-28 

voluntarily established and complied with by developers, of their own accord. The government therefore created "guidelines" regarding AI development and operation (MIC 2019, et al), called for a fundamental transformation of the governance models used by businesses (METI 2020, 2021a, 2021b), and advocated agile governance as a model (METI 2021c). The government considered that an agile governance model would help the private sector to comply with the "Social Principles of Human-Centric AI."

In this way, the Japan has engaged in international deliberations regarding the principles of human-centric AI (hereinafter referred to as "HCAI"). However, its regulatory approach in the field of information is not universal. Using personal data protection for comparison, if the EU adopts a principle-based, hard-law regulation model, and the U.S. respects the economic freedom of the private sector and expects self-regulation, Japan sets a general framework with laws and guidelines, while leaving the specific content and operation of the rules to businesses. The approach taken by Japan for the protection of personal data can be seen as an intermediate one between the EU's and the U.S.'s. For AI regulation, Japan has set agile governance as its model, relying on the unilateral efforts of private actors without imposing concrete components of HCAI. Such an approach is not assimilable with the EU's AI Act or the US's AI Risk Management Framework, for example.

The case of Japan may enrich the discussion on what legal approach to adopt—an approach that is principle-based or pragmatically oriented—to realize HCAI. Japan's approach is one of a variety of cases in this emerging period of AI regulations around the world. Examining its regulatory approach guides us also to the question of how government and private actors can contribute to HCAI.

To critically explore Japanese agile governance from an HCAI perspective, this chapter will analyze how agile governance relates to HCAI (Section II) and the Japanese government's definition of agile governance (Section III). The current state of establishing HCAI in Japan, together with the issues that are involved, will be examined in Section IV. Through this, we hope to contribute to international deliberations regarding the shape of AI governance aimed at achieving HCAI.

## WHY DOES AGILE GOVERNANCE RELATE TO HCAI?

### AI AND SOCIETY 5.0

Artificial intelligence, commonly referred to as AI, is a technology that enables advanced information processing systems to handle diverse and massive amounts of data and to perform automated decision-making without human intervention in various situations. Due to these characteristics, AI plays an important role in the future vision of a society called "Society 5.0," as advocated by the Japanese government.

Society 5.0 refers to "a human-centered society in which economic development and the resolution of social issues are compatible with each other through a highly integrated system of cyber space and physical space" (CAO 2021: 11). In such a society, it is assumed that the vast amounts of information would be exchanged between cyberspace and physical space, so that AI, as an advanced information processing technology, plays a crucial role in processing. In fact, the Japanese government's documents on the realization of Society 5.0 often mention AI.

Although AI has the potential to contribute to making society richer and better, as already pointed out in a lot of literature, there are also many drawbacks to AI, making it necessary to exercise caution in its R&D and utilization. For instance, AI, particularly when developed with technologies such as deep learning, is often criticized for perpetuating existing discrimination and disparities. Moreover, due to the so-called "black box" nature of AI, many negative outcomes can occur unintentionally, making it difficult either to identify the causes or to hold someone accountable. As a result, it is now widely recognized that the fulfillment of human centrality is crucial for the R&D and utilization of AI. AI that fulfills human centrality is also referred to as HCAI.

## Two Approaches to HCAI

To achieve the goal of developing AI that satisfies human centrality, two main approaches can be considered. The first is a direct approach: to explore the specifications that HCAI should meet *before* AI is implemented and used. If a well-defined answer can be provided to questions such as what human centrality is or what requirements HCAI should meet, the remaining issues are purely technical. By referring to that well-defined answer, AI might automatically become HCAI, perhaps with the help of mature technologies such as formal methods, which are not even called AI in modern times.

The second approach is an indirect one: instead of identifying the specification of HCAI in advance, explore and implement it *gradually* in the practice of development and utilization of AI. This approach starts with human beings coexisting with AI that may not necessarily satisfy human centrality, which then requires us to address the risks that arise from AI.

When comparing these two approaches, the first approach may appear to be the more ideal one. Indeed, there have already been some attempts to provide a clear definition of fairness, which is considered an important element of human centrality. However, for the first approach to work, it is necessary to carry out such an enterprise in a comprehensive and complete manner. As Wallach and Allen pointed out in 2009, this is an extremely difficult path to follow. Moreover, it is not a realistic option to impose a moratorium on the use of AI until such difficult attempts are successful. Since AI is rapidly developing, even more so than other information processing technologies that have been described as having "dog years," there would not be enough time to invest in identifying human centrality completely.

In contrast, the second approach may have the advantage of being a *lightweight* approach, in that it allows for progress toward achieving HCAI without the need to specify human centrality completely in advance. While this approach may have the disadvantage of being ad hoc, it can be particularly effective in situations where obtaining social consensus on what constitutes human centrality is not practical. This may be especially true in East Asia, where values and beliefs about human centrality may differ from those in the West, making it challenging to reach a consensus. Therefore, the second approach is the more practical and promising option.

## The Relationship between Agile Governance and HCAI

As noted in the preceding paragraph, in the case of the second approach, while there is no need to identify the specifications for HCAI completely, it is important to establish an *appropriate* governance framework to address risks that arise from AI that are not necessarily human-centric. "Agile governance," which is introduced in a later section, is considered to be the response from the Japanese government to this challenge.

Agile governance is a governance model that does not rely on means that strictly qualify as regulation[1]. This feature is well suited to the second approach, which aims to gradually achieve the goal of implementing HCAI, while taking into account the assumption that the full and complete requirements of HCAI cannot be obtained quickly.

In fact, agile governance is closely related to AI governance in Japan. For example, in the Japanese Ministry of Economy, Trade and Industry (METI) report of 2020, AI is positioned as an important element of Society 5.0, and it is argued that agile governance is necessary to work with the development speed of advanced technologies such as AI. In addition, the "Governance Guidelines for Implementation of AI Principles," compiled by METI to implement the social principles of human-centric AI formulated by the Japanese Cabinet Office (CAO), declares that the operation of these guidelines should be carried out in accordance with the manner of agile governance.

## Agile Software Development Analogy

Perhaps, we can learn valuable lessons that are applicable to the former discussion about the relationship between agile governance and AI by comparing two famous methodologies of software development: waterfall software development and *agile* software development.

The waterfall development generally refers to software development that follows the "waterfall model. "In the waterfall model, software development is understood as a sequence of steps, each of them depending on the preceding steps. In comparison, Agile software development, formulated in the well-known document "Agile Manifesto," derives from a group of software development methodologies called "*lightweight* software development." They were developed to overcome the shortcomings of the classical software development methodology, such as the waterfall development.

Due to the nature of dependencies of subsequent steps on preceding steps, waterfall development has two shortcomings. Firstly, it results in a *sequential* development process that does not allow for flexible revision of the outcomes of the preceding steps. Secondly, as the quality of the preceding steps directly influences the quality of subsequent steps, this leads to a *predictive* and inflexible development approach.

Given that information processing technologies are rapidly developing and the business environment that utilizes these technologies is constantly changing, such shortcomings are crucial. To avoid these shortcomings, most agile software development processes adopt two devised methods. First, they divide the software as a whole into parts while prioritizing them and then develop them gradually, rather than completing the whole software at once. Second, instead of establishing a strict development goal, they set just a rough goal and refine it gradually through development by communicating with customers (or any other stakeholders) repeatedly. Therefore, agile software development could be characterized as an *iterative* and *adaptive* approach to software development.

From the discussion above, we may derive the following lectures. First, the consideration behind agile software development could be applicable to the governance model for HCAI, because the problems faced in the governance of AI, namely the rapidity of the technologies and the business environment, are the same as those faced in software development. Second, the similarity of the conventional governance model, (which will be discussed in the following section) with waterfall development, that is, both are *predictive* and *sequential* approaches, implies the need for a new governance model. Third, as an implication of those two points, a new governance model would be more like agile governance than waterfall development.

## WHAT IS JAPANESE AGILE GOVERNANCE?

In 2019, METI established its Study Group on New Governance Models in Society 5.0. The study group has engaged in multiple discussions with the aim of creating a framework for a new governance model, based on the assumption that existing governance models will reach their limits in Society 5.0. Members of the study group include economic and legal researchers, lawyers specializing in corporate law, accountants, personnel from companies involved in AI development and fintech, and more.

This study group has produced three reports so far, suggesting that the agile governance model should replace the conventional model in AI governance because the latter would be less functional in Society 5.0. We can summarize the reports' ideas in two: (1) the primary actors of AI governance should be businesses, not government, and (2) all actors should engage in the continuous and rapid maintenance of the AI governance scheme, something that is called the "agile cycle."

### TWO PREMISES

At the outset, two premises the reports have adopted will be explained. First, three actors are key in AI governance: government, businesses, and communities/individuals. The study group describe agile governance as the reassignment of the roles those actors have. Second, it appear to understand AI governance as a process divided into three stages: rule-making, monitoring, and enforcement. During the rule-making stage, mechanisms for the governing of each actor are formulated and updated. During the monitoring stage, the behaviors of people, companies, and machines are observed to gather the information required for evaluating whether they comply with the rules formulated in the previous stage. During the enforcement stage, the actors take action to resolve the problems found in the monitoring stage. The study group points out that the conventional governance would face many obstacles at every stage (METI 2020: 15–19).

### Limits of the Conventional Governance Model

The conventional governance model is described as follows. Governments use laws to formulate detailed rules[2]. Regulatory authorities regularly monitor whether individuals and companies are complying with these rules. If any problems occur, the regulatory authorities and judiciary enforce the rules using administrative or criminal sanctions. This model is based on the following assumptions regarding society: (i) technologies and business models are slow to change, (ii) data used for monitoring can be collected by humans, (iii) all decisions are made by humans, and (iv) social activities are confined within national borders (METI 2020: 20–22).

The conventional model will cease to function within Society 5.0. First, rule-based laws and regulations will reach their limits. In the society, AI technologies are advancing so rapidly and with such complexity that it is almost impossible for the government to measure in advance the risks that AI generates. Without information about the risks, the government cannot create specific rules for deciding what is good or bad in AI use. Related to this, regulations that apply to specific industries (such as the manufacturing industry or banking industry), defined by their business models, will be difficult to apply. In a society in which business models change dynamically and AI technologies are used across business models, AI regulations enacted by every business model become dated instantly. They cannot rule actual AI practice in society.

Second, a great deal of information regarding AI is owned by companies. The analysis and interpretation of that information are difficult for anyone who lacks specialized technical knowledge. Because of this, when national governments form rules and perform monitoring, the acquisition of monitoring data and the actual process of monitoring will be difficult. Even if the governments can undertake this, it will be inefficient.

Third, problems also arise in the enforcement stage. When complex systems with embedded AI act on the physical world, whether by themselves or collaboratively with humans, it will be difficult to predict the impact of their decisions. If these decisions do have a negative impact, it will not be easy to determine where legal responsibility lies, and it will not be possible to administer sanctions or to identify to whom future preventive measures are to be mandated (the above is based on METI 2020: 22–31).

### Reassigning the Roles of Actors

The agile governance model uses the co-regulation as a standard, and redesigns the roles of government, businesses, and communities/individuals (the following is an overview of METI 2020: 38–66).

#### Government

The government plays a smaller role. It is responsible for preparing an environment in which a business-centric governance system functions effectively. During the rule-making stage, the roles of laws will change from being specific rule-based regulations to being abstract goal-based regulations that show principles and objectives to be achieved. According to the second report, the ultimate goals to be achieved by these laws are happiness and liberty. Under these goals, some core values are located: human rights, economic growth, fundamental institutions such as deliberative democracy, and sustainability (METI 2021c: 44–58). During the monitoring stage, companies are mandated by law to disclose information regarding their technologies and to be transparent and accountable. Rule-based regulations will remain in this stage. In the enforcement stage, uniform penalties will not function effectively. Given this, penalties must be imposed flexibly based on the social impact of corporate conduct. If a company's activities cause harm, the company should be required to conduct an investigation, and appropriate incentives (such as suspending prosecution based on an agreement) should be introduced so that the company will make technical improvements.

### Businesses

Businesses play a key role in the new governance model. During the rule-making stage, companies establish nonbinding guidelines and standards based on the law goals mentioned above. Individual companies then develop and utilize the in-house technologies that make up AI accordingly. The formulation of guidelines and standards is not handled by companies acting alone but must be conducted through multistakeholder dialogue. In monitoring, businesses confirm compliance using the guidelines and standards they defined themselves. They are also called on to publicly disclose and explain how they comply with the guidelines and standards. With regard to enforcement, if risks become actualized and harm is caused, businesses report to regulatory authorities of their own accord and then must cooperate in investigations.

### Communities/Individuals

Communities/individuals are positioned as actors who actively communicate their values and evaluations to society. During the rule-making stage, they participate in companies' multistakeholder dialogues. In the monitoring and enforcement stages, they evaluate companies based on corporate information disclosure. Particularly, when negatively evaluating businesses, they promote sound corporate activity and technology development/utilization through the market (such as through boycotts) and through social norms (such as through negative reputation).

## The Agile Cycle

Agile governance is also proposed as a set of specific governance processes. In Society. 5.0, AI technologies advance so rapidly that existing laws, guidelines, and social institutions regulating AI become outdated in the shorter term. Every actor relating to AI must reconsider whether the existing mechanism for regulating AI is still appropriate in the technical and social circumstances at the time. We need rapid processes to evaluate the present circumstances and improve AI regulations based on them. The agile cycle means running the cycles continuously and rapidly, as shown in the Figure 24.1 (METI 2021c: 59–60). Here, we call the outer circle Track 1 and the inner circle Track 2. We will explain this using an example of a company providing security cameras containing an AI system. Both

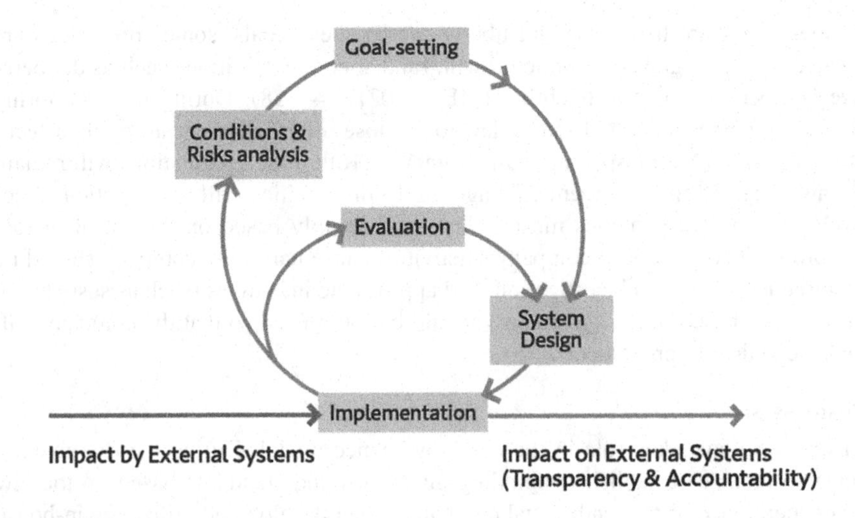


**FIGURE 24.1** The underlying idea of agile governance.

(Source: (METI 2021c: 60).)

the government and the company must run the agile cycles in each track (*see* METI 2021c: 60–69, 103–106).

Firstly, the government defines legal goal-based laws, which will adopt to all AI and other cyber-physical technologies. The laws are based on the current circumstances and the results of risk analysis of such technologies (government in Track 1). For example, the laws would contain principles such as protecting privacy and equality. The reports call this activity "governance of governance" because the laws creates a broad (but abstract) AI governance framework overlooking every area. The government organizes a monitoring and enforcement scheme to ensure that companies comply with the goal-based laws (government in Track 2).

Next, the company sets its own specific goals conforming with the goal-based regulations defined by law. It must consider concrete guidelines to protect privacy and tackle racism in operating the security cameras. Again, when setting these, a great deal of emphasis is placed on dialogue with multiple stakeholders. Supposing the company sets the specific guidelines that its security camera must erase data in one day in principle, and its AI must identify all races in an equal manner. The company organizes a monitoring team and system to conform with its own guidelines (business in Track 1). The company continuously monitors the camera's condition and improves it according to its guidelines. If necessary, the monitoring team or system will be reorganized (business in Track 2).

In parallel with Track 2, the company periodically examines whether its guidelines are appropriate given changes to the environment and risks in development. When necessary, the guidelines are modified and systems are redesigned to match the new guidelines. For example, the company recognizes that its AI unfairly detects women as presenting a greater risk than men. This is an error to be excluded. Then the company again engages in condition and risk analysis and adds to its guidelines a requirement that the AI must evaluate risk of each person regardless of their gender. In order to improve the cameras, the company redesigns its architecture and AI system according to the new guidelines. Of course, multistakeholder dialogues are also essential in this process (again, business in Track 1).

The government periodically reviews whether the goal-based laws are appropriate given changes to society's overall environment and risks. If necessary, it reconsiders the existing goals and reorganizes the monitoring and enforcement scheme (again, government in Track 1).

The dual track process is thus continuously and rapidly implemented in each sector (both the government and the company in this example here). Governance of governance is performed using these processes to maintain overall control[3].

## LOOKING INTO AGILE GOVERNANCE

Japanese agile governance has only recently been proposed. It is neither the official view of the entire Japanese government nor has it been implemented yet[4]. It attracts little attention, and there are few examples of academic debate about this topic. However, given the increasing rate of adoption of AI within society, the propositions outlined in the aforementioned reports deserve to be examined.

### Goal-based Regulations Defined by Law

The most controversial aspect of Japanese agile governance is its recommendation that the law be shifted from rule-based to goal-based regulations. If this happens, the regulatory power of the law will inevitably shrink. But, according to the reports, rule-based laws and regulations, which are a distinctive aspect of conventional governance, would cease to function in Society 5.0. In order to maintain the rule of law as restraining the arbitrary use of power, there will be an emphasis on the guidelines and standards formulated by companies, a high level of accountability, and due process with regard to the technologies involved (METI 2021c: 87–88).

A question for goal-based law is how to review the guidelines and standards created by companies to conform with the law. Without this, companies may create standards that only pursue their profit and ignore benefits for consumers or values relating to human rights. The reports do not answer this question. They mention that the government takes on the role of building trust in businesses by certifying them to be compliant with *guidelines* and *standards* (METI 2020: 43–44) but do not propose a mechanism to confirm whether law-related goals and the guidelines and standards are compatible. Ultimately, compliance with law-related goals is guaranteed by companies' self-regulation or through their social reputation.

This view seems too optimistic about companies' willingness to create and obey guidelines and standards conforming with law-related goals. Companies are likely to deviate from the law without third-party monitoring and enforcement. We think that the government still has an important role to play in supervising companies to maintain compliance with the law. However, what the reports really concern is that the government, with specified laws, hinders the innovation of AI technologies.

They suggest that as well as the law, even companies' guidelines and standards should not be strictly binding, so as not to prevent innovation (METI2020: 44). To be sure, we should avoid suppressing innovation. But the reports' suggestions might not strike a good balance between innovation and the guarantee of fundamental values in society, such as human rights.

Additionally, when the law can only create abstract goals, it is difficult for some companies to determine what behavioral standards create. They might want specified laws in AI development and implementation rather than abstract law-related goals. The reports regard multistakeholder dialogues as the solution to this problem. They envision specific guidelines and standards that conform with the goals stipulated by law as being formulated through these dialogues. The dialogue process is an essential element of agile governance, but perhaps it is burdensome for some companies to conduct this process.

## Multistakeholder Dialogues

It is important to position individuals, such as consumers and minorities, who are affected by AI, as stakeholders. When developing and operating AI, dialogues must also include representatives of groups who could, for example, suffer from statistical discrimination or be subject to unfavorable judgments at the hands of AI. For this governance model, which relativizes regulations under the law, multistakeholder dialogue could be characterized as a supplementary principle or an alternate channel of democratization. The organizers of such dialogues would likely be required to publicly disclose the reasons for their selection of dialogue participants. Rules and standards would need to be created regarding the makeup of dialogues so that people with various social backgrounds could participate in the dialogue process.

The suggestions of the reports, however, need to be more sophisticated. The major issue is that they do not show the specific criteria for identifying who the stakeholders are. The membership of a multistakeholder dialogue must be determined *before* the dialogue. It is not appropriate for companies, which are responsible for the dialogue, to create the membership rules because, if they do, the rules are likely to be made for the benefit of companies, and this would not be fair. Here, the law still has an important role to play in indicating what kind of and how many people should be included in the dialogue process. Also, unlike democratic processes, this consensus formation lacks clear rules, customs, and practices at the present. If achieving consensus took time, it would detract from the agility of the approach. Rules and practices need to be established so that the dialogue process can proceed in an agile manner.

## Realizing Value in Relation to Human Rights

Agile governance also has the potential to effectively provide society with constitutional value. In the legal world, constitutions constrain the authority of national governments at a primary level, and they are seldom directly applied to the legal relationships between private actors. However, when companies define concrete guidelines and standards, including constitutional values such as human rights, they can avoid this theoretical opposition, and directly reflect this constitutional value in their activities. For example, legal scholars and people whose human rights might be limited can take part in multistakeholder dialogues, which can produce concrete rules, thereby embedding constitutional value in AI development guidelines.

On the other hand, as the role of the law becomes more relative, the role of the courts as bastions of the protection of constitutions will be less significant. The reports make little mention of control through public law that is implemented by the judiciary. Only in exceptional cases would courts declare corporate activities to be illegal, such as when companies set standards that clearly disregard the abstract goals defined by law. We still have enough reason to doubt that agile governance serves the protection of human rights even in Society 5.0.

## CONCLUSION

This chapter has presented an overview of the Japanese version of agile governance and examined its implications for the global debate on AI. There are at least three approaches to making a legal regime to fit the AI era: (i) strictly and completely defining human centrality and teaching it to AI; (ii) scientifically denying human centrality and leaving political and legal decision-making to the autonomous judgment of AI; and (iii) not defining human centrality in advance and realizing it progressively. Both (i), which adheres strongly to the modern legal regime, and (ii), which seeks to move radically to algorithmic centrism, may lead to an ideological divide between humanism and naturalism, and even a real conflict between the two. One option is to take the pragmatic approach of (iii). If it is found to be useful, the Japanese version of agile governance may be worth listening to from a global perspective although it is just a phase of suggestion.

However, we have to confront several problems in this model. Can companies create specific rules for AI that conform with law-related goals? Are multistakeholder dialogues really functionable? Who are the stakeholders? Are human rights sufficiently protected? These questions should be answered if agile governance regards itself as the best way to actualize HCAI principles.

## NOTES

1 The report defines the governance as

> design and implementation of technical, organizational, and social systems by stakeholders, with an aim to manage risks in a society based on systems which integrates cyberspace and physical space […] at an acceptable level, while maximizing the positive impact arising from the system.
>
> (METI 2021c: 14)

2 In this chapter, "rules" means codes which specifically indicate which actions are to be taken, while "goals" do not express those actions directly but lead our evaluation in certain direction.

3 The third report explains this process more specifically (METI 2022: 22–37).

4 Yet, we can find some examples of agile governance within society. For instance, NEC Corporation, which is a comprehensive electronics manufacturer in Japan, established

its "NEC Group AI and Human Rights Principles," and has been conducting AI-related business based on these principles. These principles were formed in reference to those of the Japanese government (CAO 2019). NEC has also begun coordinating with university researchers to specify or reconsider the principles. These are practices involving the creation of specific rules regarding product development and utilization under multistakeholder dialogue, which the agile governance suggests. *See* NEC, "NEC Group AI and Human Rights Principles" (written in Japanese) https://jpn.nec.com/press/201904/images/0201-01-01.pdf; Keio Global Research Institute Project, "Review of Check Point List on Implementation of "Human Rights-By-Design" in Society: From a Legal and Technical Perspective," https://www.kgri.keio.ac.jp/en/project/kgri/2019/S19-07.html

## REFERENCES


Beck, K., et al. (2001). *Manifesto for Agile Software Development*. Retrieved from https://agilemanifesto.org/

CAO = Japan's Cabinet Office. (2019). *Social Principles of Human-Centric AI*. Retrieved from https://www8.cao.go.jp/cstp/ai/humancentricai.pdf

CAO = Japan's Cabinet Office. (2021). *6th Science, Technology, and Innovation Basic Plan*. Retrieved from https://www8.cao.go.jp/cstp/english/sti_basic_plan.pdf

METI = Japan's Ministry of Economy, Trade and Industry. (2020). *Governance Innovation: Redesigning Law and Architecture for Society 5.0*. Retrieved from https://www.meti.go.jp/press/2020/07/20200713001/20200713001-2.pdf

METI = Japan's Ministry of Economy, Trade and Industry. (2021a). *AI Governance in Japan Ver. 1.1*. (July 9, 2021). Retrieved from https://www.meti.go.jp/shingikai/mono_info_service/ai_shakai_jisso/pdf/20210709_8.pdf

METI = Japan's Ministry of Economy, Trade and Industry. (2021b). *Governance Guidelines for Implementation of AI Principles*. (July 9, 2021). Retrieved from https://www.meti.go.jp/shingikai/mono_info_service/ai_shakai_jisso/pdf/20210709_9.pdf

METI = Japan's Ministry of Economy, Trade and Industry. (2021c). *Governance Innovation Ver 2.0: A Guide to Designing and Implementing Agile Governance*. Retrieved from https://www.meti.go.jp/press/2021/07/20210730005/20210730005-2.pdf

METI = Japan's Ministry of Economy, Trade and Industry. (2022). *Agile Governance Update: How Governments, Businesses and Civil Society Can Create a Better World By Reimagining Governance*. Retrieved from https://www.meti.go.jp/press/2022/08/20220808001/20220808001-b.pdf

MIC = Japan's Ministry of Internal Affairs and Communications. (2019). *AI Utilization Guidelines* (August 9, 2019) (in Japanese). Retrieved from https://www.soumu.go.jp/main_content/000637097.pdf

Ministry of Science and Technology of the People's Republic of China. (2019). *Principles of next-generation artificial intelligence (AI) governance*, 26 September 2019.

Royce, W. W. (1970). *Managing the development of large software systems: concepts and techniques*. In *Proceedings of the 9th international conference on Software Engineering*.

Wallach, W., & Allen, C. (2009). *Moral Machines: Teaching robots right from wrong*. Oxford University Press.

# 25 A Framework for Human-Centered AI-Based Public Policies

*Jakob Kappenberger and Heiner Stuckenschmidt*
University of Mannheim, Mannheim, Germany

## INTRODUCTION

As the use of Artificial Intelligence (AI) has penetrated nearly every aspect of society, there is growing pressure on public institutions to adopt the technology frequently framed as a pathway to "responsive, efficient and fair" government (Margetts & Dorobantu, 2019). However, at the same time, AI has been linked to fostering existing inequalities and biased decision-making that disadvantages specific social groups (Saxena et al., 2021). These potential challenges are particularly relevant in the public sector, where applications of AI frequently occur in high-stake areas (Veale et al., 2018).

While this dichotomy has sparked academic interest, with multiple studies exploring the advantages and challenges associated with AI use in the public sector (e.g., de Sousa et al., 2019; Wirtz et al., 2019; Zuiderwijk et al., 2021), the focus of existing research still lies on the regulatory role of governments and other public institutions (Kuziemski & Misuraca, 2020). Consequently, as administrations around the globe are already utilizing AI, there is a need for research into frameworks that allow practitioners to analyze the potential impact of the deployment of AI in the public sector in a structured manner and examine inherent trade-offs (Harrison & Luna-Reyes, 2022; Zuiderwijk et al., 2021).

We address these research gaps by developing a framework of practical guidelines for Human-Centered AI-based public policies and applying it to a test case. Our key contributions are:

- **Theory**: Conceptualizing the findings of related work, in the remainder of Section 1, we discuss the potential advantages and challenges connected to AI use in the public sector.
- **Framework**: Combining the concept of Human-Centered Artificial Intelligence (HCAI) with policy evaluation, we present a framework of guidelines for designing Human-Centered AI-based public policies and evaluating their impact in Section 2.
- **Application**: To illustrate the framework, we apply it to a test case in Section 3: We use a generalizable simulation to model traffic and, primarily,

DOI: 10.1201/9781003320791-29

parking in inner cities. Within this simulation, as an exemplary policy, we let an AI agent set prices for the parking supply. We systematically analyze the impact of this policy on the modeled environment using our framework and conclude whether it conforms to the concept of HCAI.

## AI in the Public Sector: Applications and Challenges

Deploying AI in the public sector entails "the design, building, use, and evaluation of cognitive computing and machine learning to improve the management of public agencies, the decisions leaders make in designing and implementing public policies, and associated governance mechanisms" (Desouza, 2018).

Mirroring the general rise of AI adoption, organizations in the public sector around the globe have begun to utilize AI systems in their everyday operations.[1] For instance, various federal institutions in the US already deploy the technology. Use cases include the identification of potential violations of trading laws or adjudicating applications for social benefits (Engstrom et al., 2020). In the EU, examples encompass a Belgian agency developing a model for predicting whether daycare services require further inspections or an agricultural institution in Estonia automatically monitoring the status of grasslands (van Noordt & Misuraca, 2022). China's government is particularly progressive in terms of the deployment of AI. It is used in areas ranging from the well-publicized social credit system that entails comprehensive face recognition software (Curran & Smart, 2021) to chatbots, which are to address the citizenry's queries at the local government level (Wang et al., 2022). These examples stress the importance of both understanding the impact of AI in the public sector as well as adequately evaluating the policy outcomes that are produced supported by, or even entirely based on, AI. Following Medaglia et al. (2021), the potential benefits (and respective challenges) of this development can be structured across three areas of impact.

### Efficiency of Operations in the Public Sector

First, AI is associated with the capacity to boost the internal efficiency of public bureaucracies. It can perform mundane tasks, such as the monitoring of public grasslands in Estonia mentioned above, and thus save costs and lower the workload of employees in the public sector, which can then be deployed elsewhere. AI has already demonstrated its ability to outperform humans on selected tasks, thereby possibly lowering the error rate of internal processes of this kind in public administrations (Valle-Cruz et al., 2019).

However, apart from potentially slow adoption preventing such gains, the benefit of increased efficiency in public administrations is often also linked to potentially growing unemployment and a wide-ranging transformation of the labor market (Wirtz et al., 2019). While the extent of this problem is somewhat contested, there is generally a consensus among researchers that the rising spread of AI systems will lead to at least some jobs being lost to automation (Susar & Aquaro, 2019). Nonetheless, since AI is also bound to create new job opportunities, the outcome of the changes to the labor market it will bring are challenging to predict (Zuiderwijk et al., 2021).

### Public Decision-Making

Second, the output of AI models can support public decision-making in achieving its underlying policy goals. Due to their ability to incorporate potentially vast amounts of data into their predictions, AI, in general, and Machine Learning (ML), in particular, are frequently credited with unearthing insights humans would not be able to identify. Consequently, relying on AI input in decision-making processes may be seen as introducing more "neutrality" or "objectiveness" (Kuziemski & Misuraca, 2020). In particular, AI use in governments has been increasingly framed as a potential solution for finally supporting the attainment of the sustainable development goals of the UN (Susar & Aquaro, 2019), as it may help develop more accurate climate models and identify the most effective actions to battle climate change (Vinuesa et al., 2020).

Despite its potential advantages for the advancement of different policy goals, AI, as an example of automated decision-making (ADM), often reflects and proliferates existing biases in society (Mehrabi et al., 2022). For example, these biases can originate from the data on which the algorithm is trained, as marginal groups may lack adequate representation. This can lead to varying error rates between different social groups, as evidenced by the finding that face recognition systems trained on well-established but Western-centric open-source datasets perform significantly worse when tasked with recognizing people of color (Shankar et al., 2017). Consequently, in particular, since the deployment of AI in the public sector frequently comes with high stakes, the notion of "neutral" or "objective" AI in government appears somewhat difficult to hold as long as the data it is trained on (as one of the potential sources of bias) does not live up to these ideals (Desouza et al., 2020).

Furthermore, if AI is to support public decision-making, social acceptance, as well as the trust the public places in these systems, are essential (Wirtz et al., 2019). Experiments examining the acceptance of ADM suggest that, while purely algorithm-based decisions are perceived rather critically, a combination of human and algorithmic decision-making appears equally acceptable as decisions made exclusively by humans (Starke et al., 2022). In particular, a key role in producing legitimacy in this regard falls to the degree of transparency and the explainability of such decision systems (de Fine Licht & de Fine Licht, 2020). Designing processes that appropriately consider these points represents a crucial challenge for decision-makers in the public sector.

### Interaction between Government and Citizens

The third area in which hopes are placed on AI to advance public sector operations is that of the interaction between governments and citizens. The most popular realization of this are chatbots aimed at improving the communication between public organizations and the citizenry. On the one hand, this means lowering the strain placed on administrations by having the bot assume some of the information services usually provided by the administration's workforce. On the other hand, chatbots are tasked with automatically (and potentially faster) providing the requested information to a given user or immediately routing the user's input to the appropriate contact person (Androutsopoulou et al., 2019).

As AI systems usually require vast amounts of data to be trained, their introduction is often viewed with suspicion regarding the protection of privacy rights in the public discourse (Zhang & Dafoe, 2019). This holds particular significance since the value added by AI often rests on its ability to integrate different data sources into one data set, thus potentially allowing easier identification of specific individuals (Johnson, 2017). Furthermore, as demonstrated by the deployment of AI for social credit systems set up by the Chinese government, AI offers governments easier and deeper access to their citizens' private lives. Therefore, apart from ensuring the accessibility of AI-based public services throughout different age and social groups (Toll et al., 2019), protecting privacy as well as transparently communicating which data is used and if citizens are currently interacting with an AI is crucial to foster social trust in AI systems (Harrison & Luna-Reyes, 2022).

In summary, the deployment of AI systems appears to hold the potential to bring significant improvements to public sector operations. Nonetheless, the technology brings numerous challenges, which stress the importance of a cautious approach in introducing AI to the public sector.

## FRAMEWORK FOR HUMAN-CENTERED AI-BASED PUBLIC POLICIES

The framework that we propose to help inform such a cautious approach to AI in government is based on the concept of HCAI. HCAI emerged as an umbrella term in response to many of the challenges detailed in the previous section in general AI applications. It emphasizes constructing AI systems focusing on the human stakeholders involved (Riedl, 2019). The approach can be linked to a series of principles that constitute HCAI: *explainability*, *transparency*, *responsibility*, *ethics*, *trustworthiness*, *fairness*, and *sustainability* (Hartikainen et al., 2022). These principles guided the design of the different levels of our framework.

As a second foundation and method to determine whether a given policy abides by these principles, our framework relies on policy evaluation. There is a broad branch of research in political science concerned with the assessment of public policies, evaluating or predicting the performance of different policy programs, and relating its findings back to the policy-making process.[2] "Policy evaluation" constitutes "ascertaining the merit, that is, the worth or value of government interventions" in an evidence-based fashion (Vedung, 2020). It can either be performed ex-ante, prior to the implementation of a given policy, or ex-post after the policy in question has been enacted to inform a potential reform of that program or the design of a subsequent one (Wollmann, 2017). The framework proposed here to verify whether a given policy conforms to the concept of HCAI (Figure 25.1) is to be used in both instances as it is meant to guide both the design of new policies as well as the evaluation of implemented programs. The following remarks serve to introduce the three levels of this framework.[3]

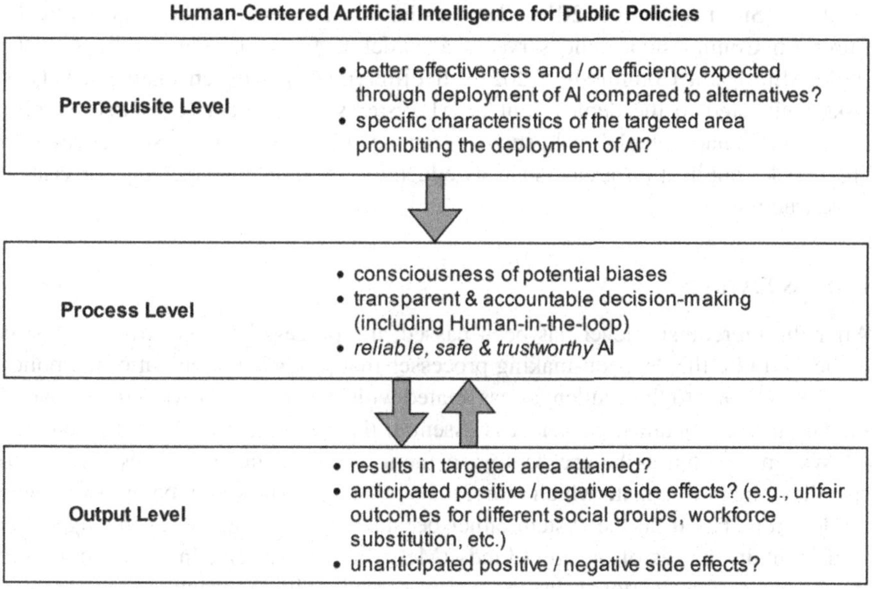

**Human-Centered Artificial Intelligence for Public Policies**

**Prerequisite Level**
- better effectiveness and / or efficiency expected through deployment of AI compared to alternatives?
- specific characteristics of the targeted area prohibiting the deployment of AI?

**Process Level**
- consciousness of potential biases
- transparent & accountable decision-making (including Human-in-the-loop)
- *reliable, safe & trustworthy* AI

**Output Level**
- results in targeted area attained?
- anticipated positive / negative side effects? (e.g., unfair outcomes for different social groups, workforce substitution, etc.)
- unanticipated positive / negative side effects?

**FIGURE 25.1** Framework for human-centered AI-based public policies.

## Prerequisite Level

To avoid the "solutionism trap" of not recognizing that technology may not represent the best remedy for a given problem (Selbst et al., 2019), before devising any policy intervention utilizing AI, policymakers should ensure that the deployment of the technology for the intended purpose is generally warranted and can be expected to bring tangible benefits (Desouza et al., 2020). The resulting policy should (with reasonable likelihood) deliver better *efficiency* and/or *effectiveness* when compared to already implemented and potential alternatives. In many instances, AI-based approaches most likely offer merely incremental performance improvements or require extensive investments in technical equipment and personnel, requiring decision-makers to make trade-offs. Crucially, this evaluation should be conducted on top of ensuring that a given policy is aimed at advancing the public good, as public institutions cannot conceptualize value based on efficiency and effectiveness alone (Mintrom & Williams, 2013).

Furthermore, some subject matters may be altogether unsuitable for the deployment of AI. This could be either due to technical causes, such as low availability of high-quality data, or the uncompetitive performance of AI approaches for the given problem (van Noordt & Misuraca, 2022). Most importantly, however, ethical or privacy considerations must inform the decision to move forward from this prerequisite stage. As alluded to in the previous section, when deployed in sensitive policy areas, such as domestic security, AI has the potential to greatly expand governments' abilities to control their citizens or negatively impact their lives with potentially biased decision-making, for example, when a face recognition suite used for searching for wanted individuals saves the data of innocent bystanders or even returns false positives (Susar & Aquaro, 2019). The regulatory framework on AI proposed by the European Commission could serve as a guideline for such considerations in the future (European Commission, 2021). It differentiates between distinct levels of risks connected to the deployment of AI systems in different scenarios, banning those with "unacceptable" risk, such as social scoring, and placing stricter rules on "high risk" application areas, such as education, asylum management, or critical infrastructure.[4]

## Process Level

After the prerequisite level has been passed, the process level governs how AI is embedded into the decision-making processes that produce the output of the policy in question. Due to the challenges associated with making use of AI in the decision-making process detailed earlier, it is essential that public actors examine potential AI systems for biases before deploying them to ensure fair outcomes. That is to say, they must investigate which individuals or social groups may be disadvantaged by decisions based on the system either because the training data or the algorithm itself contains biases of different kinds (Mehrabi et al., 2022). In doing so, due to the danger of intersectional discrimination across multiple attributes (e.g., women of color), it is not sufficient to focus on individual characteristics alone (Mann & Matzner, 2019). Crucially, as many formalizations of algorithmic fairness have been proposed, the notion of fairness suitable for a given policy is context-dependent and may be determined by the public resource being redistributed and the social groups affected (Gerdon et al., 2022).

Moreover, the decision-making process within the AI-based policy should be designed with the constructs of transparency and accountability as foundations. While these terms are connected and often used in similar contexts, they refer to distinct concepts (Williams et al., 2022). Transparency would require policymakers to provide readily available information about the role of AI in a given policy intervention and its internal functioning (de Fine Licht & de Fine Licht, 2020).[5] It is important to note that complete transparency can also have detrimental consequences in terms of the trust placed in AI systems. Studies in this area have shown that it may encourage unfounded trust in wrong outputs or mistrust in correct predictions since humans often struggle to parse confidence scores (Schmidt et al., 2020). Therefore, special care should be given to offering the appropriate information depending on the intended audience (Langer et al., 2021; Miller, 2019). For instance, practitioners within the administration or professionals in civil society organizations will most likely require more comprehensive information than the general public.

Building upon transparency, accountability must involve giving reasons and explanations regarding the design and operation of the AI system behind the policy. Most importantly, accountable decision-making processes must allow the subject of a decision to petition the enforcement of consequences should the decision in question, or its justification, prove inadequate (Binns, 2018). Consequently, AI systems utilized in government must be accompanied by the introduction of a petition mechanism that allows citizens to have the system's output checked. Furthermore, they should allow for some form of interpretability through explainable AI (XAI) to be able to provide justification. XAI methods seek to render the output of deep ML models interpretable by, for example, appropriating the learned function with a simpler, interpretable model (Samek & Müller, 2019). However, as simpler models, such as shallow decision trees or linear models, offer interpretability without XAI on top (as well as lower energy consumption in many cases), they should be preferred if lagging not too far behind in performance compared to more complex model classes.

Both transparency and accountability are imperative to detect biases in the AI system as well as to ensure public trust in the policy that is based on or supported by it (de Fine Licht & de Fine Licht, 2020; Williams et al., 2022). In particular, creating an accountable design that adheres to the concept of HCAI requires clear human responsibility (Mhlanga, 2022; Shneiderman, 2020). Consequently, the framework necessitates the implementation of the *human-in-the-loop* concept at the training and inference steps of the AI system to have a human verify the output of the algorithm and periodically provide feedback that influences the training process (Wu et al., 2022). Apart from enabling clearer accountability, this process also has the advantage that persons performing this work will gain a deeper understanding of the model's inner workings, which then should inform the external communication pertaining to the policy, thereby hopefully increasing transparency. Nonetheless, such an occupation will require extensive training and oversight as the algorithm's output may impact the specific human in the loop and their decision-making regarding intervening (Saxena et al., 2021).

Lastly, the final layer of the process level serves to introduce the best practices of *reliable*, *safe*, and *trustworthy* systems as part of the HCAI concept proposed by Shneiderman (2022) to the framework.[6] *Reliable* systems produce the output as specified by the policy they are supporting. This is ensured by following standard software engineering workflows, implementing detailed audit trails, and extensive testing. *Safety* is promoted mainly through a leadership commitment to avoid harm and a culture of openness regarding failures. Lastly, *trustworthiness* depends primarily on independent oversight. In the case of AI-based policies, this oversight should be conducted by nongovernmental and civil society organizations that, scrutinizing the information given per the transparency requirement of the framework, vet the AI use for the policy in question independently.

## Output Level

The output level of the framework features the *side-effects model* for policy evaluation proposed by Vedung (2013). The base component of this level is the goal-attainment measure, which examines whether the output of the policy in question is in accordance with the targets set during its conception.[7] For instance, this step might involve determining if a policy reconfiguring traffic lights based on the output of an AI learner improves traffic flow (although most public policies will have multiple goals informing their design). Nevertheless, as its name suggests, the side-effects model also concedes that policy interventions often have consequences outside the targeted area. Here, one can differentiate between *anticipated* and *unanticipated* side effects. For AI-based public policies, both the ex-ante collection of anticipated consequences as well as the ex-post diagnosis of unanticipated results should incorporate the known specific challenges of AI use in the public sector discussed above. Potential detrimental outcomes such as biased decisions for particular social groups or the substitution of existing labor should always feature prominently in these considerations as an extension of the analysis conducted at the process level. Conversely, as Figure 25.1 indicates, the results of applying the framework on the output level are to be used as part of a continuous loop to improve the process of the policy. This may encompass, for example, altering the composition of the training data to improve the representation of a particular social group, as the output level showed a higher error rate of the AI system's decisions for this specific group as an unanticipated side effect.

## APPLYING THE FRAMEWORK: AI-BASED PRICING FOR PARKING

In order to illustrate the core principles of our framework, this section introduces a simple test case. Our example policy intervention is aimed at improving the management of municipal parking space in a model city by utilizing AI-based pricing. This example is well-suited to demonstrate the framework, as so-called "smart cities" are one of the most prominent application areas of AI in the public sector (Ben Rjab & Mellouli, 2019). Additionally, city administrations have already experimented with using automated pricing systems for parking management (Friesen & Mingardo, 2020). Moreover, as urban parking is generally considered relatively inefficient since excessively low prices lead to high demand and adverse outcomes for urban societies, such as cruising for parking, there is pressure on policymakers to develop new solutions in this policy area (Shoup, 2011). To tackle this issue, the target of our test case policy is to manage the parking zones in our model city more efficiently by avoiding both excessive over- and underutilization. The AI system supporting this policy is governed by Reinforcement Learning (RL), a variant of ML that allows for dynamic reactions to the state of its subject without a specified ground truth.

For policies dependent in some capacity on AI, it generally appears prudent to first conduct ex-ante policy evaluation since, as detailed in the previous section, such policies may bring wide-ranging consequences that may only be partially understood beforehand. Consequently, simulations, as evidenced by their use in this research area for some time, represent a valuable tool for testing AI-based public policies in a safe environment (Desouza et al., 2020; Nespeca et al., 2022).[8] Therefore, we will employ a parking simulation written in *NetLogo* (Wilensky, 1999) to simulate the process and output level of our test policy before implementing it in the real world. The model simulates individual agents traversing a traffic grid on their way to their destination. They attempt to find a parking space close to said destination, either in one of the four curbside Controlled Parking Zones (CPZs) or supplementary garages. To determine whether they can afford a free parking space in their vicinity, the individual agents have an empirically calibrated income, which allows for classifying the model population into three income groups.[9] The visual interface of the simulation can be inspected in Figure 25.2.

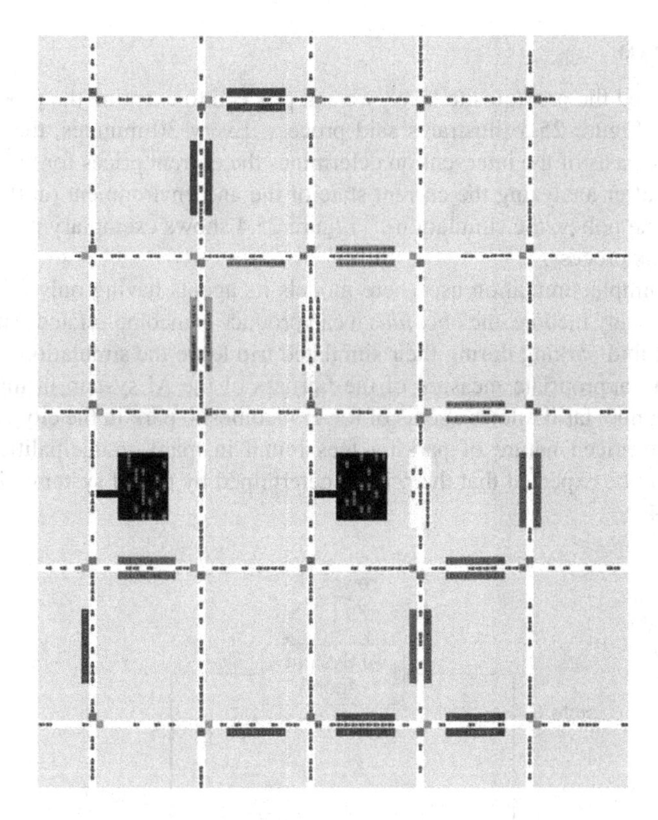

**FIGURE 25.2** Visual interface of the parking simulation.

## Prerequisite Level

On the prerequisite level, the first step is to determine whether the test policy can be expected to provide better *efficiency* and/or *effectiveness* compared to the alternatives.[10] For our test policy, effectiveness can be operationalized as the amount of time the respective systems keep the utilized capacity of the CPZs within a desired range.[11] Currently, the city administration operates a static pricing scheme corresponding to 1.8€ for the two more peripheral zones (blue and teal) and 3.6€ for the central, more popular zones (yellow and green). As Figure 25.5 shows, this strategy is relatively ineffective in keeping the utilized capacity of the different CPZs in the desired range as it cannot cope with varying demand for parking over the course of the modeled day. The AI-based pricing system proposed by the test policy, on the other hand, should prove more able to respond to changes in the environment and can therefore be expected to deliver more effectiveness regarding the policy goal.[12] Examining the *specific characteristics of the target policy area*, if one is to apply the regulatory framework proposed by the EU Commission, the pricing of public parking space would fall under "access to public services," which constitutes a high-risk application (European Commission, 2021). Thus, the application of AI in this case is permitted by the regulatory framework if special care is given to the risks inherent to the deployed system.

## Process Level

Having passed the prerequisite level, we can proceed to examine the process of our test policy. Figure 25.3 illustrates said process. Every 30 minutes, the AI system forming the basis of the intervention determines the current prices for the four curb-side CPZs after analyzing the current state of the and environment (at this stage of designing the policy, the simulation).[13] Figure 25.4 shows exemplary pricing resulting from this process.

As the simple simulation used here models its agents having only one personal attribute, namely income, the only *bias* it can produce is income-related. Since agents that do not find parking during their simulated trip leave the simulation and are not replaced, an appropriate measure of the fairness of the AI system in this instance might be in how far it allows agents of lower incomes to park in the city center. Due to the underpriced nature of parking fees found in many municipalities (Shoup, 2011), it can be expected that the pricing determined by the AI system will turn out

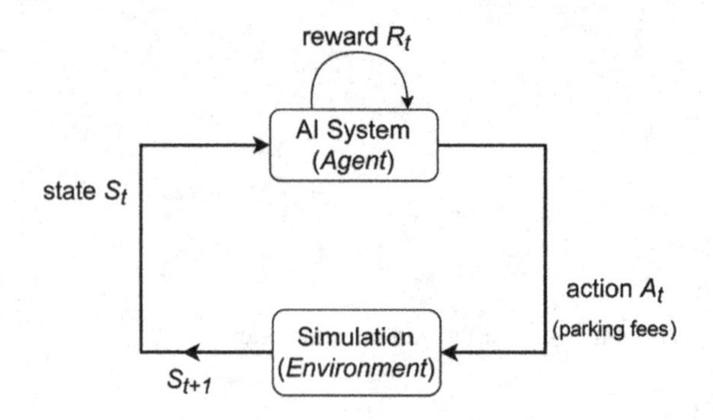


**FIGURE 25.3** Interaction between AI system and simulation.

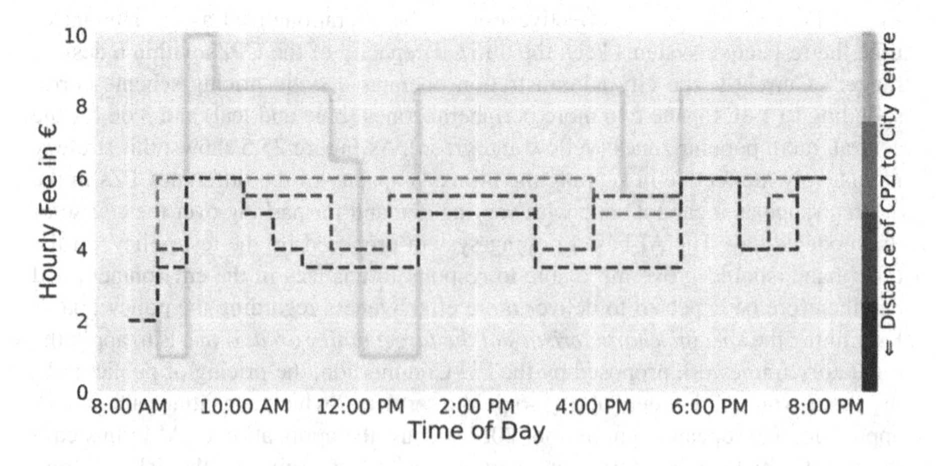


**FIGURE 25.4** Exemplary pricing as determined by AI system.

higher on average than the current scheme, thereby potentially disproportionately affecting low-income individuals.

In terms of *transparency* as well as *accountability*, applying the framework already allows for concluding that the AI system in its current form does not satisfy the requirement of XAI (or general explainability), as it provides no further insight into why the agent has chosen a particular fee for a given CPZ. Beyond that, it also lacks an implementation of the human-in-the-loop concept during the training and inference steps. As indicated by Figure 25.3, in its current form, there is no human input into the system's operation. Consequently, city administrations aiming to implement the policy would have to introduce XAI methods into the process and opt for a different RL algorithm, such as learning from human preferences (Christiano et al., 2017), that accommodates human feedback. Furthermore, as detailed in the previous section, the final policy implementing the proposed system would have to be accompanied by transparent communication regarding the deployment of the system and its functioning, as well as a petition mechanism for the public to facilitate accountability.

At this early stage of the policy cycle, the final layer of the process level serves primarily as a to-do list of how to design the software engineering workflows as well as the audit and communication processes defining the AI system. In the case of our test policy, *reliability* might be of particular interest since reliability from citizens' perspectives might be at odds with the volatile nature of the system's pricing strategy.

### Output Level

Proceeding to the output of the policy, Figure 25.5 shows the immediate effects of the pricing system on the model environment across four dimensions in comparison with the existing static scheme. Generally, the AI-based policy appears to significantly improve upon the status quo in terms of its ability to keep the utilized capacity of the curbside parking space within the desired range since its pricing succeeds in balancing over- and underutilization. Hence, the *desired results in the targeted area* are attained. Regarding *anticipated side effects*, the simulation run confirms that the policy disproportionately limits the accessibility of parking in the modeled city center, thus exhibiting a bias toward low-income individuals. Moreover, the policy also produced *unanticipated side effects* in the simulated environment.[14] Compared to the implemented baseline, the AI-based fees resulted in improved traffic flow, thus tackling the congestion caused by cruising for parking by significantly reducing the number of cars in the simulation through its aggressive pricing strategy (see Figure 25.4). Depending on the overall policy goals of the respective administration, these may constitute positive or negative side effects, which, in conjunction with the findings regarding the anticipated negative side effects, should inform subsequent iterations of the policy design.

Overall, based on this high-level analysis, the proposed AI-based policy does not conform to the concept of HCAI, as its internal process lacks both transparency and accountability, while its output shows evident bias. However, the introduced framework points to clear starting points for changes to the policy program that may result

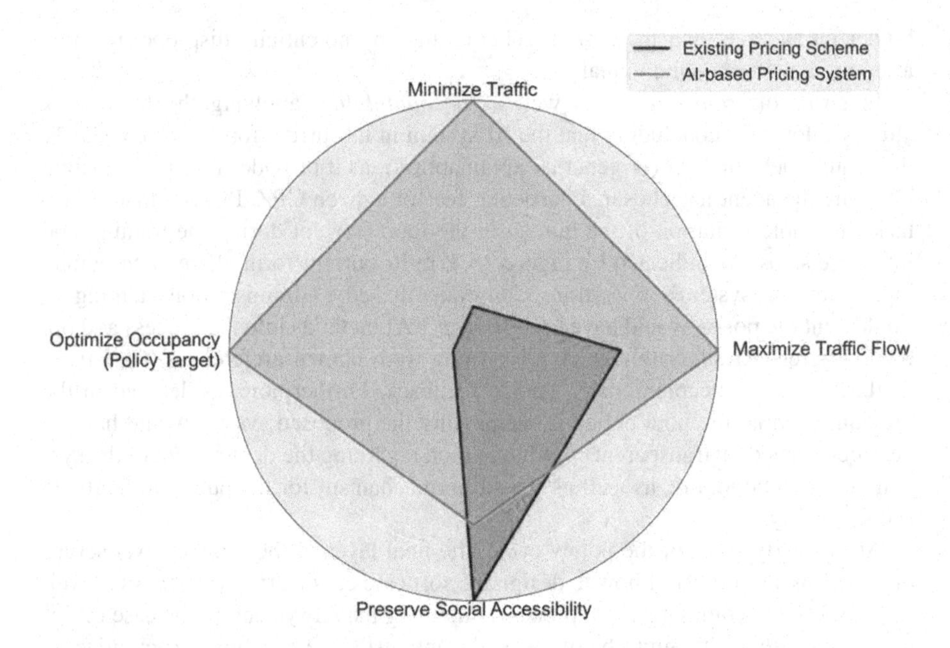


**FIGURE 25.5** Performance of the existing pricing scheme against the AI-based system across different dimensions (relative to the best value in the respective dimension).[15]

in it conforming in future iterations. Nonetheless, it is worth pointing out that a valid interpretation of these results may also conclude that a proposed policy is generally not suited for the intended purpose. In this test case, managing demand with dynamic prices equal across all income cohorts will most likely always disproportionately disadvantage those of low income. Therefore, alternative policies, such as banning car traffic in the city center entirely, may prove more attractive to policymakers.

## CONCLUSION

This chapter introduced a framework of guidelines for developing Human-Centered AI-based public policies to contribute to a structured and careful deployment of AI in the public sector. Based on existing literature on HCAI and policy evaluation, the framework allows for determining whether a policy intervention currently being designed or already implemented satisfies the core constructs that define HCAI. To demonstrate its potential use, we applied the framework to a simple model policy in a simulation setting.

Crucially, there are several limitations of our work to note. Due to space limitations, the framework covers multiple theoretical constructs and topics that are subjects of entire branches of research (e.g., AI transparency or XAI) in a high-level fashion. Instead of producing a detailed manual to be deployed in practice immediately, our aim was to offer an overview of theoretical concepts that can comprise Human-Centered AI in the public sector, pointing toward the experts in the respective fields. Therefore, our framework can merely offer general guidelines, which may serve as a foundation for more elaborate and practical processes (such as step-by-step instructions) developed together with stakeholders and policy practitioners to enable responsible deployment of AI in the public sector.

Finally, to further the research into if and how the public sector should make use of AI, we require a better understanding of the ways in which public institutions are already deploying the technology. Hence, more studies investigating current policy programs based on AI (in line with Kuziemski & Misuraca, 2020; van Noordt & Misuraca, 2022) are needed to ensure that frameworks such as the one proposed here can be refined to both better reflect the realities of public sector operations and holistically evaluate the outcomes of public policies relying on AI. Furthermore, more theoretical research into how the regulation of AI use in the public sector is to be designed is required. While there is a prevalence of proposals of voluntary standards in the literature, more severe and enforceable measures might be required, as "even seemingly trivial applications of AI by the public sector can be an instrument of exercising control over the citizens" (Kuziemski & Misuraca, 2020).[16] Lastly, the framework proposed in this chapter, as many studies on the topic, is created from an inherently Western-centric perspective. Therefore, it is crucial that the scientific and public discourse on the AI use of public actors is shaped by a more diverse set of voices in the future.

## ACKNOWLEDGEMENT

This research was conducted as part of the grant "Consequences of Artificial Intelligence for Urban Societies (CAIUS)", funded by Volkswagen Foundation. We would like to thank the editorial team, workshop participants, as well as our colleagues Lea Cohausz and Frederic Gerdon for their helpful feedback to an earlier version of this chapter.

## NOTES

1 This development can also be seen as part of the proliferation of (new) policy instruments in governments (Zehavi, 2012) in recent times.
2 The field produced a plethora of various terms for describing its methods, ranging from "policy analysis" and "policy appraisal" to "policy evaluation" (Wollmann, 2017).
3 While this framework was designed to accommodate the particular challenges of deploying AI in the public sector, it could in theory also be used in private enterprises.
4 Additionally, the public perception of AI in government may vary across different contexts, rendering some high-stake applications of AI potentially too unpopular to pursue (Kern et al., 2022).
5 This could entail if a generative parametric model is used, interpreting its parameters, or even publishing the whole training data as well as the algorithm used.
6 See Shneiderman (2022) for a detailed description. Since there is some overlap between the author's framework and the one described in this chapter, the following remarks will focus on where both complement each other.
7 Policy evaluation traditionally also focuses on establishing a causal link between the observed results and the designed policy (see Vedung (2013)).
8 Said usefulness directly depends on the quality and external validity of the model of reality used. For the model city that is the subject of this test case, it is assumed that the used simulation provides enough external validity to allow drawing conclusions based on it.

9 See Kappenberger et al. (2022) for a detailed account of the simulation.
10 In practice, this would require not only comparing against the currently implemented policy but also all considered alternatives.
11 The desired range of occupancy of the individual CPZs was defined between 75 and 90% in accordance with the goal of having "one or two spaces per block remain on average unoccupied" stated in the literature (Shoup, 2011).
12 Based on this simulation alone and without being able to account for all administrative infrastructure that would have to be created to enact AI-based parking pricing in practice, it is impossible to draw definitive conclusions regarding the efficiency of this example policy.
13 To do so, the Proximal Policy Optimization (PPO) algorithm is used (Schulman et al., 2017).
14 A limitation of combining simulations with this framework is that one is only able to diagnose unanticipated side effects on the variables, which are part of the implementation of the simulation used.
15 Dimension "optimize occupancy" shows the timeshare that CPZs are held in the desired occupancy range. "Maximize speed" is the mean value of the average normalized speed during the runs. The scores on the remaining dimensions were determined by the value of the respective indicator at the end of the day, i.e., the number of vehicles and the share of low-income agents, respectively.
16 The framework proposed here should be able to support both paradigms.

## REFERENCES

Androutsopoulou, A., Karacapilidis, N., Loukis, E., & Charalabidis, Y. (2019). Transforming the communication between citizens and government through AI-guided chatbots. *Government Information Quarterly*, *36*(2), 358–367. https://doi.org/10.1016/j.giq.2018.10.001

Ben Rjab, A., & Mellouli, S. (2019). Artificial Intelligence in Smart Cities: Systematic Literature Network Analysis. *Proceedings of the 12th International Conference on Theory and Practice of Electronic Governance*, 259–269. https://doi.org/10.1145/3326365.3326400

Binns, R. (2018). Algorithmic Accountability and Public Reason. *Philosophy & Technology*, *31*(4), 543–556. https://doi.org/10.1007/s13347-017-0263-5

Christiano, P., Leike, J., Brown, T. B., Martic, M., Legg, S., & Amodei, D. (2017). *Deep reinforcement learning from human preferences* (arXiv:1706.03741). arXiv. http://arxiv.org/abs/1706.03741

Curran, D., & Smart, A. (2021). Data-Driven Governance, Smart Urbanism and Risk-Class Inequalities: Security and Social Credit in China. *Urban Studies*, *58*(3), 487–506. https://doi.org/10.1177/0042098020927855

de Fine Licht, K., & de Fine Licht, J. (2020). Artificial Intelligence, Transparency, and Public Decision- Making: Why Explanations Are Key When Trying to Produce Perceived Legitimacy. *AI & Society*, *35*(4), 917–926. https://doi.org/10.1007/s00146-020-00960-w

Desouza, K. C. (2018). *Delivering Artificial Intelligence in Government*: IBM Center for The Business of Government. https://www.businessofgovernment.org/report/delivering-artificial-intelligence-government-challenges-and-opportunities

Desouza, K. C., Dawson, G. S., & Chenok, D. (2020). Designing, Developing, and Deploying Artificial Intelligence Systems: Lessons from and for the Public Sector. *Business Horizons*, *63*(2), 205–213. https://doi.org/10.1016/j.bushor.2019.11.004

Engstrom, D. F., Ho, D. E., Sharkey, C. M., & Cuéllar, M.-F. (2020). Government by Algorithm: Artificial Intelligence in Federal Administrative Agencies. *SSRN Electronic Journal*. https://doi.org/10.2139/ssrn.3551505

European Commission. (2021). *Proposal for a Regulation of the European Parliament and of the Council Laying down Harmonised Rules on Artificial Intelligence (Artificial Intelligence Act) and Amending Certain Union Legislative Acts*. https://digital-strategy.ec.europa.eu/en/library/proposal-regulation-laying-down-harmonised-rules-artificial-intelligence

Friesen, M., & Mingardo, G. (2020). Is Parking in Europe Ready for Dynamic Pricing? A Reality Check for the Private Sector. *Sustainability*, *12*(7), 2732. https://doi.org/10.3390/su12072732

Gerdon, F., Bach, R. L., Kern, C., & Kreuter, F. (2022). Social impacts of algorithmic decision-making: A research agenda for the social sciences. *Big Data & Society*, *9*(1). https://doi.org/10.1177/20539517221089305

Harrison, T. M., & Luna-Reyes, L. F. (2022). Cultivating Trustworthy Artificial Intelligence in Digital Government. *Social Science Computer Review*, *40*(2), 494–511. https://doi.org/10.1177/0894439320980122

Hartikainen, M., Väänänen, K., Lehtiö, A., Ala-Luopa, S., & Olsson, T. (2022). Human-Centered AI Design in Reality: A Study of Developer Companies' Practices: A study of Developer Companies' Practices. *Nordic Human-Computer Interaction Conference*, 1–11. https://doi.org/10.1145/3546155.3546677

Johnson, D. G. (2017). Ethical Issues in Big Data. In J. C. Pitt & A. Shew (Eds.), *Spaces for the Future* (1st ed., pp. 164–173). Routledge. https://doi.org/10.4324/9780203735657-16

Kappenberger, J., Theil, K., & Stuckenschmidt, H. (2022). Evaluating the Impact of AI-Based Priced Parking with Social Simulation. In F. Hopfgartner, K. Jaidka, P. Mayr, J. Jose, & J. Breitsohl (Eds.), *Social Informatics* (Vol. 13618, pp. 54–75). Springer International Publishing. https://doi.org/10.1007/978-3-031-19097-1_4

Kern, C., Gerdon, F., Bach, R. L., Keusch, F., & Kreuter, F. (2022). Humans Versus Machines: Who Is Perceived to Decide Fairer? Experimental Evidence on Attitudes Toward Automated Decision-Making. *Patterns*, *3*(10), 100591. https://doi.org/10.1016/j.patter.2022.100591

Kuziemski, M., & Misuraca, G. (2020). AI governance in the public sector: Three tales from the frontiers of automated decision-making in democratic settings. *Telecommunications Policy*, *44*(6), 101976. https://doi.org/10.1016/j.telpol.2020.101976

Langer, M., Oster, D., Speith, T., Hermanns, H., Kästner, L., Schmidt, E., Sesing, A., & Baum, K. (2021). What Do We Want from Explainable Artificial Intelligence (XAI). *Artificial Intelligence*, *296*, 103473. https://doi.org/10.1016/j.artint.2021.103473

Mann, M., & Matzner, T. (2019). Challenging Algorithmic Profiling: The Limits of Data Protection and Anti-discrimination in Responding to Emergent Discrimination. *Big Data & Society*, *6*(2). https://doi.org/10.1177/2053951719895805

Margetts, H., & Dorobantu, C. (2019). Rethink Government with AI. *Nature*, *568*(7751), 163–165. https://doi.org/10.1038/d41586-019-01099-5

Medaglia, R., Gil-Garcia, J. R., & Pardo, T. A. (2021). Artificial Intelligence in Government: Taking Stock and Moving Forward. *Social Science Computer Review*. https://doi.org/10.1177/08944393211034087

Mehrabi, N., Morstatter, F., Saxena, N., Lerman, K., & Galstyan, A. (2022). A Survey on Bias and Fairness in Machine Learning. *ACM Computing Surveys*, *54*(6), 1–35. https://doi.org/10.1145/3457607

Mhlanga, D. (2022). Human-Centered Artificial Intelligence: The Superlative Approach to Achieve Sustainable Development Goals in the Fourth Industrial Revolution. *Sustainability*, *14*(13), 7804. https://doi.org/10.3390/su14137804

Miller, T. (2019). Explanation in Artificial Intelligence: Insights from the Social Sciences. *Artificial Intelligence*, *267*, 1–38. https://doi.org/10.1016/j.artint.2018.07.007

Mintrom, M., & Williams, C. (2013). Public Policy Debate and the Rise of Policy Analysis. In E. Araral (Ed.), *Routledge Handbook of Public Policy* (pp. 3–16). Routledge.

Nespeca, V., Comes, T., & Brazier, F. (2022). A Methodology to Develop Agent-Based Models for Policy Design in Socio-Technical Systems Based on Qualitative Inquiry. In M. Czupryna & B. Kamiński (Eds.), *Advances in Social Simulation* (pp. 453–468). Springer International Publishing. https://doi.org/10.1007/978-3-030-92843-8_34

Riedl, M. O. (2019). Human-Centered Artificial Intelligence and Machine Learning. *Human Behavior and Emerging Technologies*, *1*(1), 33–36. https://doi.org/10.1002/hbe2.117

Samek, W., & Müller, K.-R. (2019). Towards Explainable Artificial Intelligence. In W. Samek, G. Montavon, A. Vedaldi, L. K. Hansen, & K.-R. Müller (Eds.), *Explainable AI: Interpreting, Explaining and Visualizing Deep Learning* (Vol. 11700, pp. 5–22). Springer International Publishing. https://doi.org/10.1007/978-3-030-28954-6_1

Saxena, D., Badillo-Urquiola, K., Wisniewski, P. J., & Guha, S. (2021). A Framework of High-Stakes Algorithmic Decision-Making for the Public Sector Developed through a Case Study of Child- Welfare. *Proceedings of the ACM on Human-Computer Interaction, 5*(CSCW2), 1–41. https://doi.org/10.1145/3476089

Schmidt, P., Biessmann, F., & Teubner, T. (2020). Transparency and trust in artificial intelligence systems. *Journal of Decision Systems, 29*(4), 260–278. https://doi.org/10.1080/12460125.2020.1819094

Schulman, J., Wolski, F., Dhariwal, P., Radford, A., & Klimov, O. (2017). Proximal Policy Optimization Algorithms. *ArXiv:1707.06347 [Cs]*. http://arxiv.org/abs/1707.06347

Selbst, A. D., Boyd, D., Friedler, S. A., Venkatasubramanian, S., & Vertesi, J. (2019). Fairness and Abstraction in Sociotechnical Systems. *Proceedings of the Conference on Fairness, Accountability, and Transparency*, 59–68. https://doi.org/10.1145/3287560.3287598

Shankar, S., Halpern, Y., Breck, E., Atwood, J., Wilson, J., & Sculley, D. (2017). *No Classification without Representation: Assessing Geodiversity Issues in Open Data Sets for the Developing World* (arXiv:1711.08536). arXiv. http://arxiv.org/abs/1711.08536

Shneiderman, B. (2020). Human-Centered Artificial Intelligence: Reliable, Safe & Trustworthy. *International Journal of Human–Computer Interaction, 36*(6), 495–504. https://doi.org/10.1080/10447318.2020.1741118

Shneiderman, B. (2022). *Human-Centered AI*. Oxford University Press.

Shoup, D. (2011). *The High Cost of Free Parking* (Updated). Planners Press, American Planning Association.

de Sousa, W. G., de Melo, E. R. P., Bermejo, P. H. D. S., Farias, R. A. S., & Gomes, A. O. (2019). How and where is artificial intelligence in the public sector going? A literature review and research agenda. *Government Information Quarterly, 36*(4), 101392. https://doi.org/10.1016/j.giq.2019.07.004

Starke, C., Baleis, J., Keller, B., & Marcinkowski, F. (2022). Fairness perceptions of algorithmic decision-making: A systematic review of the empirical literature. *Big Data & Society, 9*(2). https://doi.org/10.1177/20539517221115189

Susar, D., & Aquaro, V. (2019). Artificial Intelligence: Opportunities and Challenges for the Public Sector. *Proceedings of the 12th International Conference on Theory and Practice of Electronic Governance*, 418–426. https://doi.org/10.1145/3326365.3326420

Toll, D., Lindgren, I., Melin, U., & Madsen, C. Ø. (2019). Artificial Intelligence in Swedish Policies: Values, Benefits, Considerations and Risks. In I. Lindgren, M. Janssen, H. Lee, A. Polini, M. P. Rodríguez Bolívar, H. J. Scholl, & E. Tambouris (Eds.), *Electronic Government* (Vol. 11685, pp. 301–310). Springer International Publishing. https://doi.org/10.1007/978-3-030-27325-5_23

Valle-Cruz, D., Alejandro Ruvalcaba-Gomez, E., Sandoval-Almazan, R., & Ignacio Criado, J. (2019). A Review of Artificial Intelligence in Government and its Potential from a Public Policy Perspective. *Proceedings of the 20th Annual International Conference on Digital Government Research*, 91–99. https://doi.org/10.1145/3325112.3325242

van Noordt, C., & Misuraca, G. (2022). Exploratory Insights on Artificial Intelligence for Government in Europe. *Social Science Computer Review, 40*(2), 426–444. https://doi.org/10.1177/0894439320980449

Veale, M., Van Kleek, M., & Binns, R. (2018). Fairness and Accountability Design Needs for Algorithmic Support in High-Stakes Public Sector Decision-Making. *Proceedings of the 2018 CHI Conference on Human Factors in Computing Systems*, 1–14. https://doi.org/10.1145/3173574.3174014

Vedung, E. (2013). Six models of evaluation. In E. Araral (Ed.), *Routledge Handbook of Public Policy* (pp. 387–400). Routledge.

Vedung, E. (2020). Policy Evaluation. In D. Berg-Schlosser, B. Badie, & L. Morlino, *The SAGE Handbook of Political Science*. SAGE Publications Ltd. https://doi.org/10.4135/9781529714333

Vinuesa, R., Azizpour, H., Leite, I., Balaam, M., Dignum, V., Domisch, S., Felländer, A., Langhans, S. D., Tegmark, M., & Fuso Nerini, F. (2020). The role of artificial intelligence in achieving the Sustainable Development Goals. *Nature Communications*, *11*(1), 233. https://doi.org/10.1038/s41467-019-14108-y

Wang, Y., Zhang, N., & Zhao, X. (2022). Understanding the Determinants in the Different Government AI Adoption Stages: Evidence of Local Government Chatbots in China. *Social Science Computer Review*, *40*(2), 534–554. https://doi.org/10.1177/0894439320980132

Wilensky, U. (1999). *NetLogo*. Center for Connected Learning and Computer-Based Modeling, Northwestern University. http://ccl.northwestern.edu/netlogo/

Williams, R., Cloete, R., Cobbe, J., Cottrill, C., Edwards, P., Markovic, M., Naja, I., Ryan, F., Singh, J., & Pang, W. (2022). From transparency to accountability of intelligent systems: Moving beyond aspirations. *Data & Policy*, *4*, e7. https://doi.org/10.1017/dap.2021.37

Wirtz, B. W., Weyerer, J. C., & Geyer, C. (2019). Artificial Intelligence and the Public Sector—Applications and Challenges. *International Journal of Public Administration*, *42*(7), 596–615. https://doi.org/10.1080/01900692.2018.1498103

Wollmann, H. (2017). Policy Evaluation and Evaluation Research. In F. Fischer & G. J. Miller (Eds.), *Handbook of Public Policy Analysis: Theory, Politics, and Methods* (pp. 393–404). Routledge. https://doi.org/10.4324/9781315093192

Wu, X., Xiao, L., Sun, Y., Zhang, J., Ma, T., & He, L. (2022). A survey of human-in-the-loop for machine learning. *Future Generation Computer Systems*, *135*, 364–381. https://doi.org/10.1016/j.future.2022.05.014

Zehavi, A. (2012). *New Governance and Policy Instruments: Are Governments Going "Soft"?* Oxford University Press. https://doi.org/10.1093/oxfordhb/9780199560530.013.0017

Zhang, B., & Dafoe, A. (2019). Artificial Intelligence: American Attitudes and Trends. *SSRN Electronic Journal*. https://doi.org/10.2139/ssrn.3312874

Zuiderwijk, A., Chen, Y.-C., & Salem, F. (2021). Implications of the use of artificial intelligence in public governance: A systematic literature review and a research agenda. *Government Information Quarterly*, *38*(3), 101577. https://doi.org/10.1016/j.giq.2021.101577

# 26 Three Implementation Gaps to Harnessing Inclusive AI in Organizations

*Clementine Collett*

Oxford Internet Institute, University of Oxford, Oxford, UK

*Gina Neff*

Minderoo Centre for Technology and Democracy, University of Cambridge, Cambridge, UK

*Maria Axente*

Universiity of Cambridge, Oxford, UK

## INCLUSIVE AI

Nowadays, companies feel there's a business imperative to use AI systems and tools in many of their processes. This includes, for example, recruitment, talent management, decision-support systems, data analysis, predictive analytics, or customer interaction (Black & van Esch, 2020, 2021; Desouza et al., 2020). AI systems are thought to be key to both reducing running costs and also helping companies gain efficiencies and value (Desouza et al., 2020; Forman et al., 2020; Reynolds, 2021; Seiler, 2021), but they come with implementation challenges, especially for those companies that want to ensure that their systems are fair and not perpetuating gender and racial discrimination. This has been the focus of much academic research, which has emphasized the ability of AI to misgender (Keyes, 2018), to oppress (Browne, 2015; Noble, 2018; Woods, 2018), to exclude (Buolamwini & Gebru, 2018), and to stereotype (Kay et al., 2015).

Separately, there has been no widespread consensus about how we define "inclusivity" in organizations (Podsiadlowski, 2014). While some scholars have explored the importance of dynamics such as power and belonging to organizational inclusion (Bryer, 2020; Syed & Özbilgin, 2009) others have highlighted the context-dependent nature of the inclusion (Dobusch, 2014; Podsiadlowski, 2014). But most consistently, organizational inclusion has been linked to the actual diversity of employees and the way that organizations then allow this diversity to fulfill its potential in terms of


DOI: 10.1201/9781003320791-30

opportunity, interaction, communication, respect, and understanding (Dobusch, 2014; Podsiadlowski, 2014; Thompson, 2017; Woods, 2002). Furthermore, there has scarcely been any academic work that aims to combine work on inclusivity within organizations with what it means for AI to be inclusive.

Inclusive organizations should be considered when we speak of inclusive AI. Inclusive AI, as we define it, works toward the goal of inclusive organizations: it should encourage diversity of employees and embrace differences to help the workforce fulfill their potential and to be aware, communicative, respectful, collaborative, and understanding of one another.

In this chapter, we use the definition of inclusive AI that the Women's Forum for Economy and Society includes in their "Action Toolkit on Inclusive AI," a definition that encapsulates the objective and vision of inclusive organizations: "Inclusive AI addresses the issue of bias and discrimination with the aim of reducing inequalities, including representation, accessibility, and interpretability. Inclusive AI is non-discriminatory in its production, unbiased in its consequences, and accessible to all" (WFES, 2021, p. 9).

Inclusive AI goes beyond mere technological advancements; it places humans firmly at the forefront, emphasizing the significance of considering the well-being and empowerment of all individuals. On the other hand, Human-Centric AI (HCAI) fundamentally calls for a shift from algorithm-centric perspectives to human-centric perspectives, emphasizing the importance of designing AI systems that cater to human needs. The aims of HCAI and inclusive AI are inherently aligned, working hand in hand to create an AI-driven future that serves and uplifts our society. Therefore, it is crucial that inclusive AI remains central to the discourse surrounding HCAI efforts. By prioritizing inclusivity, AI can foster care, collaboration, sustainable communities, and environmental restoration (Shneiderman, 2021).

In order to advance the implementation of AI technologies inclusively and promote Human-Centric AI (HCAI) in industry, it is crucial to map and understand the challenges faced by organizations. This chapter aims to provide a comprehensive overview of these challenges and shed light on the specific elements of implementation that pose difficulties. By identifying and addressing these hurdles, businesses can pave the way for a more inclusive and human-centric approach to AI.

We use data from 3 focus group interviews with 11 high-level C-suite, lead data or technology officers, and industry experts on AI strategy, conducted by the authors in late 2020. From the thematic analysis of the transcripts of these interviews, we identify three implementation gaps that companies face: the engagement gap; the translation gap; and the dialogue gap. These gaps arise when companies have difficulty navigating the space between implementing AI in their organization and implementing AI inclusively.

Framing the challenges that companies face through the lens of "implementation gaps" is a novel way of identifying prevalent and yet unsolved business challenges (Fisher, 2022) and one which translates well regarding organizations' experiences. The gaps that we identify often represent a lack of support, knowledge, or collaboration, all of which we conclude by suggesting strategic solutions that could be used to narrow these implementation gaps and ultimately make AI implementation more inclusive. Overall, we argue that implementing inclusive AI requires, first and foremost, an organizational pivot to embrace inclusivity at the core of business culture and practice.

## BACKGROUND

Many companies feel the business incentive to implement AI systems. Not only this, but the COVID-19 pandemic has recently prompted more companies to implement these technologies; organizations need, more than ever, to think strategically about their resources and business decisions in order to survive (Forman et al., 2020; Reynolds, 2021; Seiler, 2021).

But businesses face many challenges when it comes to implementing AI systems. Research shows that few companies have the foundational building blocks in place for AI to generate value at scale (Chui & Malhotra, 2018). For example, businesses may struggle, practically and legally, to make use of new data analytic tools (Tambe et al., 2019) or they may not possess the necessary technological competence or data skills (EY, 2021; O'Reilly, 2021; Pan et al., 2021). Some organizations end up not implementing systems at all for lack of readiness, knowledge, guidance, or infrastructure, and those that do might not think they're gaining the full potential from these systems. In a global survey by O'Reilly of more than 3,000 people who work with AI in some way across a variety of industries, it was found that only around a quarter (26%) would describe their AI as "mature" meaning that they had revenue-bearing AI products (O'Reilly, 2021).

In the companies which do implement AI systems, employees may not fully understand how to use them, or indeed, not want to use them. It's unlikely that AI which is implemented without want from the end-user will add value or inclusivity to the business practices and outputs (Makarius et al., 2020). Even though these new ways of working can help organizations to reduce costs and increase revenues, end-users often resist adopting AI tools to guide decision-making because they see few benefits for themselves; new tools may require additional work and a perceived loss of autonomy (Kellogg et al., 2022; also see WEF, 2019).

There are clearly practical barriers in implementing AI systems which organizations are now working to overcome. More importantly, though, companies struggle to implement AI systems inclusively, sustainably, and responsibly (IBM, 2022; O'Reilly, 2021). In this chapter, we explore the nature of the struggle to implement inclusive AI and reflect upon ways that this might be overcome.

## RESEARCH PROCESS

The data used in this article was collected from 3 online focus group interviews with 11 high-level C-suite and lead data or technology officers at Fortune 500 companies and industry experts on AI strategy in late 2020.

All participants were in managerial, middle-managerial, or development positions and have an oversight on AI strategies, including anyone working across different functions utilizing AI, from Operations to IT and spanned roles from C-Suite, data scientists, modelers, specialists, and engineers. Participants came from organizations in sectors such as financial services, technology companies, oil and gas, education, and law. These focus groups were held on Zoom, lasted for one hour each, and were recorded. In this study, all participants are anonymized, and referred to by their participant number only.

The aim of the focus groups was to convene AI leaders within organizations in order to explore how they prioritize resources and achieve inclusive AI development and deployment, as well as to assess how they articulate and implement inclusive AI strategy. While this chapter focuses mainly on companies implementing AI systems, both internally and externally sourced and designed, it also touches on implementation gaps that might occur within AI vendor companies that design and develop AI systems.

The focus groups were conducted with the Women's Forum for Economy and Society through their initiative, the Women4AI Daring Circle, and they were facilitated by Professor Gina Neff. The Women4AI Daring Circle was set up by the Women's Forum for the Economy and Society in order to bring together an ecosystem of partners and advocate for making the design and development of AI truly inclusive. Its aim is to encourage organizations of all types and sizes to take tangible action to create an environment where women are empowering AI to its full potential, and AI is empowering women to their full potential.[1]

For thematic analysis, the qualitative data from the focus groups was used in addition to a broad literature review on the topic of challenges implementing AI in organizations and inclusive AI in order to identify particular themes and patterns (Braun & Clarke, 2006; Lapadat, 2010).

In our methodology, we present the results of the focus group discussions and establish a dialogue between these findings and the existing scholarly works in the field. By combining the insights and perspectives gathered from the participants with the knowledge and research already available, we aim to provide a comprehensive understanding of the challenges at hand. This integration of empirical data and scholarly literature allows us to offer a nuanced and well-informed analysis of the identified challenges, enriching the discussion with a broader context and theoretical insights.

## THREE IMPLEMENTATION GAPS TO HARNESSING INCLUSIVE AI

In this section, we discuss three implementation gaps to harnessing inclusive AI in the workplace that emerged from the analysis of the focus group data.

### The Engagement Gap

The engagement gap refers to the challenge of effectively engaging leaders and employees on issues surrounding bias, discrimination, or inclusivity in AI systems.

*Effective* engagement regarding bias and discrimination from companies would, of course, require engagement with these issues in the first place. Some companies don't engage whatsoever with issues of bias and discrimination or inclusivity because they might fear the risk that engagement with these topics exposes other issues. One participant said:

> Companies believe actually that they don't need to do a lot, actually, it's a misconception at the board level, at the top level, and that's where decisions are made. So, they don't necessarily believe they have an issue in this respect, so it's kind of lack of awareness in some companies.
>
> (Participant 8)

> While it may well be conscious or tactical ignorance from some companies, often the disengagement in other companies isn't purposeful. It was evident from the focus groups that while some companies have the intent to engage with issues of bias and discrimination, and in doing so, intent to implement inclusive AI, they don't know how to grasp the complexity of the task ahead and act upon it.
>
> (Participant 5)

The first step to navigating this implementation gap is having hard conversations about the underlying issues of bias and discrimination; raising awareness and then identifying where issues could manifest and how. In addition to education and dialogue about issues of bias and discrimination in AI, companies need to pursue increased transparency and communication about how AI systems are being developed and implemented. This will allow for scrutiny and dialogue aimed at improvements and harm identification and mitigation. We know from the literature that sustainable and inclusive innovation requires not just initiative from users, employees, and crucially, leadership but also from stakeholders such as communities affected by these technologies (Ghassim & Foss, 2020; Leslie, 2019; Plasschaert, 2019; Schiff et al., 2020; Stix, 2021).

Effective engagement in inclusive AI implementation necessitates a comprehensive and actionable mandate for inclusivity within the organization, coupled with a commitment to prioritize diversity and inclusion at all stages of product development, including the allocation of responsibility within management structures and active involvement of relevant stakeholders (leadership, suppliers, regulators, unions) to address biases and discrimination within AI systems.

Research shows that a good governance structure is crucial to influencing firm performance (Clark et al., 2014), organizational culture, and employee behavior in the workplace (Gottman et al., 1998). The ability of leaders to integrate and create bonds within teams is imperative for enhancing collective innovation (Jiang & Chen, 2018).

Unfortunately, there are a few barriers that stand in the way of leaders engaging with issues surrounding bias and discrimination, and therefore with issues of inclusive AI. First, many companies don't have diverse leadership teams (Eagly & Fischer, 2009). Research highlights how women and minority racial/ethnic groups are less likely to be seen as "fitting" the stereotype of successful performance in a leadership role (Heilman, 2012; Hoyt & Murphy, 2016; Lyness & Heilman, 2006; Wille et al., 2018) especially in the tech industry (Bello et al., 2021; UNESCO, 2019; Young et al., 2021). These factors could lead to issues of diversity, bias, and discrimination being less likely to be the focus of leaders' attention (Leavy, 2018, p. 14). However, it has become clear over the last few years that organizations realize diversity at all levels needs to be prioritized if they are to maintain sustainability and profitability.

Another barrier to engaging with inclusive AI is the persistent narrative of prioritizing efficiency and profit over responsible, sustainable, and inclusive innovation. Ethics, generally, can act as a constraint in achieving these goals, as it's seen as a trade-off between accuracy (the main KPI for models) and fairness, transparency, and privacy. For example, gathering the views of relevant stakeholders, resolving conflicts, and more time spent on team meetings, all of this is imperative to inclusivity and all of it costs money and resources, and therefore might be seen to collide with commercial incentives (Mittelstadt, 2019).

While companies see it as imperative to innovate and keep up with trends, they are not engaging with how issues of bias and discrimination, and therefore inclusivity, intersect with this implementation. They see issues of inclusivity in AI as separate from the most lucrative technologies, but they're not separate: "There's a real misunderstanding about how implementing responsible and ethical and inclusive AI can contribute, in this case in the private sector as one of their priorities, to economic growth" (Participant 6).

However, engaging with these issues is in the interest of the company, because "beyond the humanity and society question here, it's about taking competitive advantages" and this competitive advantage really comes from the fact that companies are "actually engaging with becoming a responsible business and having responsible AI or ethical AI or inclusive AI" (Participant 8). Therefore, this narrative that competitive and efficient company decisions are not also ones that are responsible, sustainable, ethical, and inclusive does not add up. It is exactly these types of mandates which has been shown, time and time again, to be beneficial to business performance even through the use of technology in these ways (Sahgal et al., 2020; WEF, 2022).

## The Translation Gap

The translation gap refers to the difficulty that companies have in translating inclusive AI principles into practice.

While principles and guidelines on AI abound, turning these principles into action creates a major sticking point for companies. There is generally a lack of understanding of the profound changes that are needed to act upon the ethical principles for AI. As Participant 2 said: "Companies don't really know how to do it. I mean they've all outlined concepts but taking action on those concepts is going to require significant changes to processes, to systems" (Participant 2).

Research tends to focus on principles surrounding ethical AI and responsible AI, and this "gap" in translating principles to practice has been noted. Schiff et al. (2020) note the lack of clarity on how to implement high-level principles for responsible AI into practices, naming this the "principles-to-practices gap." Two of the explanations they give for this gap are an overabundance of tools (or principles) for addressing responsible development and use of AI, and a lack of integration between roles and levels in organizations.

The same is true for implementing "ethical AI" principles. Some scholars have focused on trying to make ethical principles actionable (Floridi et al., 2018), but many ethical principles and frameworks for AI do not help concrete views on AI practices and are not relevant for their AI solutions, even if they are valuable for conceptualizing ethical issues (Canca, 2020; Clarke, 2019). As Mittelstadt (2019) argues, translation requires specification of high-level principles into mid-level norms and low-level requirements, and it is challenging to deduce these "without accounting for specific elements of the technology, application, context of use or relevant local norms" (Mittelstadt, 2019, p. 504). AI principles are only the starting point, and, often, they are too broad and high-level to guide ethics in practice

(Whittlestone et al., 2019). "Bridging the gap" between principles and practice and focusing on the tensions that arise during implementation is the main way to overcome, or close, this gap. There is a need for systematic intervention to translate the principles into practice (Mökander & Axente, 2021).

As Participant 6 talked about, this is an "institutional gap in translating these principles into practice," especially as we talk about concepts like "trust," "accountability," and "explainability"; what do these words mean in practice? Participant 6 suggested that this might be helped by more investment in incentivizing companies to develop ethical guidelines, but also measurement mechanisms to show the impact of different practices.

However, we do see examples of companies that are beginning to overcome these difficulties, for example, through developing internal policies and standards that describe the rules to those designing and developing these tools. We can see this in the cases of IBM and Microsoft as explored by the World Economic Forum (Green, Heider, et al., 2021a; Green, Lim, et al., 2021b).

Unlike the above literature, here, we focus on inclusive AI, rather than ethical or responsible AI specifically. Therefore, it's not just about navigating how to translate principles into practice but exploring whether these practices are inclusive in the first place.

Overwhelmingly, we found in the focus groups that what companies seem to be struggling with when navigating this gap of translating principles of inclusivity into practice with AI implementation, is the lack of concrete instruction. Participant 8 recognized that their clients have started to change the way they develop and apply AI principles. The way they've done this successfully is by putting in place processes from the beginning where the data is collected, right until the end user of the AI. The other reason this has happened is that the engagement gap had been navigated effectively, meaning that the boards were engaged with this issue and, so, often, the leaders of companies were ensuring that these principles were actionable and measurable. Stakeholders, and greater transparency, can play such a significant role in the translation of inclusive AI principles into action.

For some companies in particular industries, these concrete steps are much more present already. Some companies will have structures that could be used to promote inclusive AI, and those need to be identified, revamped to suit the purpose, and used. For example, financial services have strong regulations, policies, and audit practices. This means that they feel the high levels of data culture and high levels of collaboration between countries will make strategies for inclusive AI easier. The fact that the structures are already in place, even if the rules change, should make it smoother to implement than for industries or sectors without these elements (Participant 1 and Participant 3).

But navigating the translation gap is not as simple as translating principles into actions through inclusive regulation or policy; it's about more than that. It's about ensuring we are creating inclusive organizational cultures that foster an ethical behavior and prioritize inclusion, so that we constantly analyze and discuss how regulation and policy can harness inclusion more, and how they can be complemented by other means that incentivize ethical behavior. Participant 5 pointed out that while AI becomes more pervasive, existing regulatory and accountability frameworks won't work, so we need to, as corporations, continue to be in dialogue with policymakers. There is a risk, they said, that some bias (whether social or technological) might creep into these policies too. It must also be truly representative of global standards so that we don't further divide the Global North and Global South.

## The Dialogue Gap

The third, and most prominent gap which emerged from these focus groups was the dialogue gap when it comes to implementing inclusive AI. The dialogue gap refers to how organizations struggle to navigate inclusive communication, feedback, and idea-sharing, between those designing AI and those implementing or using AI. As Participant 2 pointed out, this is no small feat: "How do you create human feedback from an individual user? I think those are, you know, all very unanswered questions and they're also frankly very big questions" (Participant 2).

In terms of developing AI systems in the first place, dialogue is crucial. For example, in the context of a technology company, a technical team should be in constant dialogue with other "non-technical" roles in order to discuss corporate responsibility along with the social and ethical impacts of the systems, as well as what it might look like to embed these concepts into the design process, because "[t]echnical teams may also fail to imagine ways in which the system they are creating could be improved, or how other systems, tools, or methodologies could be applied to better safeguard and improve human well-being" (Schiff et al., 2020, p. 4). Microsoft just published the second version of their Responsible AI Standard that product development teams can use as concrete and actionable guidance for what to uphold as their principles. In other words, it's a way for nontechnical principles to be translated into technical actions (Microsoft, 2022).

Participant 7 pointed out the lack of engagement in relation to the technical side of things. They talked about how inclusive AI is about thinking: where can you make the mistake? A lot of technical people, they said, aren't aware of the fact that the way we collect data is already biased, before even starting to code. But this awareness is so important, otherwise one doesn't know where the mistakes are coming from. Participant 5 also pointed out that datasets are prepared by humans, and the challenge of humans categorizing data means that their own biases can creep in. We train the model, and this can end up furthering the damage. One of the issues, then, is that we don't have fully diverse and representative datasets to begin with. As Tambe et al. (2019) point out, formalizing the process of creating the underpinnings of AI models is important—these aren't decisions that should be made by data scientists alone.

Participant 6 spoke about the importance of remaining open for dialogue at the point of design. They said that this entails not just including a diverse range of voices in terms of gender, ethnicity, race, class, disability, and so on, but also in terms of having a diverse range of experts around the table from a range of disciplines, such as anthropology, sociology, philosophy and so on. Even if the usage and the purpose are important, if we don't pay attention to how it's created then it's going to be very complicated in the future (Participant 9) which is why it's crucial for organizations to include people from different teams, backgrounds, experiences, and with different skills (Participant 11).

This is just one example of why a mix of disciplines and backgrounds in design and development teams is crucial. If you ensure that your team is diverse, then you might have a better picture of what's going on and how to solve it. Diversity has often and vehemently been hailed as crucial for the teams designing and developing AI; if we want to ensure this tech is as inclusive as possible (Bello et al., 2021; Criado-Perez, 2019; Moore, 2019; Noble, 2018; UNESCO, 2019; Young et al., 2021). This is not just to do with the data which is used in the systems, and being able to consider whether this data is appropriate to be used to create an inclusive system, but it's also about thinking how algorithms are designed and what is considered during the design (Buolamwini & Gebru, 2018; D'Ignazio & Klein, 2020; Zou & Schiebinger, 2018). Dahlin (2021) points out that often, AI is seen as a technical object that only later, after it has been implemented it may have social consequences; however, we need to ensure that we include a social analysis at the outset of development (Dahlin, 2021).

In the same way, this dialogue is crucial throughout the implementation process. Organization and business studies literature constantly stresses the importance of cross-departmental and cross-hierarchy collaboration for innovation and for effective AI implementation (Bessant, 2009; Schiff et al., 2020).

In addition, this dialogue would ideally occur between the company implementing the AI and the company creating it. Participant 2 spoke about how AI systems are a truly socio-technical system which is interactive with not only the people building it but also the people being affected by it. They spoke about the "human-in-the-loop" concept, which is where the individual is giving feedback on the product they're using, and this is actively feeding back into the system: "From a corporate perspective, 'human-in-the-loop' is almost a buzz word or a buzz phrase at this point, but I will genuinely say that I have not seen it done well" (Participant 2). In essence, the ideal would be inclusive co-design which also includes marginalized populations when they're the ones most affected by the technology (Walsh & Wronsky, 2019).

We might refer to this as the "social shaping of AI"—which is when we consider questions surrounding how those affected by AI can also play a role in shaping it (Bondi et al., 2021). This would help to ensure that users of AI within companies, and those using AI outside of the workplace, can question, co-create, and get feedback on decisions and patterns of the system (Karami et al., 2016; Makarius et al., 2020; Sjödin et al., 2021).

## NARROWING THE IMPLEMENTATION GAPS

Overall, while organizations want to keep up their efficiency with the most innovative AI technology, they are at a loss as to how they can do this while implementing inclusive systems. Although we don't claim to have all the answers here, we argue that navigating these gaps to implementing inclusive AI requires, first and foremost, organizational change so that organizations themselves are inclusive.

Three organizational qualities emerged from the focus groups that might help organizations to successfully navigate these implementation gaps when it comes to designing and/or implementing inclusive AI.

### Inclusive Collaboration with Relevant Stakeholders

Transparency has often been spoken about as a key enabler or principle for ethical or responsible AI system design and development by tech companies (Google, 2021; IBM, 2021; Microsoft, 2021) and academic scholars (Felzmann et al., 2020; Larsson & Heintz, 2020). But what does transparency really mean for companies designing as well as those implementing AI? This is complex for companies, because not only are "black box" models highly complex, but different contexts might give rise to different explainability needs (The Royal Society, 2019).

It helps if we think about transparency, practically speaking, not just as a one-way system, but as inclusive collaboration with relevant stakeholders, including users. Participant 11 implied that the ability for users to not only have technology is explainable but also to really be able to understand the systems principles, functionality, and purpose. If organizations are engaged with the issues of bias and discrimination in AI, this will allow them to better assess their choices. Kellogg et al. (2022) discuss how end-users of the AI tools should be involved in the development at early stages and should also help to evaluate the tool. This will forestall expected user resistance and make resentment less likely if users are brought in later.

### Diverse and Inclusive Teams

The first, spoken about by many of the participants, was that organizations must pay attention to how they create teams and how they diversify them, having people from different disciplines, as well as different backgrounds and cognitive diversities, involved. This cannot just be lip service, but it must be complemented by a truly inclusive organizational culture. By embracing such diversity, organizations can tap into a wealth of perspectives, knowledge, and experiences, fostering creativity, innovation, and a holistic understanding of complex problems.

However, achieving inclusive AI goes beyond team composition. It necessitates the development of an inclusive organizational culture that enables AI users to provide practical feedback through reporting channels. This feedback mechanism empowers users to influence not only the technological function of AI systems but also their purpose and position within the organization and society at large. Nonetheless, it is vital to recognize that, as Sloane et al. (2020) point out, participatory design requires acknowledgment as work, with appropriate consent and compensation. Furthermore, the design process should be context-specific, ensuring that data is collected from the right individuals through a consultative and revisited approach.

### Education and Knowledge-Sharing

The participants also spoke about education and knowledge-sharing between organizations. Participant 5 pointed out that it's no good to just educate leaders about what works, but they also need to be educated about the risks of ignoring issues of inclusivity in AI. This is especially relevant within middle management, which is where most of the change and day-to-day decision-making is made in many organizations.

In addition to the education of leaders and knowledge-sharing between leaders, Participant 6 also pointed out that we need further education for the technical community and ML specialists on issues of ethics, gender bias, human rights, and non-discrimination. In addition, Participant 6 said that we also need to ensure we train and educate other stakeholder groups with regard to the basics of understanding AI and the impact of AI on our everyday lives: "This is something that I think is absolutely essential and there's a huge gap in digital literacy in institutional capacity building" (Participant 6). Everyone needs to be educated that technology has an impact, and that impact needs to be identified and managed accordingly.

## CONCLUSION

The findings from this small-scale focus group study establish an important theoretical contribution to academic research on working toward inclusive AI. It does this by suggesting three implementation gaps: engagement, translation, and dialogue gaps. These gaps help scholars frame the challenges companies face in implementing inclusive AI in their organizations. Our research can help practitioners frame their own challenges, suggesting how they might begin to narrow these gaps. Good, inclusive organizational practices make for good, inclusive AI. We recommend companies build and support diverse and inclusive teams, work collaboratively with a wide and diverse stakeholder group, and share best practices with other companies.

## NOTE

1 Led by Microsoft, the Women4AI Daring Circle's members are AXA, BNP Paribas, L'Oréal, Google, Publicis Groupe, Bayer and Lenovo, in collaboration with UNESCO as an Institutional Partner and HEC and the Oxford Internet Institute as Academic Partners.

## REFERENCES

Bello, A., Blowers, T., Schneegans, S., & Straza, T. (2021). To be smart, the digital revolution will need to be inclusive. In *UNESCO Science Report* (Issue June).

Bessant, J. (2009). *Innovation*. Dorling Kindersley Ltd.

Black, J. S., & van Esch, P. (2020). AI-enabled recruiting: What is it and how should a manager use it? [Article]. *Business Horizons*, *63*, 215–226. https://doi.org/10.1016/j.bushor.2019.12.001

Black, J. S., & van Esch, P. (2021). AI-enabled recruiting in the war for talent. *Business Horizons*, *64*(4), 513–524. https://doi.org/10.1016/j.bushor.2021.02.015

Bondi, E., Xu, L., Acosta-Navas, D., & Killian, J. A. (2021). Envisioning Communities: A Participatory Approach towards AI for Social Good. *AIES 2021 – Proceedings of the 2021 AAAI/ACM Conference on AI, Ethics, and Society*, 425–436. https://doi.org/10.1145/3461702.3462612

Braun, V., & Clarke, V. (2006). Using thematic analysis in psychology. *Qualitative Research in Psychology*, *3*(2), 77–101. https://doi.org/10.1191/1478088706qp063oa

Browne, S. (2015). *Dark matters: On the surveillance of blackness*.

Bryer, A. (2020). Making Organizations More Inclusive: The Work of Belonging [Article]. *Organization Studies*, *41*(5), 641–660. https://doi.org/10.1177/0170840618814576

Buolamwini, J., & Gebru, T. (2018). Gender Shades: Intersectional Accuracy Disparities in Commercial Gender Classification *. In *Proceedings of Machine Learning Research* (Vol. 81).

Canca, C. (2020). Operationalizing AI ethics principles. *Communications of the ACM*, *63*(12), 18–21. https://doi.org/10.1145/3430368

Chui, M., & Malhotra, S. (2018). Notes from the AI frontier: AI adoption advances, but foundational barriers remain. *McKinsey Analytics*, November, 11. https://www.mckinsey.com/featured-insights/artificial-intelligence/ai-adoption-advances-but-foundational-barriers-remain

Clark, J. R., Murphy, C., & Singer, S. J. (2014). When do leaders matter? Ownership, governance and the influence of CEOs on firm performance. *Leadership Quarterly*, *25*(2), 358–372. https://doi.org/10.1016/j.leaqua.2013.09.004

Clarke, R. (2019). Principles and business processes for responsible AI. *Computer Law and Security Review*, *35*(4), 410–422. https://doi.org/10.1016/j.clsr.2019.04.007

Criado-Perez, C. (2019). *Invisible women: Exposing data bias in a world designed for men*. Chatto & Windus.

D'Ignazio, C., & Klein, L. F. (2020). *Data feminism*. MIT Press.

Dahlin, E. (2021). Mind the gap! On the future of AI research. *Humanities and Social Sciences Communications*, *8*(1), 21–24. https://doi.org/10.1057/s41599-021-00750-9

Desouza, K. C., Dawson, G. S., & Chenok, D. (2020). Designing, developing, and deploying artificial intelligence systems: Lessons from and for the public sector. *Business Horizons*, *63*(2), 205–213. https://doi.org/10.1016/j.bushor.2019.11.004

Dobusch, L. (2014). How exclusive are inclusive organisations? [Article]. *Equality, Diversity and Inclusion: An International Journal*, *33*(3), 220–234. https://doi.org/10.1108/EDI-08-2012-0066

Eagly, A. H., & Fischer, A. (2009). Gender inequalities in power in organizations. In D. Tjosvold & B. Wisse (Eds.), *Power and Interdependence in Organizations* (pp. 186–204). https://doi.org/10.1017/CBO9780511626562.013

EY. (2021). *Data foundations and AI adoption in the UK private and third sectors* (Issue July).

Felzmann, H., Fosch-Villaronga, E., Lutz, C., & Tamò-Larrieux, A. (2020). Towards transparency by design for Artificial Intelligence. *Science and Engineering Ethics*, *26*(6), 3333–3361. https://doi.org/10.1007/s11948-020-00276-4

Fisher, G. (2022). Types of business horizons articles. *Business Horizons*, *65*(3), 241–243. https://doi.org/10.1016/j.bushor.2022.01.002

Floridi, L., Cowls, J., Beltrametti, M., Chatila, R., Chazerand, P., Dignum, V., Luetge, C., Madelin, R., Pagallo, U., Rossi, F., Schafer, B., Valcke, P., & Vayena, E. (2018). AI4People—An ethical framework for a good AI society: Opportunities, risks, principles, and recommendations. *Minds and Machines*, *28*(4), 689–707. https://doi.org/10.1007/s11023-018-9482-5

Forman, A., Glasser, N., & Lech, C. (2020). INSIGHT: Covid-19 May Push More Companies to Use AI as Hiring Tool. *Bloomberg Law*. https://news.bloomberglaw.com/daily-labor-report/insight-covid-19-may-push-more-companies-to-use-ai-as-hiring-tool

Ghassim, B., & Foss, L. (2020). How do leaders embrace stakeholder engagement for sustainability-oriented innovation? In N. Pfeffermann (Ed.), *New Leadership in Strategy and Communication*. https://doi.org/10.1007/978-3-030-19681-3

Google. (2021). *Our Principles – Google AI*. https://ai.google/principles/

Gottman, J. M., Coan, J., Carrere, S., Swanson, C., Gottman, J. M., Coan, J., Carrere, S., & Swanson, C. (1998). Predicting Marital Happiness and Stability from Newlywed Interactions Published by: National Council on Family Relations Predicting Marital Happiness and Stability from Newlywed Interactions. *Journal of Marriage and Family*, *60*(1), 5–22. https://doi.org/10.1002/job

Green, B., Heider, D., Firth-Butterfield, K., & Lim, D. (2021a). *Responsible Use of Technology: The IBM Case Study*. September.

Green, B., Lim, D., & Ratte, E. (2021b). *Responsible Use of Technology_The Microsoft Case Study*. February. https://www.weforum.org/whitepapers/responsible-use-of-technology-the-microsoft-case-study

Heilman, M. E. (2012). Gender stereotypes and workplace bias [Article]. *Research in Organizational Behavior*, *32*, 113–135. https://doi.org/10.1016/j.riob.2012.11.003

Hoyt, C. L., & Murphy, S. E. (2016). Managing to clear the air: Stereotype threat, women, and leadership [Article]. *The Leadership Quarterly*, *27*(3), 387–399. https://doi.org/10.1016/j.leaqua.2015.11.002

IBM. (2021). *IBM's Principles for Trust and Transparency*. https://www.ibm.com/policy/wp-content/uploads/2018/06/IBM_Principles_SHORT.V4.3.pdf

IBM. (2022). IBM Global AI Adoption Index 2022. In *IBM*. https://www.ibm.com/watson/resources/ai-adoption

Jiang, Y., & Chen, C. C. (2018). Integrating Knowledge Activities for Team Innovation: Effects of Transformational Leadership. *Journal of Management*, *44*(5), 1819–1847. https://doi.org/10.1177/0149206316628641

Karami, A. B., Fleury, A., Boonaert, J., & Lecoeuche, S. (2016). User in the loop: Adaptive smart homes exploiting user feedback-State of the art and future directions. *Information (Switzerland)*, *7*(2). https://doi.org/10.3390/info7020035

Kay, M., Matuszek, C., & Munson, S. A. (2015). Unequal representation and gender stereotypes in image search results for occupations. *Conference on Human Factors in Computing Systems - Proceedings, 2015-April*, 3819–3828. https://doi.org/10.1145/2702123.2702520

Kellogg, K. C., Sendak, M., & Balu, S. (2022). *AI on the Front Lines*. MIT Sloan Management Review.

Keyes, O. (2018). The Misgendering Machines: Trans/HCI Implications of Automatic Gender Recognition [Article]. *Proceedings of the ACM on Human-Computer Interaction, 2(CSCW)*, 1–22. https://doi.org/10.1145/3274357

Lapadat, J. C. (2010). Thematic Analysis [Article]. In Albert J. Mills, Gabrielle Durepose, & Elden Wiebe (Eds.), *Encyclopedia of case study research* (vol. 2, pp. 925–927). SAGE Publications.

Larsson, S., & Heintz, F. (2020). Transparency in artificial intelligence. *Internet Policy Review*, *9*(2), 1–16. https://doi.org/10.14763/2020.2.1469

Leavy, S. (2018). Gender bias in artificial intelligence: The need for diversity and gender theory in machine learning [Proceeding]. *GE '18: Proceedings of the 1st International Workshop on Gender Equality in Software Engineering*, 14–16. https://doi.org/10.1145/3195570.3195580

Leslie, D. (2019). *Understanding artificial intelligence ethics and safety systems in the public sector*. The Alan Turing Institute.

Lyness, K., & Heilman, M. (2006). When Fit Is Fundamental: Performance Evaluations and Promotions of Upper-Level Female and Male Managers. [Article]. *Journal of Applied Psychology*, *91*(4), 777. https://doi.org/10.1037/0021-9010.91.4.777

Makarius, E. E., Mukherjee, D., Fox, J. D., & Fox, A. K. (2020). Rising with the machines: A sociotechnical framework for bringing artificial intelligence into the organization. *Journal of Business Research*, *120*(July), 262–273. https://doi.org/10.1016/j.jbusres.2020.07.045

Microsoft. (2021). *Responsible AI principles from Microsoft*. https://www.microsoft.com/en-us/ai/responsible-ai?activetab=pivot1:primaryr6

Microsoft. (2022). *Microsoft Responsible AI Standard, v2* (Issue June).

Mittelstadt, B. (2019). Principles alone cannot guarantee ethical AI. *Nature Machine Intelligence*, *1*(11), 501–507. https://doi.org/10.1038/s42256-019-0114-4

Mökander, J., & Axente, M. (2021). Ethics-based auditing of automated decision-making systems: Intervention points and policy implications. *AI and Society*, *2021*, 1–36. https://doi.org/10.1007/s00146-021-01286-x

Moore, P. (2019). The Mirror for (Artificial) Intelligence: In Whose Reflection? *SSRN Electronic Journal*. https://doi.org/10.2139/ssrn.3423704

Noble, S. U. (2018). *Algorithms of oppression: How search engines reinforce racism*. New York University Press.

O'Reilly. (2021). *AI Adoption in the Enterprise*. https://www.oreilly.com/radar/ai-adoption-in-the-enterprise-2021/

Pan, Y., Froese, F., Liu, N., Hu, Y., & Ye, M. (2021). The adoption of artificial intelligence in employee recruitment: The influence of contextual factors. *International Journal of Human Resource Management*. https://doi.org/10.1080/09585192.2021.1879206

Plasschaert, A. (PwC). (2019). *Leaders need to take responsibility for — and action on — responsible AI practices*. PwC. https://www.pwc.com/gx/en/news-room/press-releases/2019/responsible-ai-dalian-wef.html

Podsiadlowski, J. H.A. (2014). Envisioning "inclusive organizations" [Article]. *Equality, Diversity and Inclusion an International Journal*, *33*(3). https://doi.org/10.1108/EDI-01-2014-0008

Reynolds, K. (2021). *COVID-19 increased use of AI. Here's why it's here to stay | World Economic Forum*. WEF. https://www.weforum.org/agenda/2021/02/covid-19-increased-use-of-ai-here-s-why-its-here-to-stay/

Sahgal, V., Haider, S. A., & Islan, W. (2020). *Staying ahead of the curve - The business case for responsible AI - Economist Intelligence Unit* (Issue October). https://www.eiu.com/n/staying-ahead-of-the-curve-the-business-case-for-responsible-ai/

Schiff, D., Rakova, B., Ayesh, A., Fanti, A., & Lennon, M. (2020). *Principles to Practices for Responsible AI: Closing the Gap*. http://arxiv.org/abs/2006.04707

Seiler, D. (2021). *COVID-19 digital transformation & technology|McKinsey*. https://www.mckinsey.com/business-functions/strategy-and-corporate-finance/our-insights/how-covid-19-has-pushed-companies-over-the-technology-tipping-point-and-transformed-business-forever

Shneiderman, B. (2021). Human-Centered AI: A New Synthesis [Bookitem]. *Human-Computer Interaction - INTERACT 2021, Part of the Lecture Notes in Computer Science Book Series (LNISA, Volume 12932)*, 12932. https://doi.org/10.1007/978-3-030-85623-6_1

Sjödin, D., Parida, V., Palmié, M., & Wincent, J. (2021). How AI capabilities enable business model innovation: Scaling AI through co-evolutionary processes and feedback loops. *Journal of Business Research*, *134*(April), 574–587. https://doi.org/10.1016/j.jbusres.2021.05.009

Sloane, M., Moss, E., Awomolo, O., & Forlano, L. (2020). Participation is not a Design Fix for Machine Learning. *Proceedings of the 37th International Conference on Machine Learning, Vienna, Austria, PMLR 119, 2020*. http://arxiv.org/abs/2007.02423

Stix, C. (2021). Actionable Principles for Artificial Intelligence Policy: Three Pathways. *Science and Engineering Ethics*, *27*(1), 1–17. https://doi.org/10.1007/s11948-020-00277-3

Syed, J., & Özbilgin, M. (2009). A relational framework for international transfer of diversity management practices. *International Journal of Human Resource Management*, *20*(12), 2435–2453. https://doi.org/10.1080/09585190903363755

Tambe, P., Cappelli, P., & Yakubovich, V. (2019). Artificial intelligence in human resources management: Challenges and A path forward. *California Management Review*, *61*(4), 15–42. https://doi.org/10.1177/0008125619867910

The Royal Society. (2019). *Explainable AI: The Basics* (Issue November). https://royalsociety.org/topics-policy/projects/explainable-ai/

Thompson, S. (2017). Defining and Measuring "Inclusion" Within an Organisation. In *K4D Helpdesk Report*. https://unesdoc.unesco.org/ark:/48223/pf0000367416.page=1

UNESCO. (2019). *I'd blush if I could: Closing gender divides in digital skills through education*. https://unesdoc.unesco.org/ark:/48223/pf0000367416.page=1
Walsh, G., & Wronsky, E. (2019). AI + co-design: Developing a novel computer-supported approach to inclusive design. *Proceedings of the ACM Conference on Computer Supported Cooperative Work, CSCW*, 408–412. https://doi.org/10.1145/3311957.3359456
WEF. (2019). *AI procurement in the public sector: Lessons from Brazil|World Economic Forum*. https://www.weforum.org/agenda/2022/05/the-brazilian-case-for-ai-procurement-in-a-box/
WEF. (2022). *Empowering AI Leadership: AI C-Suite Toolkit* (Issue January).
WFES. (2021). *Action Toolkit: On Inclusive AI*.
Whittlestone, J., Alexandrova, A., Nyrup, R., & Cave, S. (2019). The role and limits of principles in AI ethics: Towards a focus on tensions. *AIES 2019 - Proceedings of the 2019 AAAI/ACM Conference on AI, Ethics, and Society*, 195–200. https://doi.org/10.1145/3306618.3314289
Wille, B., Wiernik, B., Vergauwe, J., Vrijdags, A., & Trbovic, N. (2018). Personality characteristics of male and female executives: Distinct pathways to success? [Article]. *Journal of Vocational Behavior*, *106*, 220.
Woods, H. S. (2018). Asking more of Siri and Alexa: Feminine persona in service of surveillance capitalism [Article]. *Critical Studies in Media Communication*, *35*(4), 334–349. https://doi.org/10.1080/15295036.2018.1488082
Woods, S. (2002). Creating Inclusive Organizations: Aligning Systems with Diversity. *Profiles in Diversity Journal*, *4*(1), 38–39. www.diversityjournal.com
Young, E., Wajcman, J., & Sprejer, L. (2021). *Where are the women? Mapping the gender job gap in AI*. The Alan Turing Institute. https://www.turing.ac.uk/sites/default/files/2021-03/where-are-the-women_public-policy_full-report.pdf
Zou, J., & Schiebinger, L. (2018). AI can be sexist and racist – it's time to make it fair [Article]. *Nature*, *559*(7714), 324. https://doi.org/10.1038/d41586-018-05707-8

# 27 Tatiana Revilla's Commentary

*Tatiana Revilla*

Tecnológico de Monterrey, Monterrey, Mexico

Tatiana Revilla is a trained lawyer. She specializes in public policy issues with a gender or feminist perspective. Currently, she directs the Gender Program at the Government and Public Transformation School of the Tecnológico de Monterrey, Mexico. Among other things, her work and findings have contributed to including equality, diversity, and inclusion policies and mechanisms in the Mexican public sector and academia. Revilla graduated with distinction from the Instituto Tecnológico y de Estudios Superiores de Monterrey (ITESM).

**Tell us more about what you do in the field of AI.**

The program I direct is particularly concerned about the reproduction of gender, racial, and socioeconomic biases in the design and use of AI. We are very mindful of the impacts that these biases can have on society, politics, the economy, and people's lives in general. It's an area where we believe special attention should be paid.

The focus shouldn't only be on AI designers. Public policy experts also need to play an essential role in the development of AI. The discipline of public policy with a gender perspective or of public policy for equality is one of our main interests. We aspire to develop a multidisciplinary vision of AI, and we are particularly interested in examining how the logics of inequalities are reproduced in science, specifically in the creation of AI.

On a personal level, I have been involved in justice projects that deal with the collection, use, and processing of data related to violence, for example, violence against women. Despite the use of AI, many of the problems we faced haven't been resolved because they stem from epistemological and design issues.

This leads us to deeper debates and questioning from a scientific perspective. One of our gender program's objectives is to integrate this questioning across all disciplines and educate new students, so that they don't perceive AI as something that's purely technical, as has been the case so far.

Compared to Europe or the United States, I believe that in Mexico, AI is often seen as something separate from society, something that's purely technical, where

DOI: 10.1201/9781003320791-31

the majority doesn't see themselves involved. It's important to bring people closer to AI and make them understand that it is a daily part of their lives, that it's in their phones, that it's not just something for the future, or robots.

The challenge we face is to combine the social with the technical. We are working hard in our program to promote this approach. Although there is resistance in some academic and technical circles to incorporating gender perspectives and human rights into AI, it is crucial to make this connection for a more equitable and just implementation of AI.

**How is AI used in the justice systems of Latin America?**

In countries like China and the United States, legal operators are using AI systems to evaluate evidence, make decisions, and even issue judgments based on predictions. But this is something that needs to be handled with caution because the data used for making these predictions were labeled by people who might have been biased.

In Mexico, although we haven't yet implemented AI systems in our justice system at that level, we are disaggregating data that could be used in the future to feed them. In a couple of years, we will have to analyze and reconstruct this process because the data we are currently using lacks a gender perspective and is not human-centered. Most of this data benefits the prosecution and is not designed to capture, process, analyze, and disseminate human-centered information. So we should pay special attention to how we disaggregate data to create AI systems in the justice sector.

I can't speak for all of Latin America, but I believe that most countries, especially those that already have public policies around AI, are advancing in this field. These countries, including Mexico, Colombia, Peru, Brazil, and Chile, have already developed or are in the process of developing AI strategies.

I conducted an analysis of AI strategies in Latin America and compared the extent to which human rights, gender perspective, and intersectionality were incorporated. Mexico and Chile seem to be the strongest in incorporating these principles, but there is still much to be done. While there are strategies in place, they lack budget and alignment with the principles of organizations such as UNESCO and the OECD.

We need to progress from these strategies to binding public policies that involve all relevant actors and are mandatory to address biases in algorithmic audits, for example. Currently, these topics remain in the realm of principles and good intentions. I believe the new European Union's AI Act will be a significant reference for all regions, including Latin America.

**What did you think of the articles you read? What impacts did they have on your vision of AI?**

One article that I really enjoyed was Lévesque et al.'s "Crafting human-centered AI in workspaces for better work." It discusses the use of AI to improve workplaces but with an emphasis on avoiding biases. For example, at our university, we are redesigning our digital platforms to make them more accessible to people with different types of disabilities. We consider ourselves an inclusive institution, but the truth is that our platforms are not designed for people with disabilities.

Additionally, Lévesque et al.'s article highlights that technologies are too often implemented without prior consultations. We don't ask the primary users if these technologies benefit them or pose a risk. Many people are afraid that their jobs will be replaced by these new technological tools. Therefore, it is necessary to democratize their use further, and ask targeted users how they perceive these innovations.

That said, I have noticed that several themes are repeated throughout the different articles I read, which led me to reflect on certain aspects.

First, it is important for human-centered AI to move toward a collective epistemology rather than an individualistic one. Even the articles dealing with regulations highlight a disconnect between these regulations and technology users. Therefore, one key point is to start considering AI as something that should benefit all communities rather than some of them or just individuals.

For example, Malwina Wójcik's "Towards Addressing Inequality and Social Exclusion by Algorithms: Human-Centric AI Through the Lens of Ubuntu" was particularly interesting as it proposes a shift in our perspective. It suggests that we move away from the epistemologies and principles of the Global North, which often serve as our guide, and adopt a critical approach to how vulnerable populations have been excluded from the design and differential impact of these technologies.

This article also highlights how privileges associated with the use of AI tend to remain in the hands of the same groups of people. Furthermore, it raises the need to analyze and question algorithmic patterns and underlying unequal power relations.

The second point that caught my attention is that technology designers often consider that AI's weaknesses will be corrected with "simple" solutions, such as reviewing data or algorithms. Instead, we need to go beyond that and examine the sources of the algorithmic patterns I just talked about.

Another recurring theme in the essays I read is the governance of AI: who designs it, who participates in its deployment, how much involvement society as a whole has, etc. This is crucial to ensure the transparency and explainability of algorithms.

In terms of governance, the question is how do we ensure safety and accountability, in addition to transparency. Currently, there is no regulation requiring developers to explain the decisions made by their algorithms.

Furthermore, the papers I read remind us it is important to recognize that AI can influence our behavior and decisions. For example, we rely on applications like Waze or Google Maps to navigate, but in doing so, we surrender our autonomy to these tools. This can change how we remember things, use our time, and even perceive the world.

It is crucial to consider how AI can influence us, both in terms of its design and governance, and how it can affect our decision-making and autonomy. While AI is meant to serve us, it is essential to reflect on how we can make sure we don't end up serving it.

**What should we do to tackle this problem or challenge?**

We are not going to stop using useful tools like AI, for example, navigation applications that show us the routes with less traffic. However, after reading about these issues, I have become aware of how AI may be influencing my decisions. This has made me more conscious, and in the future, I may question the recommendations that AI gives me, such as suggestions about restaurants or museums or texts produced with ChatGTP or Bard.

*HAICU: Last week, I had an experience where Waze gave me some incorrect information. This made me realize that, although I often trust Waze as if it is always right, it can make mistakes. Before, I almost saw it as an infallible figure, but now I understand that it is simply an AI tool. This experience has led me to question how I consider and use AI tools.*

Exactly! In my case, I now review the routes proposed by Waze before blindly following its directions. Additionally, I believe it is crucial to educate young people about the influence of AI, as there are already applications that could lead them to change their behaviors, for example, tools that can suggest what they should eat based on their desired weight.

Going back to the essays I read, one that stood out was Marina Teller's essay on AI regulation, "Towards the Social Acceptability of Algorithms." She discussed how transparency and explainability of algorithms can be affected by regulation and how algorithms can become so powerful that they can compete with regulatory forces in the market and bureaucracy.

This essay made me think of how academia has focused on the institutional power of algorithms, and how they have been accepted without much reflection. Although it has been argued that there are no longer "black boxes" in algorithms, the reality is that we still don't fully understand how they make decisions.

Another relevant essay I read, Gentelet and Mizrahi's "A Human-Centered Approach to AI Governance: Operationalizing Human Rights Through Citizen Participation," focused on AI governance and current norms. It discussed the gap between existing regulations and the rapid advancement of AI. It discussed three types of regulations applied in different countries: ethical codes, algorithmic impact assessments, and AI laws. It shows that while ethical codes propose principles and values, they lack effective enforcement mechanisms. Algorithmic impact assessments and audits, on the other hand, need to be strengthened, and clear rules need to be established in each country. It is important to conduct these assessments and audits to ensure that AI technologies are implemented effectively and safely.

**How important is it, in your opinion, that scholars, like the ones who participated in this book, make their work and findings more accessible?**

It is important to write books like these ones, with a focus on the needs of practitioners, government officials, decision-makers, etc.

It is going to be a challenge. For example, I feel that there is still a certain lack of practical focus in the essays I read. While they delve into the theory and epistemology of AI extensively, there is a lack of discussion on the specific methodologies, processes, and indicators that could help designers.

I believe it would be beneficial to move away even more from the abstract and epistemological and talk more about how we can integrate concepts such as gender perspective, person-centeredness, and intersectionality into the concrete design and analysis of AI.

How do we take them into consideration? What indicators do we use? Who participates in the process? Answering these questions is crucial to bridge the gap between those who design, analyze, and use AI. Ultimately, we need more practical and concrete guidelines on how to translate these theories into practice.

**Based on your analysis and readings, what key recommendations would you give to governments to better implement a human-centered approach to AI?**

My first recommendation would be to make public decision-makers understand that AI is not just the business of developers and big companies like Amazon or Google. It is a social and political issue they have to tackle.

Secondly, AI needs to be democratized and made more transparent. It is necessary to seek the participation of multidisciplinary collegial bodies for the design, use, and evaluation of AI policies and systems. That said, it's important to note that AI will always have biases, just like humans, and that constant monitoring will be needed.

The third recommendation is for governments to drive these efforts through legislation and public policies.

Lastly, it's crucial to consider the impact of AI on people in vulnerable and excluded situations. Regulations must be put in place to ensure that individuals are not exploited for testing or data.

# 28 Rebecca Finlay's Commentary

*Rebecca Finlay*
Partnership on AI

Rebecca Finlay is the CEO of Partnership on AI (PAI), a nonprofit partnership of academic, civil society, industry, and media organizations creating solutions so that AI advances positive outcomes for people and society. PAI is notably known for bringing together voices from across the AI community and beyond to share insights, and for making actionable resources out of the collective knowledge that emerges from these encounters. As CEO, Rebecca oversees the organization's mission and strategy and ensures that the PAI Team works with its global community of partners to drive the adoption of human-centered AI.

**How do you define human-centered AI and what is its potential and importance?**

I believe that human-centered AI should be the way that we think *de facto* about AI because we are deploying AI into the systems and structures that are made up of people. Wouldn't it be great if we always thought about AI through the lens of humanity first instead of more techno-solutionist approaches? "Human-centered AI" speaks to the socio-technical nature of AI deployment in our world. It motivates us to think about AI, first and foremost, through the lens of humanity. For me, that also includes the environment.

**Can you give me a few examples of the work Partnership on AI does on human-centered AI and how it is implemented?**

Human-centered AI was central to the creation of the Partnership on AI and continues to be: PAI works on identifying the risks of AI for humans and the planet; trying to find ways to prevent potential harm to humanity; also shaping and developing guardrails to ensure that technology is researched, developed, deployed, used, monitored, and evaluated in ways that benefit humanity. Here are a few examples.

First, PAI works on documentation and explainable AI. We see documentation as a forcing function for the responsible, human-centered development of AI. Our "About ML Program," builds on Timnit Gebru's and Meg Mitchell's seminal work on


DOI: 10.1201/9781003320791-32

model cards (which are still *de facto* standards in the field) and that of many others, such as the Data Nutrition Project by Kasia Chmielinski and others.

The goal is to increase transparency on questions such as, "How are these systems developed?", "What are the "ingredients" within them?", "What are the structures and processes within companies to manage these systems?", etc. This work identifies key questions throughout an AI system's lifecycle to document, explain where possible, and make transparent its core elements and how they are used.

My view is that documentation is the foundation of human-centered AI and should continue to be influential even as the technology develops. Just recently, we heard this echoed by Sam Altman, CEO of OpenAI, in his call for model developers to agree on and disclose an ingredient list for generative AI models. I couldn't agree more. At PAI and elsewhere, we've been working on it for a while!

Another area is PAI's work on inclusive research and design. If we want AI to work for all people, including vulnerable and impacted communities, we need better mechanisms for citizen participation and engagement in the technological development process. At the moment, there's an acknowledgment across the sector that more needs to be done. Even though there is research underway and practices are evolving—the paper Garard et al. published on agriculture and sustainability in this book speaks to this—there isn't a clear, collective understanding and consensus on the most effective methods or approaches to do so. It remains a hard problem to address.

So, this year, we launched a program on inclusive research and design that is centered on building community understanding about what participatory design in AI development means in different settings and with different communities. Recently, the White House acknowledged this work at the Global Summit for Democracy in remarks by the Director of the White House Office of Science and Technology Policy, Dr. Arati Prabakhar, reiterating the importance of centering citizens' voices and perspectives in AI development.

**You talked about one of the articles you read. Tell me how these articles possibly changed your views on AI.**

In the articles that I read: I valued the focus on design methodology and specific use cases. It is very helpful for an organization, such as PAI where our mission is all about social impact, to understand how to translate principles into processes, not in the abstract but in real-world settings. AI technologies can be applied in many different ways, with very different datasets, applications, and levels of risk. Scholarship that we can translate into practice, helps us do our work to change industry practices and inform innovations in public policy. I loved that all these articles made the case that *context matters*.

The healthcare paper by Régis et al., for instance, did an excellent job of speaking to the need to balance technology with social norms and context. It was quite powerful. The authors make a clear case for the *temporal* nature of AI technologies. They present human-centered AI as evolutionary—as a process that involves technological changes, of course, but also impacts and is impacted by changes in human systems and structures. Another aspect that I particularly liked about this paper is its focus on sense-making and reflexivity, the idea that shared sense-making has a positive influence on innovation uptake. We must work on shared understanding, shared language, definitions, and processes; and we must support mechanisms that make it possible to reflect on what works and what doesn't.

Charisi and Dignum's paper on regulatory sandboxes through the lens of AI for children is also superb in providing tangible advice for practice, technology development, and policy. One could just take it off the shelf and use it. It sparked many new ideas for me.

As the paper shows, I often find there is back and forth between two extremes—no regulation and very restrictive regulation. It is as if there are only two opposing camps. I believe there is a spectrum of policy interventions between those two extremes that regulators and policymakers should be thinking about. Regulatory sandboxes are a good example of this. The subject is not new, but this paper was helpful because it shows how different governments are implementing sandboxes. It also shows ways to design regulatory sandboxes that meet not only the needs of the regulators but also those of technology developers and impacted communities.

I also enjoyed Nishimura et al.'s paper on agile governance. I was in Japan recently, and I saw some of Japan's preparation in advance of the G7 meetings. The concept of agile governance is central. This article, though it was slightly more theoretical than the other ones, provided good information about agile governance.

**If you had recommendations to make to policymakers, auditors, things that should be done to strengthen the development and implementation of human-centered AI in our societies, what would they be?**

The first recommendation would be to lean into *all* the levers already at the disposal of policymakers. That doesn't mean regulation isn't important. I *do* think we need more regulation, and I'm very interested to see the next steps for the EU AI Act. However, I also think that several of this book's articles show that there are many innovative ways to advance policy. Policymakers should also be working with multilateral organizations such as the OECD, the Global Partnership on AI, the G7 and the UN. We have to use these mechanisms to promote standards and interoperability that will drive trade as well as responsible economic development and innovation.

My second recommendation—and Nishimura et al.'s paper on agile governance talks about this—is that there's a need for policymakers to understand the role of multistakeholder organizations such as PAI. I guess it is no surprise that I believe there is an important role for organizations that can quickly and flexibly respond to technology changes by setting collective norms and protocols that can then be married up to appropriate regulation which takes longer to come into force. I want policymakers to see that organizations such as the Partnership on AI can be a resource to them. Indeed, we can translate technology into policy context; we can connect policymakers to individuals and organizations across sectors; we can help to connect to more international, jurisdictional questions.

This book, and the articles therein, are great examples of scholarship that can mobilize knowledge to make real change happen in policy, industry, and civil society.

# Index

Pages in *italics* refer to figures, pages in **bold** refer to tables, and pages followed by "n" refer to notes.

## B



## C

## I

## N

## T

## U



## V



## W

Printed in the United States
by Baker & Taylor Publisher Services

Printed in the United States
by Baker & Taylor Publisher Services